COMPUTATIONAL FLUID DYNAMICS

COMPUTATIONAL FLUID DYNAMICS

A Practical Approach

Fourth Edition

JIYUAN TU

RMIT University, Australia
University of New South Wales, Australia
Tsinghua University, P.R. China

GUAN HENG YEOH

Professor, Mechanical Engineering (CFD)
University of New South Wales, Sydney
Australian Nuclear Science and Technology Organisation
University of New South Wales
Australia

CHAOQUN LIU

Center for Numerical Simulation and Modeling
University of Texas at Arlington, Arlington
Texas, USA

YAO TAO

Associate Professor
School of Air Transportation/Flying
Shanghai University of Engineering Science
Shanghai, China

Butterworth-Heinemann
An imprint of Elsevier

Butterworth-Heinemann is an imprint of Elsevier
The Boulevard, Langford Lane, Kidlington, Oxford OX5 1GB, United Kingdom
50 Hampshire Street, 5th Floor, Cambridge, MA 02139, United States

Notices
Knowledge and best practice in this field are constantly changing. As new research and experience broaden our understanding, changes in research methods, professional practices, or medical treatment may become necessary.

Practitioners and researchers must always rely on their own experience and knowledge in evaluating and using any information, methods, compounds, or experiments described herein. In using such information or methods they should be mindful of their own safety and the safety of others, including parties for whom they have a professional responsibility.

To the fullest extent of the law, neither the Publisher nor the authors, contributors, or editors, assume any liability for any injury and/or damage to persons or property as a matter of products liability, negligence or otherwise, or from any use or operation of any methods, products, instructions, or ideas contained in the material herein.

ISBN: 978-0-323-93938-6

For information on all Butterworth-Heinemann publications
visit our website at https://www.elsevier.com/books-and-journals

Publisher: Katey Birtcher
Acquisitions Editor: Stephen Merken
Editorial Project Manager: Mason Malloy
Production Project Manager: Haritha Dharmarajan
Cover Designer: Mark Rodgers

Typeset by STRAIVE, India

Working together
to grow libraries in
developing countries

www.elsevier.com • www.bookaid.org

Contents

Preface

This book aims to offer users of Computational Fluid Dynamics (CFD) with *a suitable text* that pitches at *the right level* of assumed knowledge. CFD is a mathematically sophisticated discipline. A text is thus needed to provide simple to understand descriptions of fundamental CFD theories, basic CFD techniques and practical guidelines. In this fourth edition, we welcome the contribution of Dr Tao as we continue to provide more updates about the significant developments and important applications of CFD.

As CFD is becoming more accessible especially through use of commercial CFD codes, such as ANSYS-CFX, ANSYS-FLUENT or STAR-CCM+, mastery of CFD and attaining competency in such a skill certainly brings about a steep learning curve for practicing engineers to come up with solutions to fluid flow and thermal problems without *a priori* knowledge of the basic concepts and fundamental understanding of fluid mechanics and heat transfer. Without overwhelming the reader with excessive mathematical and theoretical illustrations of computational techniques, every effort in the *first, second, third editions* as well as the latest offering, *fourth edition,* has been made to discuss the material in a style of to maintain practicality of understanding CFD.

The *first edition* entails introducing the basic structure of the book:

Chapter 1 – An introduction to CFD and specifically designed to provide the reader with an overview of CFD and its entailed advantages, the range of applications as a research tool on various facets of industrial problems and the future use of CFD.

Chapter 2 – For first-time users on how a CFD problem is currently handled and solved. The reader will benefit through guidance of these basic processes using any commercial, shareware and in-house CFD codes. More importantly, it serves as a guidepost for the reader to other chapters relating to fundamental knowledge of CFD.

Chapter 3 – Basic thoughts and philosophy associated with CFD, along with an extensive discussion of the governing equations of fluid dynamics and heat transfer. It is vitally important chapter that provides the reader to fully appreciate, understand, and feel comfortable with, the basic physical equations and underlying principles of this discipline. By working through the worked-out examples, the reader will have a better understanding of the equations governing the conservation of mass, momentum and energy.

Chapter 4 – The first half of the chapter deals with numerical discretization. The basic numerics are illustrated with popular discretization techniques such as the finite difference and finite volume methods (adopted in the majority of commercial codes) for solving flow problems. The second half of the chapter deals with specific techniques to solve algebraic equations. The pressure-velocity coupling scheme (SIMPLE and its derivatives) forms the core information of the book. This scheme invariably constitutes the basis of most commercial CFD codes through which simulations of complex industrial problems have been successfully made.

Chapter 5 – Discussion on numerical concepts of stability, convergence, consistency, and accuracy. As the core understanding of the fundamental equations of fluid flow and heat transfer is the essence of CFD, it follows that the understanding of the techniques of achieving a CFD solution is the resultant substantive. This will enable the reader to better assess the results produced when different numerical methodologies are applied.

Chapter 6 – Since *real-world applications of CFD* are turbulent in nature after all, we devised some practical guidelines for the reader to better comprehend turbulence modeling and other models commonly applied. Carefully designed worked-out examples are provided to assist students in understanding the complex modeling concept.

Chapter 7 – Illustrations on the power of CFD through a set of industrially relevant applications on a significant range of engineering disciplines. Special efforts have been made to stimulate the inquisitive minds of the reader through exposition of some pioneering applications.

Chapter 8 – A general introduction of the basic concepts of advanced CFD techniques. The aim is to expose readers to the evolutionary use of CFD in any new emerging areas of science and engineering.

The *second edition* entails illustrating the other applications of CFD in areas of metallurgy as well as nuclear safety; extending an understanding to other discretization methods such as the finite element and spectral methods by providing a summary of the basic ideas underpinning the utilization of these methods to solve the fluid flow equations; describing the multigrid method to reflect the iterative approach that is also commonly being adopted to solve the systems of discretized equations in commercial CFD codes; and the proper handling of complex geometries.

The *third edition* entails further treatment of the fundamental physics of fluid flows; a new chapter focusing on mesh generation for CFD focusing on the key aspects relating to the use of structured, body-fitted and unstructured meshes and key practical guidelines of grid generation for CFD as well as introducing an advanced grid generation approach based on moving meshes; derivation and the generic form of the governing equations for compressible flows; providing extended discussions on transportiveness and boundedness; and descriptions of, hybrid CPU– Graphical Processing Unit (GPU) systems with the purpose of exploiting all available processing power and CPU memory resources for CFD calculations.

The *fourth edition* entails updating Chapter 8 with illustration of solar-induced natural convection and dynamic mesh modelling of human motion, and new results demonstrating the air/particle flow in a human respiratory system. Herein, CFD can be seen to play a key role in assisting building designers for energy-saving measures through identifying the flow and heat transfer characteristics in the presence of radiation and natural convection and enabling the evaluation of natural ventilation rates being enhanced by solar thermal energy. For the turbulent flow over a moving manikin, CFD has the propensity of further comprehending the indoor aerodynamics associated with the movement of occupants and their consequent impacts on contaminant control in different types of environments. In Chapter 9, more recent references are provided to keep abreast with the fast-paced development of CFD approaches and methodologies and further discussions on the development of CFD in the CUDA environment. Two new topics, which are reduced-order modelling approaches and machine learning accelerated CFD, are also introduced and incorporated in this chapter to demonstrate the novel and innovative applications of CFD into different facets of engineering problems.

Jiyuan Tu
Guan Heng Yeoh
Chaoqun Liu
Yao Tao

Qualified instructors can find teaching ancillaries for the book, including solutions manual, lab guides, lecture notes, and more, by visiting https://educate.elsevier.com/book/details/9780323939386 and following the instructions for obtaining access. For all users, a bonus online case studies chapter can be downloaded from the book's companion site at https://www.elsevier.com/books-and-journals/book-companion/9780323939386.

CHAPTER 1

Introduction

1.1 WHAT IS COMPUTATIONAL FLUID DYNAMICS

Computational fluid dynamics has certainly come of age in industrial applications and academia research. In the very beginning this popular field of study, accredited by its infamous acronym CFD, was only renowned in high-technology engineering areas of aeronautics and astronautics, but now, it is becoming a rapidly adopted household methodology in solving complex problems in modern engineering practice. CFD, derived from different disciplines of fluid mechanics and heat transfer, is also finding its way into other important uncharted areas especially in process, chemical, civil, and environmental engineering. Construction of new and better-improved system designs and optimization carried out on existing equipment through computational simulations are resulting in enhanced efficiency and lower operating costs. With the concerns of global warming and the increasing world population, engineers in power generation industries are heavily relying on CFD to reduce development and retrofitting costs. These computational studies are currently being performed to address pertinent issues relating to technologies for clean and renewable power as well as meeting strict regulation challenges of emissions control and substantial reduction of environmental pollutants.

Nevertheless, the basic question remains: What is actually *computational fluid dynamics*? In retrospect, it has certainly become a new branch integrating the disciplines of fluid mechanics not only with mathematics but also with computer science, as illustrated in Fig. 1.1. Let us briefly discuss each of these individual disciplines. Fluid mechanics is essentially the study of fluids either in motion (*fluid in dynamic mode*) or at rest (*fluid in stationary mode*). CFD is particularly dedicated to the former, fluids that are in motion, and how the fluid flow behavior influences processes that may include heat transfer and possibly chemical reactions in combusting flows. This directly infers to the "*fluid dynamics*" description appearing in the terminology. Additionally, the physical characteristics of the fluid motion can usually be described through fundamental *mathematical* equations, usually in partial differential form, which govern a process of interest and are often called governing equations in CFD (see Chapter 3 for more insights). In order to solve these *mathematical* equations, they are converted by *computer scientists* using high-level computer programming languages into computer programs or software packages. The "*computational*" part simply means the study of the fluid flow through numerical simulations, which involves employing computer programs or software packages performed on high-speed digital computers to attain the numerical solutions. Another question arises: Do we actually require the expertise of three

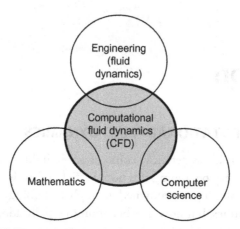

Fig. 1.1 The different disciplines contained within computational fluid dynamics.

specific people from each discipline, namely, fluids engineering, mathematics, and computer science, to come together for the development of CFD programs or even to conduct CFD simulations? The answer is obviously not and more likely it is expected that this field demands a person who will proficiently obtain more or less some subsets of the knowledge from each discipline.

CFD has also become one of the three basic methods or approaches that can be employed to solve problems in fluid dynamics and heat transfer. As demonstrated in Fig. 1.2, each approach is strongly interlinked to each other and does not lie in isolation. Traditionally, both experimental and analytical methods have been used to study the various aspects of fluid dynamics and to assist engineers in the design of equipment and industrial processes involving fluid flow and heat transfer. With the advent of digital computers, the computational (numerical) aspect has emerged as another viable approach. Although the analytical method is still practiced by many and experiments will continue

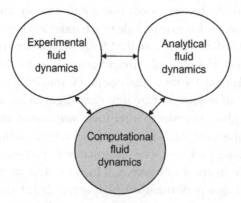

Fig. 1.2 The three basic approaches to solving problems in fluid dynamics and heat transfer.

to be significantly performed, the trend is clearly toward greater reliance on the computational approach for industrial designs, particularly when the fluid flows are very complex.

In the past, potential or novice users would probably learn CFD by investing a substantial amount of time writing their own computer programs. With the increasing demands from industries or even within academia to acquire the knowledge of CFD in a much shorter timeframe, it is not surprising that the interest in abandoning writing computer programs is escalating in favor of using more commercially available software packages. Multi-purpose CFD programs are gradually earning the approval, and with the advancement of models to better encapsulate the flow physics, these software packages are also gaining wide acceptance. There are numerous advantages to applying these computer programs. Since the mundane groundwork of the writing and testing of these computer codes has been thoroughly carried out by the "developers" of respective software companies, today's potential or novice CFD users are comforted by not having to deal with these types of issues. Such a program can be readily employed to solve numerous fluid flow problems.

Despite the well-developed methodologies within the computational codes, CFD is certainly more than just being proficient in operating these software packages. Bearing this in mind, the primary focus of this book is thus oriented to better educate potential or novice users in employing CFD in a more judicious manner, equally supplementing the understanding of underlying basic concepts and the technical know-how in better tackling fluid flow problems. For other users who are inclined to pursue a post-graduate research study or are currently undergoing research through the development of new mathematical models to solve more complex flow problems, they should consult other CFD books (e.g. Fletcher, 1991; Anderson, 1995; Versteeg and Malalasekera, 1995) and our future book, in which we intend to concentrate on presenting a step-by-step procedure of initially understanding the physics of new fluid dynamics problems at hand, developing new mathematical models to represent the flow physics, and implementing appropriate numerical techniques or methods to test these models in a CFD program.

CFD has indeed become a powerful tool to be employed either for pure/applied research or industrial applications. Computational simulations and analyses are increasingly performed in many fluid engineering applications that include: aerospace engineering (airplanes, rocket engines, etc.), automotive engineering (reducing drag coefficients for cars and trucks, improving air intake in engines, etc.), biomedical engineering (blood flow in artificial hearts and through scents, breathing, etc.), chemical engineering (fluid flowing through pumps and pipes, etc.), civil and environmental engineering (river restoration, pollutant dispersion, etc.), power engineering (improving turbine efficiency, wind farm sitting and performance prediction, etc.), and sports engineering (swimming equipment, golf swing mechanics, reducing drag in biking, etc.). Through CFD, one can gain an increased knowledge of how system components are expected to perform, so as to make the required improvements for design and optimization studies. CFD actually asks the question, "What if...?", before a commitment is undertaken to execute any design alteration. When one ponders the planet we live on, almost everything revolves in one way or another around a fluid or moves within a fluid.

CFD is also revolutionizing the teaching and learning of fluid mechanics and thermal science in higher education institutions through the visualization of complex fluid flows. The development of CFD-based software packages being more user-friendly is allowing students to visually reinforce the concepts of fluid flow and heat transfer. This software allows teachers to create their own examples or customize pre-defined existing ones. Using carefully constructed examples, students are introduced to the effective use of CFD for solving fluid flow problems and can instinctively develop an intuitive feel for the flow physics. In the next section we discuss some important advantages and further expound on how CFD has evolved and is applied in practice.

1.2 ADVANTAGES OF COMPUTATIONAL FLUID DYNAMICS

With the rapid advancement of digital computers, CFD is poised to remain at the forefront of cutting-edge research in the sciences of fluid dynamics and heat transfer. Also, the emergence of CFD as a practical tool in modern engineering practice is steadily attracting much interest and appeal.

There are many advantages to considering computational fluid dynamics. Firstly, the theoretical development of the computational sciences focuses on the construction and solution of the governing equations and the study of various approximations to these equations. CFD presents the perfect opportunity to study specific terms in the governing equations in a more detailed fashion. New paths of theoretical development are realized, which could not have been possible without the introduction of this branch of the computational approach. Secondly, CFD complements experimental and analytical approaches by providing an alternative cost-effective means of simulating real fluid flows. Particularly, CFD substantially reduces lead times and costs in designs and production compared with the experimental-based approach and offers the ability to solve a range of complicated flow problems where the analytical approach is lacking. These advantages are realized through the increasing performance power in computer hardware and its declining costs. Thirdly, CFD has the capacity to simulate flow conditions that are not reproducible in experimental tests found in geophysical and biological fluid dynamics, such as nuclear accident scenarios or scenarios that are too huge or too remote to be simulated experimentally (e.g. Indonesian Tsunami of 2004). Fourthly, CFD can provide rather detailed, visualized, and comprehensive information when compared to analytical and experimental fluid dynamics.

In practice, CFD permits alternative designs to be evaluated over a range of dimensionless parameters that may include the Reynolds number, Mach number, Rayleigh number, and flow orientation. The utilization of such an approach is usually very effective in the early stages of development for fluid system designs. It may also prove to be significantly cheaper in contrast to the ever-increasing spiraling cost of performing experiments. In many cases, where details of the fluid flow are important, CFD can provide detailed information and understanding of the flow processes to be obtained, such as the occurrence of flow separation or whether the wall temperature exceeds some maximum limit. With technological improvements and competition requiring a higher degree of optimal designs and as new high-technological applications demand precise prediction of flow behaviors, experimental development may eventually be too costly to initiate. CFD presents an alternative option.

Nevertheless, the favorable appraisal of CFD thus far does not suggest that it will soon replace experimental testing as means to gather information for design purposes. It is rather considered of being complementary in solving fluid mechanics problems. For example, wind-tunnel testing is a typical piece of experimental equipment that still provides invaluable information for the simulation of real flows at a reduced scale. For the design of engineering components especially in an aircraft that depends critically on the flow behavior, carrying out wind-tunnel experiment remains an economically viable alternative over full-scale measurement. Wind-tunnels are very effective for obtaining the global information of the complete lift and drag on a body and the surface distributions at key locations. In other applications where CFD still remains a relatively primitive state of development, an experiment-based approach remains the primary source of information especially when complex flows such as multi-phase flows, boiling, or condensation are involved.

In spite of the many upbeat assessments, the reader must also be fully aware of some inherent limitations of applying CFD. Numerical errors exist in computations and hence there will be differences between the computed results and reality. Visualization of numerical solutions using vectors, contours, or animated movies of unsteady flows are by far the most effective ways of interpreting the huge amount of data generated from the numerical calculation. However, there is a danger that an erroneous solution that may look good may not correspond to the expected flow behavior! The authors have encountered numerous incorrect numerically produced flow characteristics that could have been interpreted as acceptable physical phenomena. Wonderfully bright color pictures may provide a sense of realism of the actual fluid mechanics inside the flow system but they are worthless if they are not quantitatively correct. Numerical results obtained must always be thoroughly examined before they are believed. Hence a CFD user needs to learn how to properly analyze and make a critical judgment on the computed results. This is one of the important aims of this book.

1.3 APPLICATION OF COMPUTATIONAL FLUID DYNAMICS

1.3.1 As a Research Tool

CFD can be employed to better understand the physical events or processes that occur in the flow of fluids around and within the designated objects. These events are closely related to the action and interaction of phenomena associated with dissipation, diffusion, convection, boundary layers, and turbulence. Whether the flows are incompressible or compressible, many of the most important aspects inside these types of flows are non-linear and, as a consequence, often do not have any analytic solution. This motivates the need to seek numerical solutions for the partial differential equations and it would seem to invalidate the use of linear algebra for the classification of the numerical methods. Our experiences have nevertheless demonstrated that such is not the case.

CFD, analogous to wind-tunnel tests, can be employed as a *research tool* to perform *numerical experiments*. We examine one of these *numerical experiments*, garnered from our research investigation of Chen et al. (2004), in order to demonstrate the feasible use of CFD as a *research tool* and impart some understanding of this philosophy. Fig. 1.3A represents a "snapshot" taken for an unsteady flow past two side-by-side cylinders at a given instant of time. In Fig. 1.3B the comparative visualization of the numerical calculations based on a large eddy simulation (LES) model attests to the power of CFD modeling to capture the complex flow characteristics. This example clearly illustrates how CFD can be utilized to better understand the observed flow structures and some important physical aspects of a flow field, similar to a real laboratory experiment. For the case of three side-by-side cylinders shown in Fig. 1.4, here again is another example of how CFD simulations can work harmoniously with experiments, providing not only qualitative comparison but also a means to interpret some basic phenomenological aspects of the experimental condition. More importantly, *numerical experiments* can

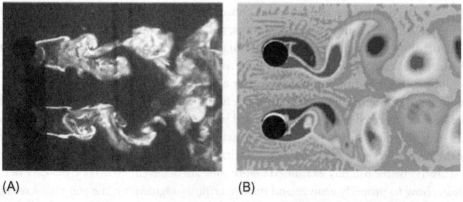

(A)   (B)

Fig. 1.3 Example of a CFD numerical experiment for a flow past two side-by-side cylinders: (A) experimental observation and (B) numerical simulation.

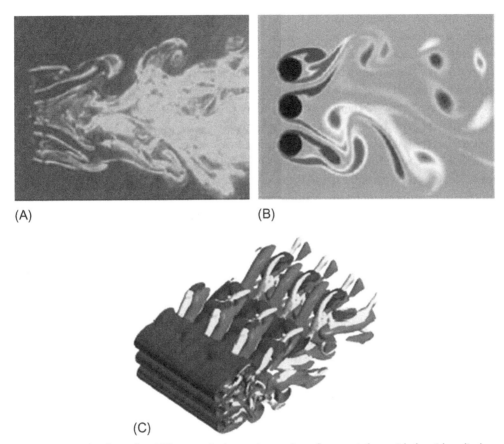

Fig. 1.4 Example of another CFD numerical experiment for a flow past three side-by-side cylinders: (A) experimental observation, (B) numerical simulation on a two-dimensional cross-sectional plane, and (C) three-dimensional representation of the fluid flow through numerical simulation.

provide more comprehensive information and details of the flow as visualized in three dimensions, as shown in Fig. 1.4C, when compared to laboratory experiments. These graphical examples clearly affirm the value of *numerical experiments* carried out within the framework of CFD.

1.3.2 As an Educational Tool

The educational spectrum of CFD generally ranges from advanced graduate course to a couple of days of software training. At the top of the spectrum, students are encouraged to develop numerical schemes and model challenging physical phenomena. Software training can be considered to be at the bottom of the spectrum, where the focus is predominantly directed toward details of the software capability. Prior knowledge of CFD is normally assumed and such a training is offered by companies for engineers to keep up with the latest advances in CFD. The philosophy of an undergraduate course at an introductory level lies in the middle of this spectrum. What is required is to expose students to both theoretical foundation of CFD, beginning from the physical descriptions rather than the mathematics behind the governing equations and with more focus on the physical interpretation of the flow physics such as boundary layer, separation, and boundary conditions. Software training is a secondary product of the learning process through which students will have to learn CFD by practice.

To meet the balance between theory and hands-on, project-based learning (PBL) can be used in the course for the software training. The main mission is to expose students to essential CFD concepts and expand the learning experience with real-world applications, which is becoming an increasingly important skill in today's job market. Nowadays, ease of use and the broad capability of commercial CFD software packages have enabled CFD to be brought down into the undergraduate classroom. With user-friendly graphical user interfaces guiding the students through stages of geometry creation and mesh generation, to be further exemplified in Chapter 2, computational simulations, and viewing the results by means of vectors, contours, or animated movies, the teaching of CFD has never been a so visually exhilarating and welcoming experience. Within these graphical user interfaces, line graphs are also provided to assist users in assessing the CFD simulations by either tracking the convergence history or monitoring the surface distribution of certain fluid forces such as the lift force through the lift coefficient. The prime derivative of these graphical user interfaces is certainly more than just introducing CFD technology to undergraduates. They are intended to arouse students' interests to not only learn about the basic fluid dynamics and heat transfer but entice them to further extend their learning experience to other transport phenomena of all kinds that may exist in practice or in nature.

The authors believe that there are two prime benefits of exposing undergraduates to CFD. Firstly, the experience with a more hands-on approach as adopted to better understand the concepts of fluid flow and heat transfer deepens greatly the students' understanding of the fluid flow as well as the thermal phenomena. Particularly, the visualization capability greatly reinforces students' intuition of the flow and heat transfer behavior. Secondly, such an approach opens the door to new classes of problems that can be solved by undergraduates who are no longer limited by the narrow range of classical solutions of engineering fluid mechanics and heat transfer.

1.3.3 As a Design Tool

CFD, likewise as research and educational tools, is also becoming an integral part of engineering design and analysis environment in prominent industries. Companies are progressively seeking industrial solutions through the extensive use of CFD for the optimization of product development and processes and/or to predict the performance of new designs even before they are manufactured or implemented. Software applications can now provide numerical analyses and solutions to pertinent flow problems through the employment of common desktop computers. As a viable design tool, CFD has assisted by providing significant and substantial insights into the flow characteristics within the equipment and processes required to increase production, improve longevity, and decrease waste. The increasing computer processing power is certainly revolutionizing the use of CFD in new and existing industries. These industrial solutions are expounded in the proceeding sections to further demonstrate the wide application of this specific technology in practice.

1.3.4 Aerospace

Computational fluid dynamics has certainly enjoyed a long and illustrious history of development and application in the aerospace and defense industries. To maintain an edge in a very competitive environment, CFD is playing a crucial role in overcoming many challenges faced by these industries in improving flight and in solving a diverse array of designs. Indeed, many engineers would associate CFD with its well-known application to aerodynamics by the calculation of the lifting force on an aircraft wingspan. Nevertheless, as methods and resources have augmented in power and ease of use, practitioners have expanded the scope of application beyond the calculation of lift. Today, CFD is being applied to many greater difficult operational problems that were too unwieldy to analyze or solve with computational tools in the past.

The simulation of fluid path lines in the vicinity of an F18 jet (left) and prediction of pressure coefficient contours at a $10°$ angle of attack around a supersonic missile system with grid fins (right) are illustrated in Fig. 1.5. These examples are just a small sampling of numerous applications in aerodynamic design and military application.

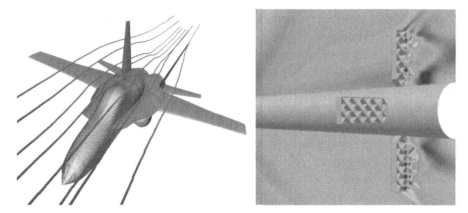

Fig. 1.5 Example of CFD results for applications in aerospace and defense industries. *(Courtesy of ANSYS-FLUENT and US Army Research Laboratory and ANSYS-FLUENT.)*

There are also other applications that CFD has been employed in resolving a number of complex operational problems in aircraft designs such as studying the impact of trailing vortices on the safe operation of successive aircraft taking-off and landing on the runway as well as enhancing the passenger and crew comfort by improving cabin ventilation, heating, and cooling. Efforts to better understand and suppress the noise produced by heavy artillery and the safe operation of a military helicopter upon firing a missile that could impinge on the airframe or the tail rotor are just some other operational problems that CFD is increasingly being employed for in military applications. As a versatile, robust, and powerful tool, CFD is rigorously meeting the broad physical modeling demands to investigate relevant complex phenomena in aerospace and defense-related designs. CFD is undoubtedly becoming a household simulation tool within these industries since the need to save time and money by reducing development costs and accelerate time to market and the need to improve the overall performance of system configurations are becoming more prevalent.

1.3.5 Automotive Engineering

Automobile engineers are increasingly relying on more simulation techniques to bring forth new vehicle design concepts to fruition. Computer-aided engineering has been at the forefront of creating innovative new internal systems that will better enhance the overall driving experience, improve driver and passenger comfort and safety, and advance fuel economy. Computational fluid dynamics has long been an essential element in automotive design and manufacture. Besides the aerospace and aerodynamics industries, this branch of engineering has also embraced much of this technology in research as well as in practice; the use of CFD is thus well entrenched in many disciplines as the engineering simulation tool for even the most difficult challenges.

CFD in automotive engineering has provided many advantages. This technology has delivered the ability to shorten cycles, optimize existing engineering components and systems to improve energy efficiency and meet strict standards and specifications, improve the in-car environment, and study the important external aerodynamics, as illustrated in Fig. 1.6. Specifically, CFD has shown measurable results by decreasing emissions with powertrain and engine analyses, increasing fuel economy, durability, and performance through aerodynamic investigations and increasing the reliability of brake components. Like in the aerospace industry, CFD has been used to determine the effects of local geometry changes on the aerodynamic forces and provides a significant capability to directly compare a multitude of vehicle designs. This reduces the dependence on time-consuming, expensive clay models and wind-tunnel experiments and delivers quicker design turn-around.

More importantly, CFD modeling has provided insights into features of in-cylinder flows that would otherwise be too difficult and expensive to obtain experimentally.

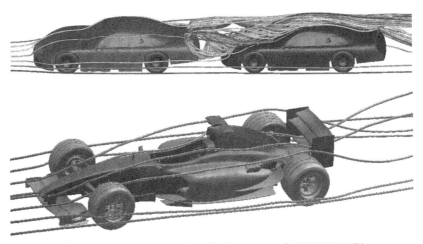

Fig. 1.6 Examples of automotive aerodynamics. *(Both courtesy of ANSYS-FLUENT.)*

Numerical simulations allow the ease of investigating different valve and port designs that can lead to improved engine performance through better breathing and more induction change distribution. Within the cylinder itself, moving and deforming grids permit the means of simulating the piston and valve motion, such as described by the example of Fig. 1.7 for a diesel internal combustion engine. The intake runner (blue) is on the left, and the exhaust runner (yellow) is on the right. Here, CFD allows engineers to understand how changes in port and combustion chamber design affect the engine performance such as volume efficiency or swirl and tumble characteristics. The flow physics described by the piston movement from the top dead center down to the bottom dead center and

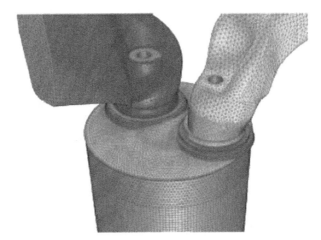

Fig. 1.7 Example of CFD applications in a diesel internal combustion engine. *(Courtesy of ANSYS-FLUENT, Internal Combustion Engine Design.)*

subsequently returning to its original position within the combustion chamber allows engineers to examine in detail the transient flow patterns inside the cylinder during the complete engine cycle. Cold flow simulations such as this address many challenging problems faced by engineers in the automotive industry. Through the application of dynamic mesh adaptation, CFD simulations of internal combustion engines are being performed with greater speed and the ease-of-use meeting the competitive progressive demands in the automotive industry.

1.3.6 Biomedical Science and Engineering

Medical researchers are nowadays relying on simulation tools to assist in predicting the behavior of circulatory blood flow inside the human body. Computational simulations can provide invaluable information that is extremely difficult to be obtained experimentally and they also allow many variations of fluid dynamics problems to be parametrically studied. Figs. 1.8 and 1.9 illustrate just one of the many sample applications of CFD in the biomedical area where the blood flows through original stenosed and virtually stented arteries have been predicted herein. With the breadth of physical models and advances in areas of fluid-structure interaction, particle tracking, turbulence modeling, and better meshing facilities, rigorous CFD analysis is increasingly performed to study the fluid phenomena inside the human vascular system. Medical simulations of circulatory functions

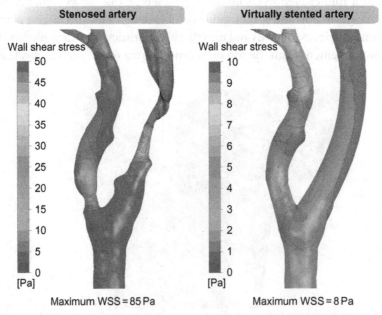

Fig. 1.8 Example of CFD prediction of wall shear stress (WSS) for original stenosed and virtually stented arteries.

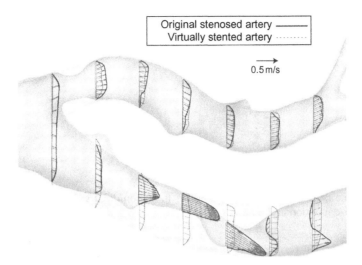

Fig. 1.9 Example of predicted velocity profiles for original stenosed and virtually stented arteries.

offer many benefits. They can lower the chances of post-operative complications, assist in developing better surgical procedures, and deliver a good understanding of biological processes as well as more efficient and less destructive medical equipment such as blood pumps. For example, CFD is being increasingly employed via virtual prototyping to recommend the best design for surgical reconstructions such as carotid endarterectomy and to better understand the blood flow through an aneurysm in the abdominal artery.

CFD is also gaining an enormous interest in the pharmaceutical industries. With the ever-increasing pollution levels causing respiratory problems and frequent asthma attacks, the need has never been greater to predict and optimize inhaled therapies. CFD can provide essential insights by simulating the entire drug delivery process for particulate, aerosol, and gaseous drug types, from the inception of device design, through the airways and down into the lungs of the pulmonary system. Through sophisticated multi-phase models, the motion of aerosol droplets and drug particles and their transport/deposition characteristics within the airways are predicted to ascertain the drug concentration in the lungs, as shown by the example exemplifying the modeling of particle formation/dispersion from nasal sprayers and particle transport/deposition in the nasal cavity in Fig. 1.10. In order to better emulate the actual fluid flow through the airways of the bronchial tree of the pulmonary system such as shown in Fig. 1.11, medical images obtained from accurate CT or MRI scans are converted into a geometrical model, which are subsequently used for advanced CFD flow simulations. Through these CFD simulations, deposition and uptake can thus be customized in a manner targeting particular drugs, delivery devices, diseases, and even individual patients. For example, Fig. 1.12 illustrates the predictions of pressure coefficient values for specific patient cases with and without asthma.

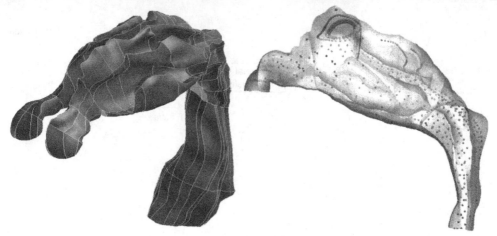

Fig. 1.10 Example of CFD application for particle formation/dispersion from nasal sprayers and particle transport/deposition in the nasal cavity.

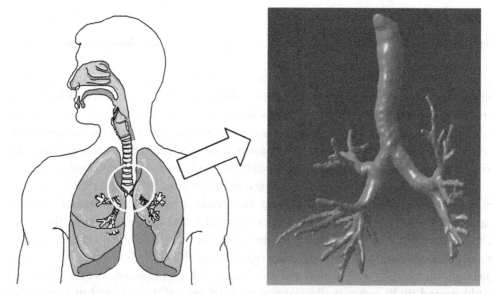

Fig. 1.11 Example of bronchial tree geometry created from CT data for CFD simulation.

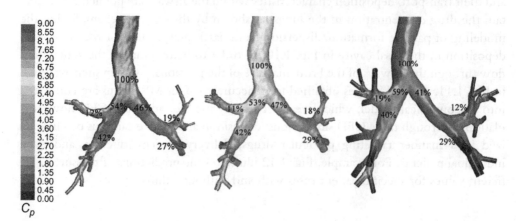

Fig. 1.12 Example of CFD prediction of pressure coefficient (C_p) values for specific patient cases with and without asthma.

As demands for new and better healthcare products continue to swell at a rapid pace, "health" companies will be required to perform more research and develop promising new products. Within these diverse healthcare sectors that may include the aforementioned biomedical, pharmaceuticals, and other areas associated with medical equipment and "general-health" personal products, CFD is becoming an important tool in the identification and improvement of new products to meet the surging market aided primarily by the aging population.

1.3.7 Chemical and Mineral Processing

Many world necessities revolve around the chemical and mineral processing industries. By applying large quantities of heat and energy to physically or chemically deform raw materials, these industries have certainly helped to mold essential products for food and health as well as vital advanced technological equipment in computing and biotechnology. In the face of increasing industrial competitiveness, these industries are confronted with major challenges to meet the world's demands and needs for the present without compromising for the future. This translates into making the operational processes becoming more energy efficient, safer, and flexible while better containing and reducing emissions.

For example, improving the performance of gas-sparged stirred tank reactors is considered to be of paramount importance in the chemical industry. Fig. 1.13 presents contours of different bubble size distribution within the stirred tank accompanied by the local flow behavior, indicated by velocity vectors, around one of the rotating blades. The detailed information on the transport of liquid and gases through the use of CFD and population balance approaches ensures that engineers have the best available possible data to work with in order to increase yield by improving fluid flows, thereby reducing operating costs and increasing system efficiency.

Also, in the world of manufacturing no industry is more important than minerals, particularly in Australia. Minerals processing span a wide range of activities and many of these processes involve complex fluid flow, heat, and mass transfer phenomena inside aggressive and hostile environments. By modeling, optimizing, and improving processes such as *classification*, *separation*, and *filtration*, CFD is at the forefront of providing designs with greater efficiencies and significant production outputs. Fig. 1.14 illustrates a *separation* process in minerals processing that involves the use of gas cyclones and hydro-cyclones. A gas cyclone is a commonly used apparatus that utilizes gravity and centrifugal force to separate solid particles from a gas stream. The hydro-cyclone is a similar device; however, the operating fluid is liquid rather than gas. Here, the generated centrifugal forces are strong enough to cause the solids to separate from the liquid, upon which these separated solids fall down under gravity into the accumulator vessel situated beneath the hydro-cyclone. CFD simulations through the application of advanced

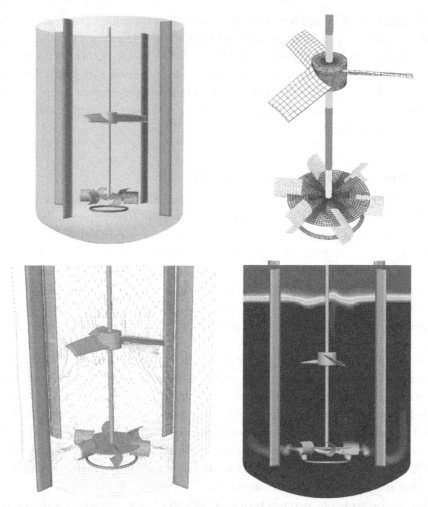

Fig. 1.13 Example of CFD application in the simulation of gas-sparged stirred tank reactor. *(Courtesy of ANSYS-FLUENT.)*

turbulence models are performed to ascertain the different flow physics within the cyclone geometry for the correct prediction of the fluid flows. The behavior of the particulates can then be simulated using suitable particle transport models, either employing the Eulerian or Lagrangian multi-phase approach.

The above examples represent merely a tip of the iceberg of the many CFD applications in the processing industries. Advanced CFD is also heavily involved in other important manufacturing sectors such as extraction metallurgy as well as oil, gas, and petrochemical industries. The recent advances in the modeling capabilities of CFD have indeed opened up many opportunities to achieve significant technological strides in the complex and demanding world of the chemical and process industries. Primarily, CFD is

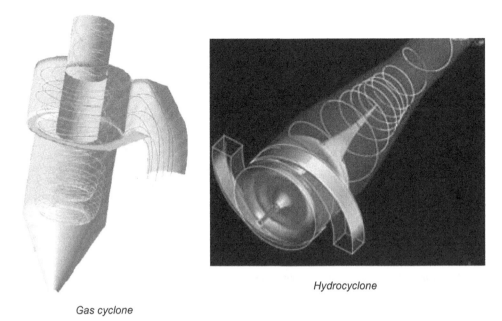

Hydrocyclone

Gas cyclone

Fig. 1.14 Example of CFD application in the simulation of gas cyclone and hydro-cyclone. *(Courtesy of ANSYS-FLUENT and ESSS and Petrobras, ANSYS-CFX.)*

playing a crucial role in extrapolating a process from the laboratory and pilot plant scale to the industrial plant scale by combining different processes into smaller, compact, and efficient units instead of treating them individually. This has produced an upgrade in the efficiency of a plant with the same existing constraints.

1.3.8 Civil and Environmental Engineering

Governments, research institutes, and corporations are actively seeking ways to meet environmental legislations and guarantees by decreasing waste while maintaining acceptable production levels fueled by increasing market demands. In many cases CFD simulations have been at the heart of resolving many environmental issues. For instance, CFD has been used to predict the pollutant plume being dispersed from a cooling tower subject to wind conditions, as shown in Fig. 1.15.

In addition, CFD can assist especially in ensuring compliance with strict regulations during the early design stages of construction. Fig. 1.16 represents the pre-construction for a new 22 m tank at a water treatment plant. Owing to the huge construction cost, which may exceed millions of dollars, virtual computer-aided models can be built and analyzed at a fraction of the time and cost by exploring all aspects of the design before construction is commenced. To determine the feasibility of such a construction, flow modeling is also performed (also shown in Fig. 1.16), which provides insights into the

Fig. 1.15 Example of CFD application to plume dispersion from a cooling tower. *(Courtesy of ANSYS-FLUENT.)*

Fig. 1.16 Example of CFD application to the construction of a new tank at a water treatment plant. The top right-hand corner figure describes the CFD simulation of the water tank that will be installed within the excavated construction site. *(Courtesy of MMI Engineering.)*

flow behavior for the proposed tank that would not have been possible through physical modeling. The added understanding gained from CFD simulation provides confidence in the design proposal, thus avoiding the added costs of over-sizing and over-specification while reducing risk.

Concerns about an architectural structure exposed to environmental elements have recently brought about an important study on the flow of air and water around the

Fig. 1.17 Example of CFD application to the flow of air and water around the Itsukushima Torii (Gate) located in the sea near Hiroshima, Japan. *(Courtesy of ANSYS-FLUENT.)*

Itsukushima Torii (Gate), a large 17 m high wooden structure, located in the sea near Hiroshima, Japan, as illustrated in Fig. 1.17. By using the dynamic mesh model to periodically move a wall so as to act as a wave generator and a volume of the fluid model to track the air-water interface to replicate the motion of the sea waves, significant insights into the flow characteristics around the architectural structure were realized. The ability to capture and better understand all the associated flow processes has certainly allowed environmental engineers the foundation whereby to better assess the environmental impacts on the exiting structure.

1.3.9 Metallurgy

Metallurgical processes involve materials flow in different forms (solid, liquids, gases, and their mixtures) from part or one of the equipment to another. Some of the typical processes include the extraction of metals from various types of iron ore to hot metal and from hot metal to steel; from copper concentrates to pure copper; and from aluminum scrap to pure aluminum or its alloys. These processes are generally complicated due to its multi-phase, high temperature, and highly reactive nature. The complex phenomena of flow, heat and mass transport, and heterogeneous chemical reactions play very important roles in determining the overall performance of large-scale reactors that accommodate such processes.

In recent times, CFD has been found very useful in studying various metallurgical processes and various aspects of a specific process. Through CFD, it has been shown to provide an insightful understanding of modification and optimization of the operation and design in existing processes and new process developments. It is worthwhile to mention that metallurgical processes are by nature challenging for CFD modeling since a lot of the phenomena have not yet been properly described or incorporated into the general CFD framework. Nevertheless, many new and significant developments of multi-physics

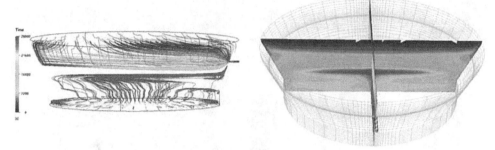

Fig. 1.18 Example of CFD application to predict molten iron flow (left, timeline) and carbon dissolution (right, concentration) in the blast furnace hearth. *(After Yang et al., 2006.)*

models are taking place to aptly simulate increasingly complicated industrial processes involving flow and transport of mass, momentum, energy, and chemical species in multi-phase and high-temperature reactive systems.

The use of CFD in simulating heterogeneous and slow dissolution of packed bed code particles is illustrated in Fig. 1.18. In this example the predicted flow pattern of molten iron and carbon dissolution in the blast furnace hearth is obtained through the population balance model coupled with the flow model. By tracking the local changes in size distribution and bed porosity, many aspects related to different initial coke size distributions, the use of inert versus reactive coke types, and the use of a mix of different code types can be investigated. Another example is the application of CFD for the redesigning of off-gas systems at different Ferro-Silicon production plants in order to assess operating conditions and the qualitative effects of different process parameters. A typical simulation is depicted in Fig. 1.19, where the colored surface represents the interface of the flames and the colors represent the flame temperature. Such prediction assists in better understanding the off-gas combustion, gas collection in the hood, transport and burn-out of soot and charge particles, carbon loss, heat transfer, and particle deposition on channel walls.

1.3.10 Nuclear Safety

During the last decade, the need for more accurate computational models for relevant safety analyzes of nuclear facilities has sparked an escalating interest in CFD to feasibly predict a number of important flow phenomena which otherwise may not have been possible through other simplified approaches. Some specific problems such as those arising from pressurized thermal shock, coolant mixing, and thermal stripping, as well as containment issues in nuclear reactors, have certainly motivated enormous research activities for the application of CFD to analyze such problems.

CFD calculations have been performed for coolant mixing in pressurized water reactors. This problem has significant interest to the nuclear community, particularly in

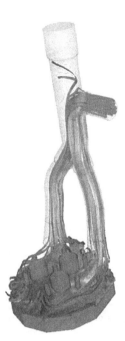

Fig. 1.19 Example of CFD application to the flow of gas in an off-gas system at Furnace 1 at Elkem Thamshavn, Norway. *(After Johansen, 2003.)*

attempting to understand the stationary and transient mixing of coolant in stream line break and boron dilution scenarios where the mixing phenomena have a tremendous impact on the economical operation and structural integrity of such facilities. Fig. 1.20 illustrates the case of a pump start-up due to a strong impulse-driven flow at the inlet nozzle where the horizontal part of the flow dominates in the downcomer in a pressurized water reactor. CFD predictions gave in–depth insights into the injection, which is distributed into two main jets where the maximum of the tracer concentration at the core inlet appears at the opposite part of the loop the tracer is injected. In relation to small break loss of coolant accident, slug flow that can occur in the cold leg of a pressurized water reactor is potentially hazardous. Strong oscillating pressure levels formed behind the liquid slugs can significantly affect and subsequently weaken the structure of the system. The calculated slug in a horizontal channel determined through the inhomogeneous multi-phase model is in good agreement with the experiment, as exemplified in Fig. 1.21. Although the entrainment of small bubbles in the front of the slug could not be observed, the characteristic of the slug front wave rolling over and breaking is clearly identified. Based on the impending need for analysis of long, large-scale, and transient problems, the capability of CFD for performing full containment analysis is still currently out of reach with the available computational power and resources. Nevertheless, the use of CFD in addressing the transport of gases in multi-compartment geometry remains

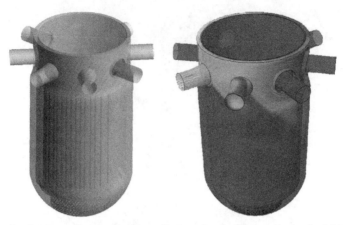

Fig. 1.20 Example of CFD application to the prediction of turbulent mixing in the ROCOM test facility during boron dilution transients (start-up of the first coolant pump). *(After Höhne et al., 2010.)*

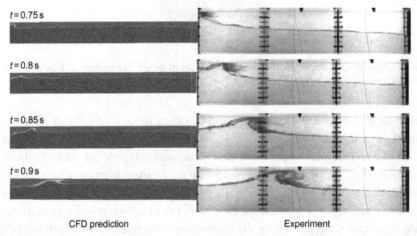

Fig. 1.21 Example of CFD application to predicting stratified flow in the cold leg of pressurized water reactor. *(After Höhne et al., 2010.)*

realistically achievable and CFD with a detailed sufficient mesh can give reliable answers on issues relevant to containment simulation.

1.3.11 Power Generation and Renewable Energy

In an increasingly competitive energy market utilities and equipment manufacturers are turning to CFD to provide a technological edge through a better understanding of the equipment and processes within these industries. Although traditional electric power generation sources are still widely used, a renewable power source such as wind energy is emerging as a potential alternative to power generation. To maximize return on

Fig. 1.22 Example of CFD application in predicting the airflow over a wind turbine in the vicinity of a proposed wind farm. *(Courtesy of Siemens PLM.)*

investment, CFD is being employed to optimize the turbine blades for generating constant power under varying wind conditions, as demonstrated by a typical three-dimensional simulation of the hydraulics in a complete wind turbine depicted in Fig. 1.22. CFD is also the only technology that has proven to accurately model wind farm resource distribution, especially for highly complex terrain with steep inclines, as shown in the same figure. Importantly, CFD has allowed the positioning of turbines throughout an area to achieve efficient wind capture and to minimize wake interaction. Wind resource assessments through CFD have allowed engineers to better study the economic viability of wind farms where accurate results are needed in order to reduce the financial risk.

The abundance of coal in many parts of the world such as Australia has made this raw mineral a popular fuel over many years in the power generation industries (see Fig. 1.23). Within these industries, the burning of coal in large furnaces has largely been associated with the release of environmentally harmful and hazardous pollutants such as CO, NOx, SOx, and mercury. To meet strict state, national, and international regulations, CFD simulations have greatly assisted engineers in identifying areas where deficiencies in design occurred. More importantly, the causes of these ineffective operations can be established in order to reduce emissions in a cost-effective manner. With this information, design improvements or operating strategies can be implemented to maintain satisfactory levels of residual carbon-in-ash as well as achieve better flexibility in plant operation. For example, a boiler operation is optimized through the reduction of fuel consumption, pollutant emissions, slagging, and degradation of the tubes, while downtime is reduced through prolonged component life. For clean power technologies, gasification offers the promise

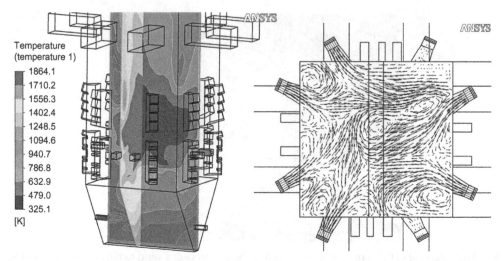

Fig. 1.23 Example of CFD simulation on an industrial pulverized coal furnace.

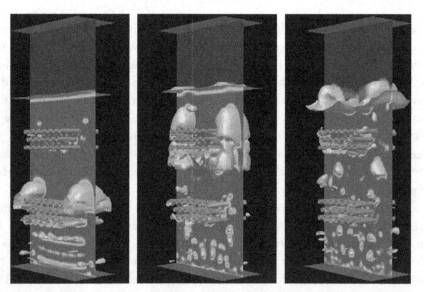

Fig. 1.24 Example of CFD simulation of bubbles in fluidized coal bed. *(Courtesy of ANSYS-FLUENT, Coal Gasification.)*

of increased generation efficiency and reduced emissions. Simulations can be performed to meet the modeling needs of this important technology, as typified by the combusting fluidized coal bed in addition to other relevant unit operations, as demonstrated in Fig. 1.24. By combining complete system simulations and detailed three-dimensional component analysis, engineers are at the liberty of making better design decisions and, ultimately, products of improved quality at low manufacturing costs.

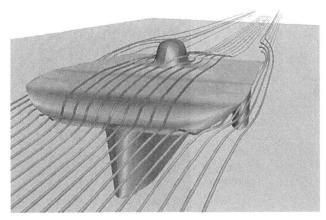

Fig. 1.25 Example of CFD simulation of airflow over SolarFox 3. *(Courtesy of http://www.ucl.ac.uk/solarfox/vehicles_main/SolarFox_3)*

SolarFox 3, a highly competitive, fast car, shown in Fig. 1.25, is the latest vehicle being developed by the University College London Team, which has been created to run solely on solar energy. CFD simulations have proven to be useful in designing the outer body, wheel fairings, and canopy. The shape of the vehicle has been specifically crafted to significantly reduce the drag, and the teardrop curvature of the wheel fairings and canopy contribute to its aerodynamic profile. All materials on the vehicle have been carefully chosen to ensure that the overall weight is kept to a minimum. By minimizing the weight, a low rolling resistance is thus achieved, which promotes a higher acceleration of the vehicle. The monocrystalline solar cells are divided into three arrays in accordance with the curvature of the upper body, which allows for maximum solar energy absorption.

1.3.12 Sports

Very recently, one of the most innovative uses of CFD in the sports arena is to "design" the optimum stroke to achieve peak propulsive performance for elite swimmers, as demonstrated by the example in Fig. 1.26. In aspiring to attain the extra edge USA Swimming, the national governing body for competitive swimming in the United States, commissioned CFD investigations to evaluate the flow around the hand and forearm of a swimmer during the propulsion phases of the freestyle and butterfly strokes. By applying CFD, steady-state lift and drag forces for the hand and arm are determined by applying sophisticated turbulence model and adaptive meshing. For this example, the force coefficients, evaluated at angles of attack ranging from $-15°$ to $195°$, and for various states of water turbulence, were found to compare very well with coefficients developed experimentally in a wind tunnel, a tow tank, and a flume. The successful comparison of the simulated results with experimental data thus validates the chosen CFD modeling technique.

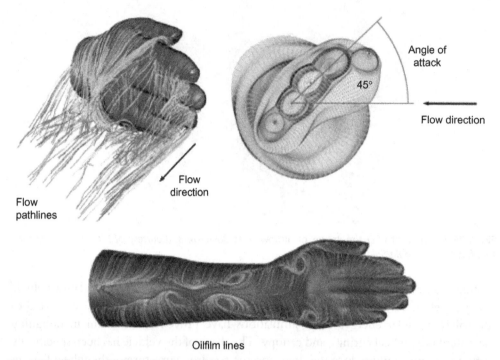

Fig. 1.26 Example of CFD application for designing the optimum stroke. *(Courtesy of USA Swimming, Honeywell Engines and Systems and ANSYS-FLUENT.)*

In the swimming literature the hand is often compared to an aerofoil; a polar diagram developed from CFD analyses shows otherwise, where the aerodynamic efficiency of the hand has been ascertained to be significantly less than that of an aerofoil of a similar aspect ratio. At velocities reached by swimmers in competitive racing, flow pathlines predicted through CFD reveal a highly three-dimensional flow with significant boundary layer separation. Large vortices form on the downstream side of the hand, and smaller tip vortices twirl off the fingertips in a manner similar to those flowing from aircraft wings or turbine blades. By undertaking further assessments, CFD analyses can further study the effects of arm and hand acceleration and deceleration on the swimmer's ability to generate propulsive forces.

Cycling presents another sports arena where CFD has played a major role in designing the best possible bike to shave crucial milliseconds off the athletes' times. The Sports Engineering Research Group (SERG) at the University of Sheffield, working closely with British Cycling, have provided the British cycling team with the means of constructing the fastest "legal" bike in the world. During the Olympic Games, they won a total of four medals – two gold, one silver, and one bronze – a fantastic achievement for British cyclists. Their winning edge was solely attributed to SERG's aerodynamic optimization performed on the bike. Through CFD, SERG redesigned the Olympic

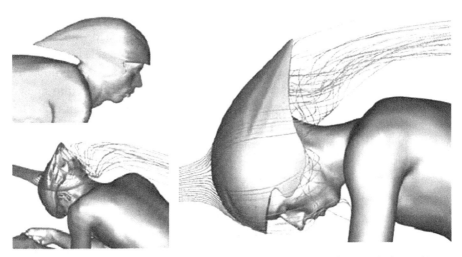

Fig. 1.27 Example of CFD simulation on designing the ultimate aerodynamic helmet. *(Courtesy of Sports Engineering at CSES, Sheffield Hallam University and ANSYS-FLUENT.)*

bikes' forks and handlebar arrangement. CFD also helped the team choose the most streamlined design for the aerodynamic helmet, as exemplified in Fig. 1.27. By better understanding the flow pathlines over the aerodynamic helmet, a range of helmet designs were manufactured to accommodate different head styles, achieving the ultimate cycling efficiency. SERG's recommendations ensured the British cycling team had the competitive advantage in their quest for gold.

1.4 THE FUTURE OF COMPUTATIONAL FLUID DYNAMICS

We are witnessing a renaissance of computer simulation technology in many industrial applications. This changing landscape is partly attributed to the rapid evolution of CFD techniques and models. For example, state-of-the-art models for simulating complex fluid mechanics problems such as jet flames, buoyant fires, and multi-phase and/or multi-component flows are now being progressively applied, especially through the availability of multi-purpose commercial CFD computer programs. The increasing usage of these types of codes in industries is a clear testimony whereby very demanding practical problems are now being analyzed by CFD. With decreasing hardware costs and rapid computing times, engineers are increasingly relying on this reliable yet easy-to-use CFD tool for delivering accurate results, as already described by the examples given in the previous sections.

Additionally, significant advances in virtual technology and electronic reporting are allowing engineers to swiftly view and interrogate the CFD predictions and make necessary assessments and judgments on a given engineering design. In industries CFD will eventually be so entrenched in the design process that new product development will evolve toward "zero prototype engineering." Such a conceptual design approach is not a mere flight of the imagination but rather a reality in the foreseeable future especially in the automotive industry. Looking ahead, full-vehicle CFD models with underhood, climate control and external aerodynamics will eventually be assembled into one comprehensive model to solve and analyze vehicle designs in hours instead of days. Time-dependent simulations will be routinely performed to investigate every possible design aspect. Other related "co-simulation" areas in ascertaining the structural integrity as well as aero-acoustics of the vehicle will also be computed concurrently alongside the CFD models. Engineering judgment will be consistently exercised on the spot through *real-time* assessments of proposed customized design simulations in selecting the most optimum vehicle.

In the area of research the advances in computational resources are establishing large eddy simulation (LES) as the preferred methodology for many turbulence investigations of fundamental fluid dynamics problems. Since all real-world flows are inherently unsteady, LES provides the means of obtaining such solutions and is gradually replacing traditional two-equation models in academic research. The demand for LES modeling is steadily growing. LES has made significant in-roads especially for single-phase fluid flows. In combustion research LES has also gained much respectability, particularly in capturing the complex flame characteristics due to better accommodating the unsteadiness of the large-scale turbulence structure affecting the combustion process. Although much effort has been focused on developing more robust CFD models to predict complicated multi-phase physics involving gas-liquid, gas-solid, liquid-solid, or gas-liquid-solid flows, LES remains in its infant stage of application to cater to these types of flow problems. Instead, two-equation turbulence models are still very prevalent in order to account for the turbulence within such flows. LES may be adopted as the preferred turbulence model for multi-phase flows in the future but in the meantime, the immediate need is to further develop more sophisticated two-equation turbulence models to resolve such kinds of flows.

Based on current computational resources, numerical calculations performed through LES can be rather long and arduous due to a large number of grid nodal points required for computations. However, the ever-escalating trend of fast computing in the foreseeable future will permit such calculations to be performed at a more regular frequency. Also, with the model gradually moving away from the confines of academic research into the industry environment, it is not entirely surprising that LES will eventually become a household methodology for investigating many physical aspects of practical industrial flows. There are many challenging prospects for the use of CFD in industry and research. Perhaps, we can ponder toward the day when all turbulence flow problems can be resolved directly without the consideration of any models. Direct numerical simulation

(DNS) of turbulent flows in academia research and possibly in some facets of industrial applications may well become a *distinct certainty* instead of a *distinct impossibility*.

1.5 SUMMARY

The array of examples illustrated in this introductory chapter clearly depicts how computational fluid dynamics has evolved through decades through the rigorous development of numerical techniques. Despite the long-standing usage within academic circles, computational fluid dynamics, better known by its acronym CFD, is flourishing in many industrial sectors. This unprecedented occurrence is partly due to the increasing availability and accessibility of multi-purpose commercial computer programs. These codes have certainly fueled much of the swelling demand for industrial CFD applications. CFD research is also at the crossroads of progressive usage. Escalating computational power has permitted the ability to incorporate more sophisticated models to better resolve increasingly complex flow transport phenomena. As multi-purpose commercial codes become more commonplace in many educational institutions, they are revolutionizing the way by which CFD is being taught. Lately, simpler versions of these codes are being developed into dedicated educational tools for teaching and learning purposes. They are greatly assisting potential and novice users to learn CFD without them *reinventing the wheel* (i.e. duplicating their efforts in writing and debugging their own computational codes).

CFD computation usually involves the generation of a set of numbers or digits that will hopefully provide a realistic approximation of a real-life fluid system. Nevertheless, the main outcome of any CFD exercise is that the reader acquires an improved understanding of the flow behavior for the system in question. The main ingredients of learning CFD are to gain experience and a thorough understanding of the flow physics and the fundamentals of the numerical techniques and models. Additionally, practical guidelines for good operating practice are ever more needed to increase competency in the use of this powerful tool. The intention of this book is therefore specifically written to address all of the aforementioned issues and to better equip the reader with the necessary background material for a good understanding of the internal workings inside a CFD code and its successful operation.

In the next chapter we begin by discovering and coming to grips with how a CFD code works through the various elements that constitute a complete CFD analysis. Since a greater emphasis has been placed within this book on the practicality of CFD, many practical steps and aspects exemplifying the important elements of CFD analysis will be highlighted. The primary aim is to expose the reader to numerous operations that occur behind many existing commercial and shareware computer codes. With the surging demand for CFD to solve a spectrum of fluid-related problems, the aptitude to properly employ and apply this methodology has never been so important in the current climate.

REVIEW QUESTIONS

1.1 From which industry has CFD emerged?

1.2 Which three disciplines is CFD derived from?

1.3 Traditionally, CFD has been used to solve aerospace and automotive engineering applications such as drag and lift for airplanes and cars. What examples can you think of where CFD is being used within non-traditional fluid engineering applications?

1.4 What are some of the advantages of using CFD?

1.5 What are the limitations and disadvantages of using CFD?

1.6 What CFD measurements can be obtained to assist the designs for safety and comfort in passenger airplanes?

1.7 What detailed results can be achieved in the modeling of F18 jets that are difficult to achieve in experiments?

1.8 How is CFD being used as a research tool, a design tool, and an educational tool in academic fields, such as thermal fluids?

1.9 What aspect of CFD provides an advantage to allow a better understanding of fluid flow and heat transfer?

1.10 How can CFD reduce development costs in time to market during the design stage?

1.11 How would CFD be used to reduce drag in automotive engineering?

1.12 Why would parametric studies using CFD be more effective than large-scale wind tunnel experiments?

1.13 Can CFD be used to investigate piston engines in automotive engineering? How would this be possible, computationally?

1.14 The biomedical science field is turning to CFD to resolve flows within the human airway and vascular systems. What advantages does CFD hold over experiments in obtaining these numerical results?

1.15 What can CFD provide for evaluating surgical reconstructions for blood flows in arteries?

1.16 What other technologies are involved so that CFD can be performed inside the human body?

1.17 What details can CFD capture in the simulation of hydro-cyclones, a process commonly used in the minerals industry?

1.18 What types of flow can CFD investigate in the separation process in minerals processing?

1.19 How is CFD being used in the civil and environmental industry?

1.20 How is CFD being employed in understanding metallurgical processes?

1.21 In metallurgy studies CFD is able to model many phases at once (solid, liquid, gas), known as multi-phase modeling. Can you think of other multi-phase problems that CFD could possibly be used for?

1.22 What kind of problems has CFD been applied to in nuclear safety analyses?

1.23 What type of two applications is CFD being used for in the power generation industry? What kind of data is collected and how is this useful in increasing the efficiency in power generation?

1.24 What important parameters would you look for when modeling wind farms using CFD?

1.25 How can CFD assist swimmers to improve their swimming strokes?

1.26 What competitive edge can CFD give to a cycling team?

1.27 In the future to what extent will CFD be involved in the product development process in manufacturing?

1.28 What is the future of CFD?

CHAPTER 2

CFD Solution Procedure – A Beginning

2.1 INTRODUCTION

With the widespread availability and ease of accessibility of many commercial, shareware, and in-house computer codes, today's CFD users will most probably acquire their necessary skills and knowledge rather differently from yesterday's CFD users. Such codes have become very prevalent, prominent, and widely used in many fields of academia, industrial sectors, and major research centers. It is therefore not surprising that potential or novice CFD users would be more inclined nowadays to resort to these available codes for learning CFD. The evolution of commercial codes toward more user-friendly environments and applications certainly reflects such demand.

Applying ready-developed CFD codes has certain advantages. Potential or novice users can initially treat these codes as black boxes (from the perspective of a student setting and not in a professional setting) and operate them with the main intentions of just practicing and familiarizing the many important features of these codes. Without any prior basic CFD knowledge, a first-time user can perform the necessary operations that involve setting up a fluid problem, solving the numerical problem, and managing some graphical representation of the results attained. Although the process can be regarded as very mechanically and laboriously driven, the exposure for a first-time user to CFD is not an intimidating or daunting experience. Interested users are generally curious and they are usually motivated to further investigate the mysterious aspects that are contained within these black boxes.

Fostering and cultivating the keenness and eagerness of interested users in learning CFD are important ingredients for maturing these users as eventual expert CFD users. The acquired skills and expertise fulfill the requirements of being competent and knowledgeable in analyzing and interpreting the computational solution, whether it is physically realistic or numerically accurate. There have been many pitfalls in the past where the over-exposure of hardcore mathematical formulations of the fluid flow equations and their numerical representations has certainly caused much angst among first-time users. More often than not, these users have become rather frustrated, disillusioned, and lamented at the prospect of learning CFD. However, CFD lies in the core comprehension of the fundamental principles concerning the fluid flow processes and analyzing the computational solutions. The integration of practical experience with theoretical knowledge to learning CFD is therefore tantamount. The practical side of CFD, in arousing and sustaining the level of enthusiasm of new users, shares the same foundational

Computational Fluid Dynamics
https://doi.org/10.1016/B978-0-323-93938-6.00004-X

importance with the theoretical component in equipping these new users with the required level of CFD knowledge for better tackling fluid flow problems.

This book attempts to reconcile the two diametrically opposing approaches to teaching CFD. We begin by introducing the reader's perspective as a first-time user to the salient features that are common in many commercial and possibly in some shareware CFD codes. These codes are usually structured around the robust numerical algorithms that can tackle fluid flow problems. In order that easy access is granted to their solving power, almost all current commercial and possibly some shareware CFD packages include user-friendly Graphical User Interface (GUI) applications and environments to input problem parameters and to examine the computed results. Hence these codes provide a complete CFD analysis consisting of three main elements:

- *pre-processor*
- *solver*
- *post-processor*

This chapter, in retrospect, is intended to address the many practical steps and aspects that are exemplified in these three important elements in the attempt to uncover the numerous operations beneath many of the shareware and commercial CFD codes. Fig. 2.1 presents a framework that illustrates the interconnectivity of the three aforementioned

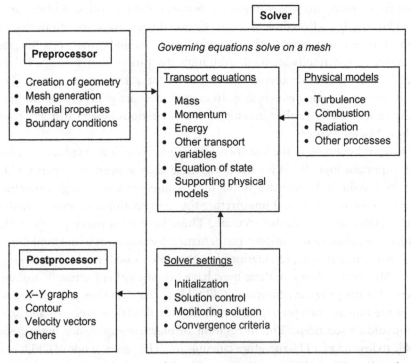

Fig. 2.1 The interconnectivity functions of the three main elements within a CFD analysis framework.

elements within the CFD analysis. The functions of these three elements will be examined in more detail in the subsequent sections within this chapter.

2.1.1 Shareware CFD

Today's CFD users have the luxury to download possible shareware or freeware CFD codes from certain websites through the internet. Under the website of http://www.cfd-online.com/Links/soft.htm/, the reader is offered, under the software category, a catalog of CFD codes listed as shareware products through the *shareware, freeware* option link.

Nevertheless, first-time CFD users may wish to search the internet to gain immediate access to an interactive CFD code. First-time users may be required to register themselves in order to freely access the interactive CFD code. The website provides simple CFD flow problems for first-time users to solve and allows a colorful graphical representation of the computed results. More advanced users may wish to visit an excellent web portal to acquire all the necessary aspects related to CFD at http://www.cfd-online.com/.

2.1.2 Commercial CFD

Under the software category of the web link http://www.cfd-online.com/Links/soft.htm/, the reader may wish to uncover the list of commercial codes that are currently available in the market through the *commercial* option link. Table 2.1 presents the internet links of some of the popular commercial CFD packages; it is by no means an exhaustive list. Commercial CFD vendors have invested much time, effort, and expense in the concerted development of user-friendly GUIs to make CFD very accessible and facilitate its usage and application in handling very complex fluid flow problems. We present some typical GUIs that a first-time user may experience and encounter in the course of employing a number of the commercial CFD packages tabulated in Table 2.1. The two interface fronts chosen are those that have been developed by ANSYS-CFX and ANSYS-FLUENT and they are illustrated in Figs. 2.2 and 2.3. These interface fronts given below are not to be construed as an endorsement of these specific products but

Table 2.1 Internet Links to Some Popular Commercial CFD Packages.

Developer	Code	Distributor Web Address
ANSYS, Inc.	CFX	http://www.ansys.com/
ANSYS, Inc.	FLUENT	http://www.fluent.com/
CD–adapco	STAR–CCM+	http://www.cd–adapco.com/
CHAM	PHOENICS	http://www.cham.co.uk/
COMSOL, Inc.	COMSOL	http://www.comsol.com/
ESI Group	CFD–ACE+	http://www.esi–group.com/
Flow Science	FLOW3D	http://www.flow3d.com/

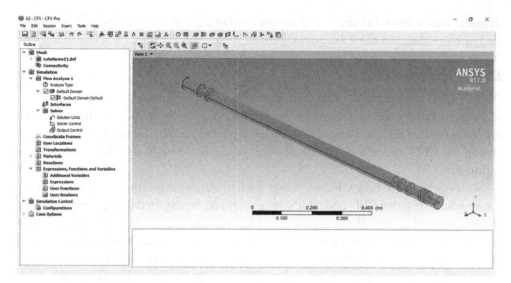

Fig. 2.2 A typical ANSYS-CFX graphical user interface.

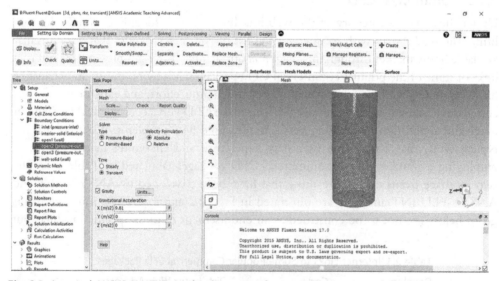

Fig. 2.3 A typical ANSYS-FLUENT graphical user interface.

we simply treat them as GUI examples for illustration purposes. Other commercial packages tabulated in Table 2.1 would most likely have similar user-friendly interface fronts of appealing graphical appearances and features such as those possessed by the two afore-mentioned commercial packages. While using these packages, the user will attempt to navigate through these interface fronts by accomplishing a certain number of basic steps to set up and solve the flow problem and obtain a CFD solution. These steps will be described and discussed in the next section.

2.2 PROBLEM SETUP – PRE-PROCESS

2.2.1 Creation of Geometry – Step 1

The first step in any CFD analysis is the definition and creation of the geometry of the flow region, that is, the *computational domain* for the CFD calculations. Let us consider two flow cases: a fluid flowing between two stationary parallel plates and a fluid passing through two cylinders in an open surrounding. It is important that the reader should always acknowledge the real physical flow representation of the problem that is to be solved, as demonstrated by the respective *physical domains* in Figs. 2.4 and 2.5. For the purpose of illustration, we designate the former and latter cases as Case 1 and Case 2, respectively. We shall assume that the two cylinders in Case 2 have the same length as the width W of the overall domain. We shall also assume that the width W of both flows within the three-dimensional *physical domains* is sufficiently large in order that the flows can be taken to be invariant along this transverse direction. Hence Case 1 and Case 2 can simply be considered two-dimensional *computational domains* for CFD calculations. These two flow cases will be repeatedly taken as illustrative examples to demonstrate the various basic steps that are involved in the pre-process, solver, and post-process stages.

There are certain distinct dissimilarities in the nature of these two flow problems. Case 1 represents an *internal flow* problem, while Case 2 is taken typically as an *external flow* scenario. In both cases the fluid enters from the left boundary and exits at the right boundary of the *computational domains*. The main difference between these two flows is

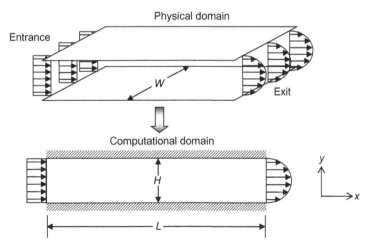

Fig. 2.4 Case 1: fluid flowing between two stationary parallel plates.

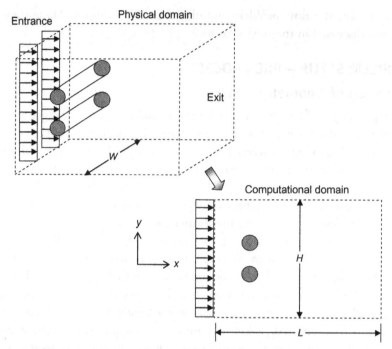

Fig. 2.5 Case 2: fluid passing over two cylinders in an open surrounding.

accentuated by the top and bottom boundaries, which brings about the classification of *internal* and *external flows*. In Case 1 the fluid flow is bounded within a domain of rigid walls as represented by the horizontal external walls of the two stationary parallel plates. That is not the characteristic of the fluid flow in Case 2, as the fluid can either take the inflow or outflow boundary conditions at the top and bottom boundaries.

One important aspect that the reader should always take note of in the creation of the geometry for CFD calculations is to allow the flow dynamics to be sufficiently developed across the length L of these computational domains. For Case 1, we require the flow to be *fully developed* as it exits the domain. The physical interpretation and meaning of the concept of *fully developed* will be expounded in Chapter 3. For Case 2, we require to encapsulate the occurrence of complex *wake-making* development that persists behind the two cylinders as the flow passes over these cylinders. This phenomenon is analogous to the formation and shedding of vortices that are commonly experienced for flow past a cylinder. In this particular case the top and bottom boundary effects may influence the flow passing over these two cylinders; the height H of the domain needs to be prescribed at a distance to sufficiently remove any of these boundary effects on the fluid flow surrounding the two cylinders while still being manageable for CFD calculations. More practical guidance on this issue will be addressed in Chapter 7.

2.2.2 Mesh Generation – Step 2

The second step, that is, mesh generation, constitutes one of the most important steps during the pre-process stage after the definition of the domain geometry. CFD requires the sub-division of the domain into a number of smaller, non-overlapping sub-domains in order to solve the flow physics within the domain geometry that has been created; this results in the generation of a *mesh* (or grid) of *cells* (elements or control volumes) over-laying the whole domain geometry. The essential fluid flows that are described in each of these cells are usually solved numerically so that the discrete values of the flow properties such as the velocity, pressure, temperature, and other transport parameters of interest are determined. This yields the CFD solution to the flow problem that is being solved. The accuracy of a CFD solution is strongly influenced by the number of cells in the mesh within the computational domain. While, in general, increasing the number of cells will improve the accuracy of the solutions, it is also influenced by many other factors, such as the type of mesh, the order of accuracy of the numerical method, and the adequacy of the techniques chosen to the physics of the problem. However, the accuracy of a solution is strongly dependent on the imposed limitations dominated by the computational costs and calculation turnover times. Issues concerning numerical accuracy and computational efficiency will be explored in Chapter 6.

The majority of the time spent in industry on a CFD project is usually devoted to successfully generating a mesh for the domain geometry. Most commercial CFD codes have developed their own dedicated CAD-style interface and/or facilities to import data from solid modeler packages such as PARASOLID, PRO/ENGINEER, SOLID EDGE, SOLIDWORKS, and UNIGRAPHICS to maximize productivity and allow the ease for geometry creation. The mesh for the created geometry can be subsequently realized through in-built powerful mesh generators that reside within these respective codes. Nevertheless, the reader should be well aware that it is still up to the skills of the CFD user to design a mesh that is a suitable compromise between the desired accuracy and solution cost. The generation of an appropriate mesh for CFD calculations will be further discussed in Chapter 7.

The mesh generation step is illustrated for Case 1 and Case 2. For relatively simple geometries such as the created geometry domain for Case 1, an overlay mesh of *structured* cells that generally comprises a regular distribution of rectangular cells can be readily realized. Fig. 2.6 shows a mesh of 20 (*L*) × 20 (*H*) cells resulting in a total number of 400 cells

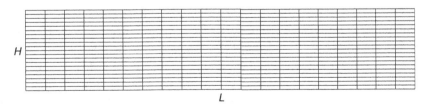

Fig. 2.6 Structured meshing for fluid flowing between two stationary parallel plates.

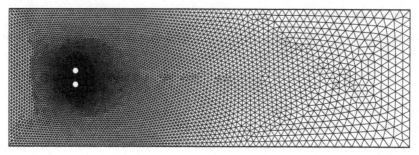

Fig. 2.7 Unstructured meshing for fluid passing over two cylinders in an open surrounding.

allocated for the Case 1 geometry. For more complex geometries, the meshing by triangular cells allows the flexibility of mesh generation for geometries having complicated shape boundaries. Fig. 2.7 illustrates a typical distribution of triangular cells within the computational domain for Case 2 geometry with a mesh totaling 16,637 cells mapping the whole flow domain.

The specific type of meshes represented in Case 1 and Case 2 domain geometries should be construed as just illustrative examples for the intended purpose of demonstrating the mesh generation step. It is not unprecedented and rather legitimate to interchangeably use an *unstructured* mesh in place of a *structured* mesh for Case 1 and vice versa for Case 2. It is also not uncommon and in some practices a requirement to embrace a combination of *structured* and *unstructured* meshes for more realistic simulations within flow domains that may include many inherent complex geometrical intricacies. More practical guidance on mesh generation will be provided in Chapter 7.

2.2.3 Selection of Physics and Fluid Properties – Step 3

Many industrial CFD flow problems may require solutions to very complex physical flow processes such as the accommodation of complicated chemical reactions in combusting fluid flows. The inclusion of combustion and possibly radiation models in the CFD calculations are generally prerequisites to the successful modeling of these types of flows. Combustion and radiation processes have the tendency to strongly influence local and global heat transport, which consequently affects the overall fluid dynamics within the flow domain. It is therefore imperative that a CFD user carefully identifies the underlying flow physics that is unique to the particular fluid flow system.

For clarity and ease of reference, a flowchart highlighting the various flow physics that may be encountered within the framework of CFD and heat transfer processes is presented in Fig. 2.8. Under the main banner of *Computational Fluid Dynamics & Heat Transfer*, a CFD user declares initially whether simulations of the fluid flow system are to be attained for *transient/unsteady* or *steady* solutions. He/she subsequently defines which class of fluids the flows may belong to: *inviscid* or *viscous*. *Inviscid* fluid flows are generally

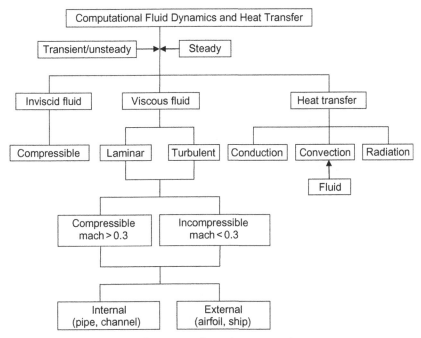

Fig. 2.8 A flowchart encapsulating the various flow physics in CFD.

compressible and the consideration of fluid compressibility in the flow physics can usually be handled through the *Panel Method*. *Viscous* fluid flows can, however, exist in their *laminar* or *turbulent* state. Under these two flow conditions, prior knowledge of whether the fluids are *compressible* or *incompressible* is required. The classification of *internal* and *external* flows for viscous fluids allows the user to appropriately treat these types of flow problems in a manner that has been discussed in Section 2.2.1. Also, the transport of heat may contribute significantly to the fluid flow process, which comprises three heat transfer modes: *conduction*, *convection*, and *radiation*. For convection, the dominant mode of *heat transfer* will more likely be driven by the *convective* fluid flow rather than by other modes of *conduction* and *radiation*. Nevertheless, there are circumstances where *radiation* and *convection* can co-exist and dominate the *heat transfer* especially in the expansion of fires.

Let us review the various basic steps that have been described thus far for the pre-process stage. In Step 1 we have created two-dimensional computational domains for Case 1 and Case 2. Step 2 illustrates the generation of an overlay mesh of structured cells for Case 1 and unstructured cells of different sizes for Case 2. Step 3, the current step in the pre-process stage, consists of the identification and formulation of the flow problems in terms of the physical phenomena. A CFD user must select the appropriate flow physics, as discussed above, in order to correctly simulate the characteristics of the fluid flow. For simplicity, a steady CFD solution is considered for Case 1 and Case 2 and the fluid flows in both cases can be taken to be viscous, laminar, incompressible, and isothermal (without heat transfer). It is rather important that setting up the flow physics is also accompanied by ascertaining what fluid is used within the flow domain. For example, air or water has its own unique fluid and thermal properties. Appropriate properties are therefore required to be assigned to correctly define the particular fluid in this pre-process step. Fluid properties such as *density* and *viscosity* (dynamic) can usually be imposed through the GUIs in many commercial CFD codes.

2.2.4 Specification of Boundary Conditions – Step 4

The complex nature of many fluid flow behaviors has important implications in which boundary conditions are prescribed for the flow problem. A CFD user needs to define appropriate conditions that mimic the real physical representation of the fluid flow into a solvable CFD problem.

The fourth step in the pre-process stage deals with the specification of permissible boundary conditions that are available for impending simulations. Evidently, where there exist *inflow* and *outflow* boundaries within the flow domain, suitable fluid flow boundary conditions are required to accommodate the fluid behavior entering and leaving the flow domain. The flow domain may also be bounded by *open* boundaries. Although the intricacies of *open* boundary conditions are still subject to much theoretical debate, this boundary condition remains the simplest and cheapest form to prescribe when compared with other theoretically more satisfying selections in CFD. Appropriate boundary conditions are also required to be assigned for external stationary *solid wall* boundaries that bound the flow geometry and the surrounding walls of possible internal obstacles within the flow domain.

We illustrate the applications of the aforementioned boundary conditions to Case 1 and Case 2 for CFD calculations. Schematic descriptions of the boundary conditions are demonstrated in Fig. 2.9 for Case 1 and Fig. 2.10 for Case 2.

In Subsection 2.2.3 the considerations of a viscous fluid, laminar, incompressible, and isothermal assumed for the purpose of illustration in both Case 1 and Case 2 require the prescription of only the fluid velocity on all the bounding walls of the computational domain. By definition, the velocities are zero for the external stationary *solid walls* bounding the flow domain in Case 1 (see Fig. 2.9) and the two cylindrical motionless *solid walls* in Case 2 (see Fig. 2.10). For the *inflow* boundary conditions, the user is required to ascertain the inlet fluid velocity in order to stipulate the fluid entering into both of these flow domains. At the *outflow* boundaries indicating the fluid departure, only one outlet condition, typically a specified relative pressure, is imposed. A far-field flow boundary condition can be imposed for the open boundaries that either applies the inflow or outflow boundary conditions. Care should always be exercised in handling these *open* boundaries, where they have to be defined far enough away from the region of interest within the solution domain in order to obtain physically meaningful results.

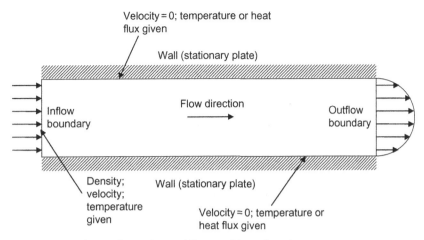

Fig. 2.9 Boundary conditions for an internal flow problem: Case 1.

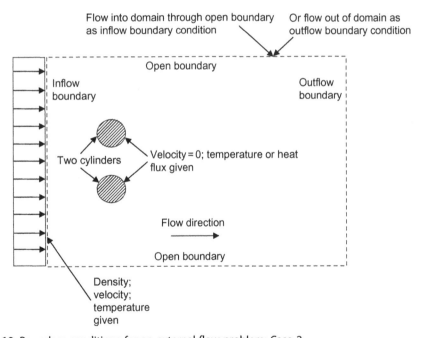

Fig. 2.10 Boundary conditions for an external flow problem: Case 2.

Nevertheless, for a general fluid flow case where the transport of heat is dominant within the flow domain, as may be experienced by the flows in Case 1 and Case 2, the user is obliged to prescribe the surface boundaries of the flow domains and obstacles either by the given temperature or heat flux distributions, as shown in Figs. 2.9 and 2.10. In compressible-like flows such as buoyancy-driven flows and combusting flows such as fire where there exists a strong density variation with temperature, the density emerges as part of the solution everywhere within the flow domains. The density corresponding to the particular fluid flowing into these domains at the inflow boundaries needs to be specified and is usually determined directly from the imposed temperature and pressure. However, no boundary conditions are required for the density at the outflow boundaries since the fluid density replicates the outgoing fluid characteristics through these boundaries. For the open boundaries such as in Case 2, the density can either be represented by the inflow or outflow boundary conditions.

General purpose CFD codes also often allow the prescription of inflow and outflow pressure or mass flow rate boundary conditions. By setting fixed pressure values, sources and sinks of mass placed at the boundaries ensure the correct mass flow into and out of the solution zone across the constant pressure boundaries. It is also feasible to allocate directly sources and sinks of mass at the boundaries by the mass flow rates instead of pressures to retain the overall mass balance for the flow domain. To take advantage of special geometrical features that the solution region may possess, *symmetric* and *cyclic* boundary conditions can be employed to speed up the computations and enhance the computational accuracy by placing an additional number of cells to the simplified geometry. Fig. 2.11 shows the boundary geometry for which symmetry boundary condition can be imposed for Case 1, while Fig. 2.12 illustrates a generic geometry where cyclic boundary conditions may be useful. The physical meanings of the various boundary conditions that have been described herein to close the fluid flow system will be expounded in Chapter 3.

Thus far, we have concentrated on the application of various boundary conditions that only pertain to *subsonic* fluid flows, that is, flows below the speed of sound. There is another broad range of fluid flows that can possibly achieve speeds near and above the speed of sound, which are generally classified as *hypersonic, transonic,* and *supersonic* fluid flows. The prevalence of such complex flows is evident in many aerodynamic investigations. At these high speeds, the viscous regions in the flow are usually exceedingly thin and the enveloping flow in a large part of the solution domain behaves effectively inviscid. Applying boundary conditions predominantly for viscous flow to a largely

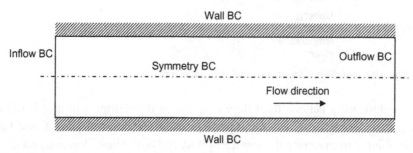

Fig. 2.11 Definition of symmetry boundary condition for Case 1.

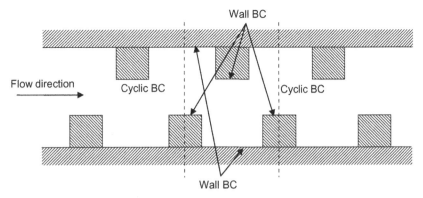

Fig. 2.12 Definition of cyclic boundary condition for a generic geometry.

dominated inviscid flow may yield unphysical behavior of the shockwaves propagating through the flow domain. Physical description and application of boundary conditions for *supersonic* flows will be covered in Chapters 3 and 8. For additional details, interested readers are advised to refer to Fletcher (1991) and Anderson (1995) for more discussions and prescriptions of the boundary conditions for fluid flows that are *hypersonic* and *transonic*. The majority of the commercial codes make claims and boast the capability to resolve all fluid flow regimes: *subsonic, transonic,* and/or *supersonic*. They usually, however, perform most effectively below the speed of sound, that is, *subsonic,* as a consequence of the boundary conditions outlined above.

2.3 NUMERICAL SOLUTION – CFD SOLVER

The appropriate usage of either an in-house or a commercial CFD code commands the core understanding of the underlying numerical aspects inside the *CFD solver*. This section focuses on the treatment of the solver element. A *CFD solver* can usually be described and envisaged by the solution procedure presented in Fig. 2.13. The prerequisite processes in the solution procedure that have implications on the computational solution are: *initialization, solution control, monitoring solution, CFD calculation,* and *checking for convergence*. A CFD user, whether applying in-house or commercial codes, needs to gain the necessary insights and knowledge pertaining to the workings of these prerequisite processes in order to skillfully utilize the many solver features and better navigate the underlying "black box" operations that reside in many of these codes.

2.3.1 Initialization and Solution Control – Step 5

The fifth step of the CFD analysis encompasses two prerequisite processes within the CFD solver, which are *initialization* and *solution control*.

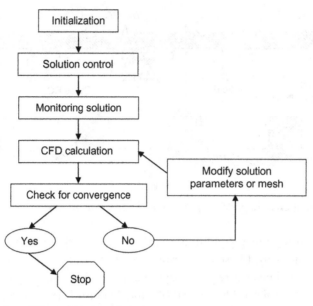

Fig. 2.13 An overview of the solution procedure.

Firstly, the underlying physical phenomena in real fluid flows that are generally complex and non-linear within such flows usually require the treatment of the key phenomena to be resolved through an iterative solution approach. The iterative procedure generally involves all the discrete values of the flow properties such as the velocity, pressure, temperature, and other transport parameters of interest to be *initialized* before calculating a solution. In theory, initial conditions can be purely arbitrary. However, in practice, there are certain advantages to imposing initial conditions *intelligently*. Good initial conditions are crucial to the iterative procedure; two reasons why the reader as a CFD user should undertake the appropriate selection of initial conditions are:

- If the initial conditions are close to the final steady-state solution, the iterative procedure will converge quicker and results will be calculated in a shorter computational time.
- If the initial conditions are far away from reality, the computations will result in longer computational efforts to reach the desired convergence. Also, improper initial conditions may lead to the iterative procedure misbehaving and possibly "blowing up" or diverging.

Secondly, the setting up of appropriate parameters in the *solution control* usually entails the specification of appropriate *discretization (interpolation) schemes* and the selection of suitable *iterative solvers*.

Almost all well-established and thoroughly validated general-purpose commercial codes adopt the *finite volume method* (Chapter 5) as their standard numerical solution technique. The algebraic forms of equations governing the fluid flow within these codes are usually approximated by the application of finite-difference-type approximations to a finite volume cell in space. At each face of the cell volume, surface fluxes of the transport variables that are required can be determined through different interpolation schemes. Some of the common interpolation schemes are: *First-Order Upwind, Second-Order Upwind, Second-Order Central,* and *Quadratic Upstream Interpolation Convective Kinetics (QUICK).* The inability of the *Central* scheme to identify the flow direction results in the formulation of other schemes such as *Upwind* or *QUICK,* which means the interpolation methods are biased on the upstream occurrence of the fluid flow and thus account for the flow direction. The choice of a higher-order interpolation scheme may achieve the desired level of accuracy that needs to be evaluated at the cell faces. Solution procedures such as SIMPLE, SIMPLEC, or PISO algorithm are popular in many commercial codes. The method is geared toward guaranteeing correct linkage between the pressure and velocity, which predominantly accounts for the mass conservation within the flow domain. At present, our intention is not to dwell on the many underlying numerical properties but rather to present simply the choice of interpolation schemes and pressure-velocity coupling methods that are offered as standard options in many CFD codes. It is imperative that some background knowledge on the appropriate selection of these options is acquired before any CFD calculation is performed. More discussions and practical guidance on the many numerical issues pertaining to the application of interpolation schemes and pressure-velocity coupling methods will be provided in Chapters 5 and 7.

Iterative solvers, so-called number-crunching engines for numerical calculations, are employed to resolve the algebraic equations. Nowadays, robust solvers such as the algebraic multi-grid (AMG) algorithm and conjugate gradient methods are standard features in many commercial codes. Other popular solvers such as strongly implicit procedure (SIP) or Stone's method and TDMA line-by-line procedure are also prominently employed by many users in the CFD community. More descriptions of these solvers are provided in Chapter 6. Solver controlling parameters that exist inside these solvers in commercial codes tend to be optimally configured for efficient matrix calculations. The desired performance can usually be achieved through the default settings prescribed within these codes.

Step 5 hereby completes the specification of various relevant physical features pertaining to the intended fluid flow process of a CFD problem. To exemplify the *CFD calculation* process of the solution procedure in Fig. 2.13, Case 1 and Case 2 fluid flow problems described in Section 2.1.1 are revisited to demonstrate the iterative operations that are typically visualized through the ANSYS-CFX and ANSYS-FLUENT GUIs in the next proceeding step below. We conveniently employ the default settings of the interpolation schemes and pressure-velocity coupling methods and retain the optimal controlling parameters that govern the performance of the iterative solvers to simply illustrate the important features of the numerical computations.

2.3.2 Monitoring Convergence – Step 6

The sixth step of the CFD solver involves the interlinking operations of three separate processes: *solution monitoring, CFD calculation,* and *checking for convergence.* Two aspects that characterize a successful CFD computational solution are *convergence* (Chapters 5 and 6) of the iterative process and *grid independence* (Chapters 6).

Convergence can usually be assessed by progressively tracking the *imbalances* that are accumulated by the advancement of the numerical calculations of the algebraic equations through each iteration step. These *imbalances* measure the overall conservation of the flow properties; they are also commonly known as the so-called *residuals* (Chapters 5 and 6) that are generally viewed through commercial code GUIs. Examples of these GUIs by ANSYS-CFX and ANSYS-FLUENT that represent the downward trends of the *residuals* for Case 1 are illustrated in Figs. 2.14 and 2.15, respectively. These downward tendencies clearly point to the *continual removal* as opposed to *possible accumulation* of any unwanted *imbalances,* thereby causing the iterative process to *converge* rather than to *diverge.* A *converged* solution is achieved when the residuals fall below some *convergence criteria* or *tolerance* (Chapter 6) that are preset inside the solver controlling parameters of the iterative solvers. We indicate different default settings of the *tolerance values* (Chapter 6) that are used by ANSYS-CFX and ANSYS-FLUENT to terminate the iterative process. They are prescribed at values of 1×10^{-4} and 1×10^{-3}, respectively, as shown in Figs. 2.14 and 2.15. A converged solution is obtained rather quickly from either ANSYS-CFX or ANSYS-FLUENT because of the sheer nature of the flow in Case 1 being rather straightforward. Besides examining the *residuals,* the user may use other monitoring variables such as the lift, drag, or moment force to ascertain the convergence of the numerical computations. Fig. 2.16 shows the lift coefficient (Cl) history for Case 2 employing the commercial code ANSYS-FLUENT. It is not surprising that the fluid

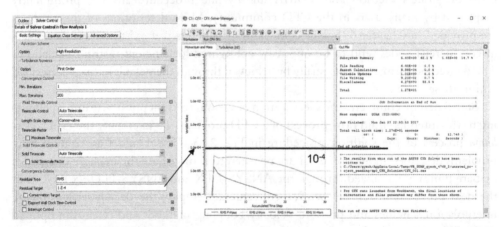

Fig. 2.14 Typical ANSYS-CFX GUIs for monitoring convergence corresponding to the prescribed convergence criteria.

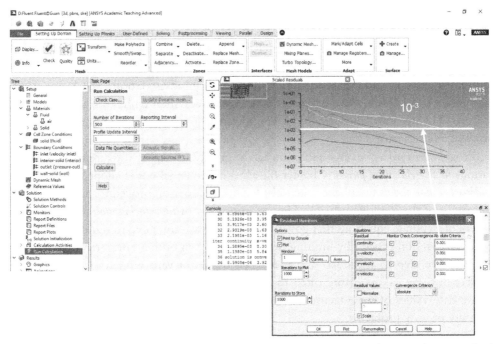

Fig. 2.15 Typical FLUENT GUIs for monitoring convergence corresponding to the prescribed convergence criteria.

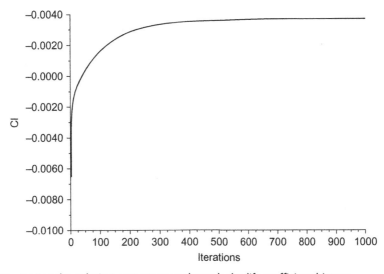

Fig. 2.16 Monitoring the solution convergence through the lift coefficient history.

flow process in Case 2, being more complicated than Case 1, requires more iterations to reach *convergence*. No appreciable change to the lift coefficient is observed after 700 iterative steps and the computational solution is thus deemed to be converged at about 1000 iterations as the lift coefficient plateaus to a fixed value of around 0.0035. In addition to monitoring residual and variable histories, the user is well advised to also check the overall mass balance and possibly heat balance for the fluid flow system within the computational domain. The net imbalance should be minimized as low as possible to ensure adequate property conservation.

Progress toward a converged solution can be greatly assisted by the careful selection of various *under-relaxation factors* (Chapter 6). Most commercial codes adopt some form of *under-relaxation factors* to enhance the *stability* of the numerical procedure and ensure the *convergence* of the iterative process. The incorporation of *under-relaxation factors* into the system of algebraic equations that govern the fluid flow is intended to significantly moderate the iteration process by limiting the change in each of the transport variables from one iterative step to the next. There are no straightforward guidelines for pertinent choices of these factors. More often than not, in-depth experience in the selection of appropriate values of these factors can only be gained by the extensive investigation of a variety of flow problems.

Estimating the *errors* (Chapter 6) introduced by an inadequate mesh design for a general flow can be rather arduous. Preparing a good initial mesh design usually requires preknowledge of or insight into the expected properties of the flow. In dealing with the coarseness of a mesh, the only way to eliminate the *errors* is to embrace the procedure of successive refinement of an initially coarse mesh until certain key results exhibit no appreciable changes. This process forms part of the extended arm within the solution procedure, as described in Fig. 2.17. A systematic search of *grid-independent* results generally leads to the accomplishment of high-quality CFD solutions.

The reader should take note that the many aspects such as *convergence*, *convergence criteria or tolerance values*, *residuals*, *stability*, *errors*, *under-relaxation factors*, and *grid independence* greatly underpin the many numerical considerations for the simulation of a CFD problem. These conceptual terminologies may be rather difficult to fathom at this juncture. Nevertheless, supplementary discussions will be provided later in other chapters to adjunct the knowledge of the basics already attained herein.

2.4 RESULT REPORT AND VISUALIZATION – POST-PROCESS

CFD has a reputation for generating vivid graphical images and, while some of them are promotional that are usually displayed in stunning and superb colorful outputs, the ability to present the computational results effectively is an invaluable design tool. In this section we concentrate on some essential computer graphic techniques frequently encountered in the presentation of CFD data. The majority of ways that the CFD results are

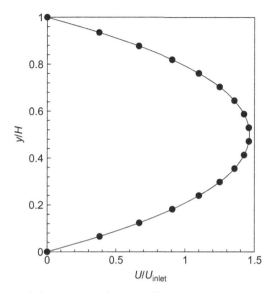

Fig. 2.17 X-Y plot of a parabolic laminar velocity profile at the fully developed region for Case 1.

emphasized graphically can be classified under different categories. Each of these categories, to be discussed below, assists the CFD user to better analyze and visualize the many relevant physical characteristics within the fluid flow problem.

Commercial CFD codes such as ANSYS-CFX, ANSYS-FLUENT, STAR-CCM+, and others often incorporate impressive visualization tools within their user-friendly GUIs to allow users to graphically view the results of a CFD calculation at the end of a computational simulation. There also exist, however, many excellent stand-alone applications of independent computer graphic software packages that the reader may opt to utilize for his/her CFD applications. Table 2.2 presents a list of some currently available graphical packages. The majority of these packages cater to various computer platforms, although some may only operate primarily in UNIX systems. Some of the commercial packages such as FIELDVIEW, TECPLOT, and ENSIGHT have been specifically developed for post-processing CFD results, while others are more general-purpose visualization tools that may be equally applied for CFD applications. For simple graphical representations, the use of an open-source plotting package such as GNUPLOT is popular among many researchers in the CFD community. It can be obtained freely through the internet website: http://www.gnuplot.info/.

For the remainder of this section, we demonstrate the different categories that can be applied to illustrate a CFD solution through a popular, versatile computer graphic software package, namely, TECPLOT. These categories will be mainly exemplified in the context of the solutions obtained from our previously defined flow cases: Case 1 – a fluid flowing between two stationary parallel plates – and Case 2 – a fluid passing through two

Table 2.2 Internet Links to Some Popular Computer Graphic Software Packages.

Developer	Code	Distributor Web Address
Advanced Visual Systems	AVS, Gsharp, Toolmaster	http://www.avs.com/
Amtec Engineering	Tecplot	http://www.amtec.com/
CEI	EnSight	http://www.ceintl.com/
IBM (free, apparently)	OpenDx	http://www.opendx.org/
Intelligent Light	FieldView	http://www.ilight.com
Numerical Algorithms Group (NAG)	Iris Explorer	http://www.nag.co.uk/
Visual Numerics	PV-Wave	http://www.roguewave.com/
Kitware	ParaView	http://www.paraview.org/
Department of Energy (DOE) Advanced Simulation and Computing Initiative (ASCI)	VisIt	https://wcl.llnl.gov/

cylinders in an open surrounding. An overview of these categories may also assist the reader in becoming more accustomed to how a CFD solution can be processed and visualized within a commercial code at the end of a simulation.

The use of TECPLOT for our graphical representations of the CFD results should not be interpreted as being skewed toward the endorsement of a specific product. Rather, we aim to simply provide the reader with an example of a standard graphics software approach. With the rapid evolution of new graphical techniques, probing the flow behavior inside a CFD problem may well be better achieved through other software packages. On that occasion, it is entirely up to the reader's choice to employ the appropriate graphics software package for their perusal.

2.4.1 X-Y Plots

These plots are mainly two-dimensional graphs that represent the variation of one dependent transport variable against another independent variable. They can usually be drawn by hand or more conveniently by many plotting packages. Such X-Y plots are the most precise and quantitative way to present the numerical data. Often, laboratory data is gathered by straight-line traverses. These graphs are therefore a popular way of directly comparing the numerical data with the experimental measured values. Also, logarithmic scales allow the identification of important flow effects occurring especially in the vicinity of solid boundaries. These graphs are widely used for presenting line profiles of velocity and for plots of surface quantities such as pressure and skin-friction coefficient. They are usually meant to be very easily identifiable; the reader can readily read the results without resorting to any mental or arithmetic interpolation.

An X-Y plot of a laminar velocity profile at the *fully developed* region for Case 1 is shown in Fig. 2.17. The significance of a parabolic profile characterizes the flow physics typically experienced for a fluid flowing within a parallel-plate channel. Another possible way of visualizing the development of the fluid flow is through the use of successive two-dimensional graphical profiles, as shown for Case 1 in Fig. 2.18. The flow distribution gradually changes from a uniform profile specified at the entrance boundary (left) to a parabolic profile as it travels downstream toward the channel exit boundary (right). More discussions on the physical aspects of this simple flow will be further expounded in Chapter 3. For the more complex flow structure of Case 2, the normalized horizontal velocity profile along the entire length of the computational domain, as shown in Fig. 2.19, can be represented at the mid-height location to better illustrate the deceleration and acceleration characteristics of the flow behavior as the fluid passes over the two cylinders. Further downstream, the fluid flow subsequently diffuses and recovers partially toward the free-stream condition near the exit boundary (right). The observed physical aspects of this particular flow will be further explored in Chapter 3.

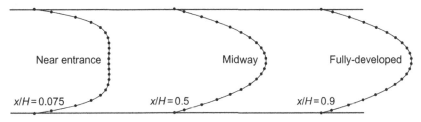

Fig. 2.18 Successive two-dimensional velocity profiles of a developing flow for Case 1.

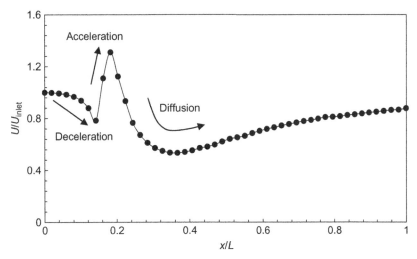

Fig. 2.19 X-Y plot of the normalized horizontal velocity along the length L midway between two cylinders for Case 2.

2.4.2 Vector Plots

A vector plot provides the means whereby a vector quantity (usually velocity) is displayed at discrete points with an arrow whose orientation indicates direction and whose size indicates magnitude. It generally presents a perspective view of the flow field in two dimensions. In a three-dimensional flow field different slices of two-dimensional planes containing the vector quantities can be generated in different orientations to better scrutinize the global flow phenomena. If the mesh densities are considerably high, the CFD user can either interpolate or reduce the numbers of output locations to prevent the clustering of these arrows from "obliterating" the graphical plot.

Fig. 2.20 illustrates a typical velocity vector plot representing the fluid flowing along the parallel-plate channel. This plot gives an alternative view of the developing flow previously envisaged in Fig. 2.17. The different colors represented in the vector plot depict the composite association of the velocity vectors with another dependent transport variable. For this particular flow case, we have arbitrarily chosen the distribution of dynamic pressure within the flow domain that impels the fluid flow. The range of colors displayed in the plot indicates the distribution between high (red) and low (blue) pressures that are effective inside the fluid flow process. Nevertheless, the CFD user can also freely select other transport variables that better emphasize significant physical aspects of the flow phenomena. For example, where heat transfer may be important, the velocity vectors can be coupled with the temperature distribution to illustrate the transport of hot fluid within the flow domain.

For the complex wake-developing flow of Case 2, the velocity vectors, as presented in Fig. 2.21, emphasize the presence of localized recirculation vortices as the fluid passes over and separates behind the proximity of the two cylinders. At the mid-height location, the varying sizes of the velocity vectors illustrating the approaching flow upstream, transitional flow in the vicinity of the two cylinders, and departing flow downstream clearly demonstrate the three distinct flow characteristics of deceleration, acceleration, and diffusion observed in Fig. 2.19, respectively. Here again like in Case 1, the array of the color spectrum that is displayed here illustrates the spatial distribution of the dynamic pressure.

2.4.3 Contour Plots

Contour plotting presents another useful and effective graphical technique that is frequently utilized in viewing CFD results. It is not surprising to imagine the proliferation

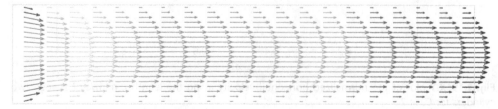

Fig. 2.20 Velocity vectors showing the flow development along the parallel-plate channel for Case 1.

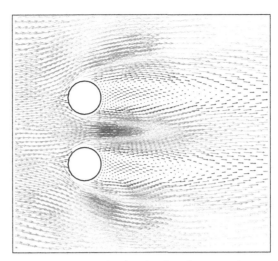

Fig. 2.21 Velocity vectors accentuating the localized wake recirculation zones behind the two cylinders for Case 2.

of contour plots ever since the advent of the computer. In CFD contour plots are one of the most commonly found graphical representations of data. A contour line (also known as *isoline*) can be described as a line indicative of some property that is constant in space. The equivalent representation in three dimensions is an *isosurface*. In contrast to X–Y plots, contour plots like vector plots provide a global description of the fluid flow encapsulated in one view. Generally, contours are plotted such that the difference between the numerical value of the dependent transport variable from one contour line to an adjacent contour line is held constant. The use of contour plots is usually not targeted for the precision evaluation of the numerical values between contour lines. Although some mental and/or numerical interpolation can be performed between the contour lines in space, it is to say the very least an imprecise process. The actual numerical values represented by the *isolines* of these plots are sometimes less important than their overall disposition. In practice, the contours are usually linearly scaled. However, to better capture the hidden details in some small regions within the flow field, the reader may be required to intrepidly employ other types of scaling choices to reveal these isolated flow behaviors. For the contour plots where the intervals are the same, the clustering of these lines indicates rapid changes in the flow quantities. Such plots are particularly useful especially in locating propagating shocks and discontinuities.

Figs. 2.22 and 2.23 are examples of *flooded* contour plots for Case 1. Here, a constant flow-field property of any transport variable is denoted by the constant intensity of the color shading. In these two plots the so-called "rainbow-scale" color map is employed to illustrate the distributions of the dimensionless resultant velocity normalized with respect to the inlet velocity and the dynamic pressure within the flow domain. The changing flooded contours near the entrance (left boundary) in Fig. 2.22 further confirms the

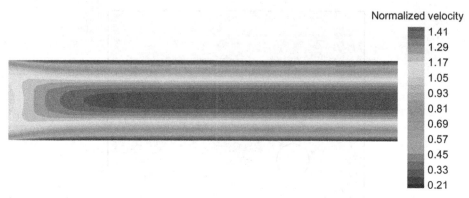

Fig. 2.22 Flooded contours on a rainbow-scale color map for the distribution of Case 1 normalized velocity.

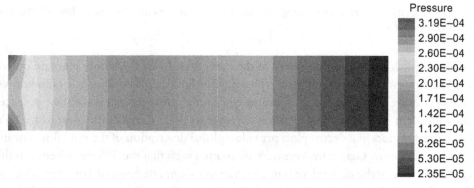

Fig. 2.23 Flooded contours on a rainbow-scale color map for the distribution of Case 1 dynamic pressure.

development of the fluid flow as previously observed by the successive velocity profiles in the X–Y plot in Fig. 2.18 and velocity vector plot in Fig. 2.20. In contrast, no appreciable change in the velocity is observed near the exit (right boundary). The successive reduction of the pressure, as indicated by the contour plot in Fig. 2.23, demonstrates the pressure gradient driving the fluid flow from the source imposed at the left boundary toward the sink located at the right boundary of the parallel-plate channel. At this point, we would like to draw the reader's attention to the color map represented by the pressure contour plot that is associated with the array of colors represented by the velocity vectors in Fig. 2.20.

Figs. 2.24 and 2.25 further exemplify another type of contour plots for the flow-field situation characterizing the flow-field situation of Case 2. The former illustrates the line contour representation of the pressure coefficient, while the latter demonstrates the dynamic pressure distribution by a "gray-scaled" color map flooded contour for the

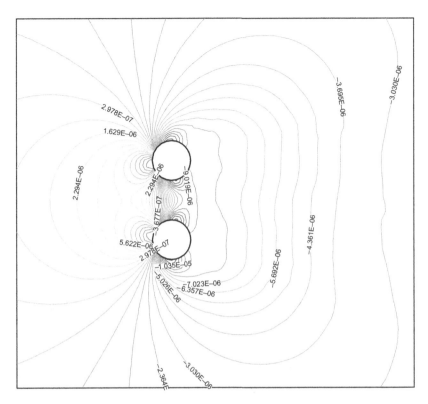

Fig. 2.24 Line contours on a rainbow-scale color map for the distribution of Case 2 pressure coefficient.

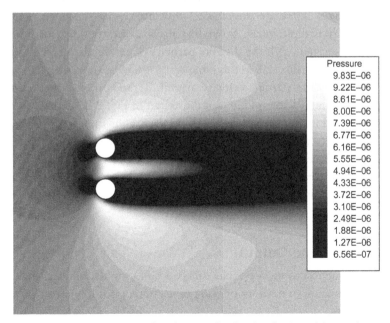

Fig. 2.25 Flooded contours on a gray-scale color map for the distribution of Case 2 dynamic pressure.

complex flow around the two cylinders. In both figures contours that are tightly clustered around the two cylinders clearly indicate the presence of vigorous flow activity as the fluid passes over the top and bottom cylinders. The manifestation of positive and negative pressure coefficients accompanied by the high and low dynamic pressures in the vicinity of the two cylindrical surfaces causes the flow stream velocities to change significantly across the curved surfaces. If the pressure *decreases* in the downstream direction, then the boundary layer thickness reduces; this case is termed a *favorable pressure gradient*. If, however, the pressure *increases* in the downstream direction, then the boundary layer thickens rapidly; this case is termed an *adverse pressure gradient*. This adverse pressure gradient, together with the action of the shear forces acting at a sufficient length, causes the boundary layer to come to rest. The flow separates from the surface, leading to the formation of reversed flow eddies as represented by the apparent recirculation vortices (wakes) seen in the velocity vector plot of Fig. 2.21. More physical explanations of these flow behaviors will be further discussed in Chapter 3.

2.4.4 Other Plots

The application of *streamlines* in the post-processing CFD stage, as in all aspects of fluid dynamics, is another exceptional tool for examining the nature of a flow either in two or three dimensions. By definition, *streamlines* are parallel to the mean velocity vector, where they trace the flow pattern using *massless* particles. For example, they can generally be obtained by integrating the spatial three velocity components expressed in a three-dimensional Cartesian frame: $dx / dt = u$, $dy / dt = v$, and $dz / dt = w$. For edification purposes, the reader may well benefit at this point of time to be accustomed to other existing terminologies synonymous with *streamlines* that are widely used in many graphical software packages. They are *streamtraces*, *streaklines*, or *path lines*. In more complex flow problems such as multi-phase flows that involve the transport of solid particles, the *particle tracks* associated with the discrete particles of certain diameter and mass being injected inside the parent fluid fall in this same category. Here, important information on the particle residence time, particle velocity magnitude, and other properties can be duly extracted. An example of a streamline plot illustrated in Fig. 2.26 defines the basic flow topology of localized recirculation zones behind the two cylinders, as previously identified from the velocity vector plot in Fig. 2.26. This tool can often reveal important features that could be obscured in some isolated flow regimes, which is clearly demonstrated by a more definitive representation of the observed wake-developing vortices through the streamline plot. Like in the velocity vector plot, the color spectrum corresponds to the spatial distribution of the dynamic pressure.

2.4.5 Data Report and Output

It is generally impractical to view the *raw data* of a CFD simulation especially on a mesh that may entail thousands or millions of grid points, except possibly on a reasonably small

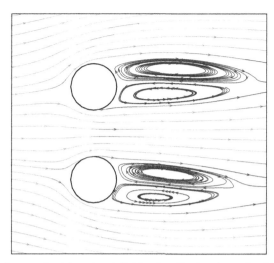

Fig. 2.26 Streamline plot emphasizing the definitive localized wake recirculation zones behind the two cylinders for Case 2.

"mapped" grid. Other alphanumeric reporting approaches need, however, to be adopted, of which such reports may be helpful to qualitatively check the attained numerical solution and/or extract the quantitative results for post-analysis purposes. Important variables such as surface fluxes, forces, and integrals can be evaluated at each respective boundary encompassing the computational domain. Some relevant data of significant interest are the evaluation of the mass flux in/out of each boundary that provides a clear indication of whether mass is conserved within the flow domain and the determination of the components of forces/moments on a surface such as wall shear stresses that may reveal important features connected with the flow physics of separation/reattachment/impingement. In some fluid mechanics problems the transient/unsteady nature of a flow may be important. Data of a number of pertinent transport variables can be tracked and dumped into format/unformatted files to adequately describe the time-development response of the fluid flow process.

2.4.6 Animation

CFD data lends itself very neatly to the animation category – *moving pictures* of the data produced through the CFD simulation. Animation, as with other graphical display tools, represents not only a technical record of quantitative results; it is also an artistic work of art. Some examples in which animation has assisted in enhancing the physical representation of the fluid flow processes are the movement of particles with the fluid in multiphase flows, moving geometries such as mixing tanks, and shock propagation in high-speed flows. Nowadays, a collection of short-length animated videos or movies bristling with multiple frames of brightly colorful representations of CFD simulations can be found on many internet websites. It is undoubtedly an effective visualization tool for education and marketing purposes. The ever-increasing availability of these *moving pictures* is indeed a tribute to the tireless efforts of computer graphic developers for *bringing the fluid flow to life*. Example of animated CFD results of a developing free-standing fire within a flickering period can be found in Chapter 8; the reader may well choose to advance to Chapter 8 to view those results that illustrate the puffing behavior and other physical aspects of the fire dynamics in better appreciating the pivotal role of animation in CFD.

2.5 SUMMARY

Chapter 2 has been purposefully written to expose potential or novice users to a powerful tool or technique for tackling problems associated with fluid flows. We have intended to capture the quintessence of how the CFD discipline has evolved; the prevalence of many commercial, shareware, and in-house computer codes is unquestionably a true testimonial of the dynamically evolving discipline. The widespread availability of these codes has certainly provided a favorable environment for students or new users learning CFD. Our intention throughout this chapter has been to provide proper guidance and possible supplementary knowledge emphasizing the many practical aspects within the framework of CFD analysis.

Although reference has been made to a number of commercial codes, this chapter is not intended to replace the user manuals of the CFD codes. The descriptions and explanations of many basic practical steps that may be akin to extracting the necessary parts or sections of a user manual are predominantly geared to assist the reader in achieving the goal of attaining the eventual computational solution. At the completion of this chapter, we strongly encourage the reader to begin applying any available CFD codes even without any *prior* knowledge of the fundamentals in order for he/she to be familiarized with the many facets of CFD.

A complete CFD analysis consists of *pre-processor*, *solver*, and *post-processor*. It simply encompasses the procedures of appropriately setting up the flow problem, solving and monitoring the solution, and analyzing the CFD results at the end of the simulation. Nevertheless, amid these rather mechanically driven "black-box" operations, there are many underlying fundamental principles beneath each of these three elements. The reader at this moment may wish to carefully review the many basic practical steps that have been carried out in handling Case 1 and Case 2 flow problems within this chapter. While performing the review, a list of questions presented below may assist the reader in contemplating and perhaps being more aware of the numerous theoretical and numerical issues that are profoundly embedded inside the CFD analysis, which are:

- What are the physical flow processes of the CFD problem? (Chapter 3)
- How is the flow physics described in mathematical equations? (Chapter 3)
- What are the equations governing the fluid flow and heat transfer? (Chapter 3)

- Why are boundary conditions important and how are they applied? (Chapter 3)
- What are the physical meanings of the boundary conditions? (Chapter 3)
- Why does a flow domain require to be sub-divided into many smaller non-overlapping sub-domains or a computational mesh/grid? (Chapter 4)
- What types of meshes are commonly encountered in CFD? (Chapter 4)
- How are the mathematical equations solved? (Chapter 5)
- How are computational methods/techniques employed? (Chapter 5)
- What is the meaning of monitoring curves? (Chapter 6)
- How is the numerical procedure terminated? (Chapter 6)
- What are solution errors? (Chapter 6)
- How is a computational solution assessed to be correct, numerically accurate, and physically meaningful? (Chapter 6)
- When dealing with more complex flow problems, are there any other available methods/techniques or practical experiences or general guidelines that can assist in overcoming convergence difficulties? (Chapter 7)
- Are there any additional illustrative examples using CFD and how can the solution be better analyzed? (Chapter 8)
- What are the future advancements in CFD? (Chapter 9)

Various chapters indicated above at the end of each question are primarily intended to guide the reader where the theoretical and numerical considerations of CFD will be established in other chapters of this book. These chapters are aimed to comprehensively answer these questions while acknowledging the many practical aspects that have been covered in this chapter. Needless to say, applying CFD in practice goes hand in hand with the understanding of the basic equations governing the fluid flow process. These equations are based on conservation laws and theories. Discussions of their physical significance and implications as well as respective formulations are addressed in the next chapter.

REVIEW QUESTIONS

2.1 How are commercial codes allowing CFD analyses to be carried out with ease for the novice user?
2.2 What are the main elements involved in a complete CFD analysis?
2.3 Why is it important to correctly define the computational domain for the fluid flow problem? Give an example of this.
2.4 What is the consequence of using a very fine mesh (i.e. a very large number of cells) compared with using a coarse mesh (i.e. a small number of cells)?
2.5 What is the main difference between a structured and unstructured mesh and when are they applied to physical domains?
2.6 What types of boundary conditions can be imposed on the computational domain?

2.7 What type of boundary can be used for a computational boundary that represents an open physical boundary?

2.8 What advantages can a *symmetry* boundary condition and a *cyclic* boundary condition provide and when can they be applied?

2.9 What is the main purpose of a *CFD solver*?

2.10 What are the advantages of providing intelligent values for the initial solution?

2.11 What is an iteration process and how is it performed?

2.12 What does the convergence criterion control?

2.13 What is the main purpose of the *post-processing* stage?

2.14 What are the advantages of using X–Y plots? Give examples of what CFD results X–Y plots can capture.

2.15 Why are contour plots best used to display the distribution of a variable?

2.16 What is the meaning of a streamline? What advantages do they have over other plot types?

CHAPTER 3

Governing Equations for CFD – Fundamentals

3.1 INTRODUCTION

CFD is fundamentally based on the *governing equations* of fluid dynamics. They represent mathematical statements of the *conservation laws of physics*. The purpose of this chapter is to introduce the derivation and discussion of these equations, where the following physical laws are adopted:

- Mass is conserved for the fluid.
- Newton's second law, the rate of change of momentum equals the sum of forces acting on the fluid.
- First law of thermodynamics, the rate of change of energy equals the sum of rate of heat addition to and the rate of work done on the fluid.

It is important that anyone concerned with CFD possesses some understanding of the physical phenomena of fluid motion, as it is these phenomena that CFD analyzes and predicts. All of CFD is based on these equations; we must therefore begin our understanding at the most basic description of the fluid flow processes and the meaning and significance of each of the *terms* within them. After these equations are obtained, forms particularly suited for use in formulating CFD solutions will be delineated. The physical aspects of the boundary conditions and their appropriate mathematical statements will also be developed since the appropriate *numerical* form of the physical boundary condition is strongly dependent on the particular mathematical form of the governing equations and numerical algorithm used. At the completion of this chapter, it is the authors' endeavor to remove some of the underlying mysteries surrounding the prediction of a fluid in motion through computer-based tools and to replace them with a solid understanding of the equations governing the fluid transport.

3.2 THE CONTINUITY EQUATION

3.2.1 Mass Conservation

One conservation law that is pertinent to fluid flow is that *matter may neither be created nor destroyed*. Consider an arbitrary control volume V fixed in space and time in Fig. 3.1. The fluid moves through the fixed control volume, flowing across the control surface.

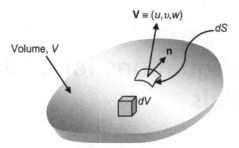

Fig. 3.1 Finite control volume fixed in space.

Mass conservation requires that the rate of change of mass within the control volume is equivalent to the mass flux crossing the surface S of volume V. In *integral form*:

$$\frac{d}{dt}\int_V \rho\, dV = -\int_S \rho \mathbf{V} \cdot \mathbf{n}\, dS \tag{3.1}$$

where $\mathbf{n}$ is the unit normal vector. We can apply Gauss's divergence theorem that equates the volume integral of a divergence of a vector to an area integral over the surface that defines the volume. This is stated as

$$\int_V div\,\rho\,\mathbf{V}\, dV = -\int_S \rho \mathbf{V} \cdot \mathbf{n}\, dS \tag{3.2}$$

Using the above theorem, the surface integral in Eq. (3.1) may be replaced by a volume integral; hence the equation becomes

$$\int_V \left[\frac{\partial \rho}{\partial t} + \nabla \cdot (\rho \mathbf{V})\right] dV = 0 \tag{3.3}$$

where $\nabla \cdot (\rho \mathbf{V}) \equiv div\,\rho \mathbf{V}$. Since Eq. (3.3) is valid for any size of volume V, the implication is that

$$\frac{\partial \rho}{\partial t} + \nabla \cdot (\rho \mathbf{V}) = 0 \tag{3.4}$$

Eq. (3.4) is the *mass conservation*. In the Cartesian coordinate system it can be expressed as

$$\frac{\partial \rho}{\partial t} + \frac{\partial(\rho u)}{\partial x} + \frac{\partial(\rho v)}{\partial y} + \frac{\partial(\rho w)}{\partial z} = 0 \tag{3.5}$$

where the fluid velocity $\mathbf{V}$ at any point in the flow field is described by the local velocity components u, v, and w, which are, in general, functions of location (x,y,z) and time (t).

Alternatively, consider the scenario of a fluid flowing between two stationary parallel plates, as illustrated in Fig. 3.2. An infinitesimal small control volume $\Delta x\, \Delta y\, \Delta z$ fixed in space

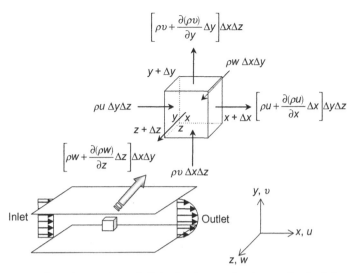

Fig. 3.2 The conservation of mass in an infinitesimal control volume of a fluid flow between two stationary parallel plates.

(enlarged to the right of the figure) is analyzed where the *mass conservation* statement applies to the (u,v,w) flow field. Transport due to such motion is often referred to as *advection*. The conservation law requires that, for unsteady flow, *the rate of increase of mass within the fluid element equals the net rate at which mass enters the control volume* (inflow – outflow); in other words,

$$\frac{dm}{dt} = \sum_{in} \dot{m} - \sum_{out} \dot{m} \tag{3.6}$$

The rate at which mass enters the control volume through the surface perpendicular to x may be expressed as $(\rho u)\, \Delta y\, \Delta z$, where ρ is the local density of the fluid and similarly through the surfaces perpendicular to y and z as $(\rho v)\, \Delta x\, \Delta z$ and $(\rho w)\, \Delta x\, \Delta y$, respectively. The rate at which the mass leaves the surface at $x + \Delta x$ may be expressed through a truncated Taylor series as

$$\left[(\rho u) + \frac{\partial(\rho u)}{\partial x}\Delta x \right]\Delta y \Delta z + O(\Delta x, \Delta V) \quad \text{where} \quad \Delta V = \Delta x\, \Delta y\, \Delta z \tag{3.7}$$

Similarly, the rate at which mass leaves the surfaces at $y + \Delta y$ and $z + \Delta z$ may also be expressed as

$$\left[(\rho v) + \frac{\partial(\rho v)}{\partial y}\Delta y \right]\Delta x \Delta z + O(\Delta y, \Delta V) \quad \text{and}$$

$$\left[(\rho w) + \frac{\partial(\rho w)}{\partial z}\Delta z \right]\Delta x \Delta y + O(\Delta z, \Delta V) \tag{3.8}$$

Since the mass of the fluid element m is given by $\rho\,\Delta x\,\Delta y\,\Delta z$, Eq. (3.6) becomes

$$\frac{\partial(\rho\Delta x\Delta y\Delta z)}{\partial t} = (\rho u)\Delta y\Delta z + (\rho v)\Delta x\Delta z + (\rho w)\Delta x\Delta y - \left[(\rho u) + \frac{\partial(\rho u)}{\partial x}\Delta x\right]\Delta y\Delta z$$

$$-\left[(\rho v) + \frac{\partial(\rho v)}{\partial y}\Delta y\right]\Delta x\Delta z - \left[(\rho w) + \frac{\partial(\rho w)}{\partial z}\Delta z\right]\Delta x\Delta y$$

$$+ \Delta V\ O(\Delta x, \Delta y, \Delta z)$$

$$(3.9)$$

In the limit, canceling terms and dividing by the constant-size $\Delta x\,\Delta y\,\Delta z$, we obtain

$$\frac{\partial\rho}{\partial t} + \frac{\partial(\rho u)}{\partial x} + \frac{\partial(\rho v)}{\partial y} + \frac{\partial(\rho w)}{\partial z} = 0 \qquad (3.10)$$

Eq. (3.10) is exactly the same form as derived in Eq. (3.5). This equation is precisely the **partial differential form** of the *continuity equation*. We have shown that the integral form in Eq. (3.1) can, after some manipulation, yield the partial differential form. This specific differential form is usually called the *conservation form*. Both Eqs (3.1) and (3.10) are in conservation form; the manipulation performed does not alter the situation.

In Chapter 2 a two-dimensional CFD analysis is performed for a channel flow that is described by the fluid flow between two stationary parallel plates (see Fig. 3.2). This is made possible by the assumption that the dimension in the z coordinate direction is sufficiently large such that the flow remains invariant along this coordinate direction. For a variable fluid property – varying density ρ – the continuity equation for a compressible flow in two dimensions is given by

$$\frac{\partial\rho}{\partial t} + \frac{\partial(\rho u)}{\partial x} + \frac{\partial(\rho v)}{\partial y} + \underbrace{\frac{\partial(\rho w)}{\partial z}}_{= 0} = 0 \qquad \text{or} \qquad \boxed{\frac{\partial\rho}{\partial t} + \frac{\partial(\rho u)}{\partial x} + \frac{\partial(\rho v)}{\partial y} = 0}$$

Flow invariant in the z direction

$$(3.11)$$

If the fluid is taken to be incompressible, the density ρ is constant, that is, the spatial and temporal variations in density are neglected relative to those velocity components of u, v, and w. We can obtain the continuity equation in two dimensions for an incompressible flow as

$$\underbrace{\frac{\partial\rho}{\partial t}}_{\text{constant density}} + \rho\left(\frac{\partial u}{\partial x} + \frac{\partial v}{\partial y} + \underbrace{\frac{\partial w}{\partial z}}_{= 0}\right) = 0 \qquad \text{or} \qquad \boxed{\frac{\partial u}{\partial x} + \frac{\partial v}{\partial y} = 0}$$

Flow invariant in the z direction

$$(3.12)$$

3.2.2 Physical Interpretation

Let us examine the physical meaning of the continuity equation of Eq. (3.12) as applied to an infinitesimal small control volume for the two-dimensional case of the fluid flow between two parallel plates to illustrate the fundamental physical principle. Two situations are considered.

Consider the first situation, if $\partial u/\partial x > 0$ then the velocity at the surface at $x + \Delta x$ is greater than the velocity at the surface x, that is, $u(x + \Delta x) > u(x)$. Since more fluid is physically *leaving* the control volume than *entering* along the x direction, there should be more fluid *entering* than *leaving* along the y direction. Here, $\partial v/\partial y < 0$ and the velocity at the surface $y + \Delta y$ is less than the velocity at the surface y, that is, $v(y + \Delta y) < v(y)$.

Alternatively, for the second situation, if $\partial u/\partial x < 0$ then the velocity at the surface at $x + \Delta x$ is less than the velocity at the surface x, that is, $u(x + \Delta x) < u(x)$. Since more fluid is physically *entering* the control volume than *leaving* along the x direction, there should be more fluid *leaving* than *entering* along the y direction. Here, $\partial v/\partial y > 0$ and the velocity at the surface $y + \Delta y$ is greater than the velocity at the surface y, that is, $v(y + \Delta y) > v(y)$.

Both situations satisfy the continuity equation: $\partial u/\partial x + \partial v/\partial y = 0$ (*mass conservation*).

EXAMPLES

Example 3.1

Consider a laminar boundary layer that can be approximated as having a velocity profile $u(x) = U_\infty y/\delta$, where $\delta = cx^{1/2}$, c is a constant, U_∞ is the free-stream velocity, and δ is the boundary layer thickness. With reference to the two-dimensional fluid flow over a flat plate, as shown in Fig. 3.1.1 below, determine the velocity v (vertical component) inside the boundary layer.

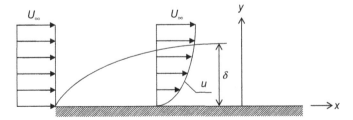

Fig. 3.1.1 Two-dimensional flow over a flat plate.

Solution: As the boundary layer grows downstream, the horizontal velocity u is gradually slowed down due to viscous effect and the no-slip condition at the surface of the flat plate. In order to satisfy the continuity equation, the vertical velocity v should be positive and acting to remove the fluid away from the boundary layer.

We begin the analysis by substituting the velocity profile $u(x)$ into Eq. (3.13), yielding

$$\frac{\partial}{\partial x}\left(\frac{U_\infty y}{cx^{1/2}}\right) + \frac{\partial v}{\partial y} = 0 \Rightarrow -\frac{U_\infty y}{2cx^{3/2}} = -\frac{\partial v}{\partial y}$$

Integrating the vertical velocity v with respect to y, the equation becomes

$$v = \frac{U_\infty y^2}{4cx^{3/2}} = \left(\frac{U_\infty y}{cx^{1/2}}\right)\left(\frac{y}{4x}\right) = \frac{uy}{4x}.$$

Discussion: This physically means that the velocity ratio $v/u = y/4x$ increases away from the surface at a fixed x location, that is, it decreases further downstream at a fixed y location. At the edge of the boundary layer, $y = \delta = cx^{1/2}$, the velocity ratio v/u equals $c/4x^{1/2}$. If the constant c is assumed unity, the boundary layer thickness δ and the velocity ratio v/u as a function of the horizontal distance x from the leading edge of the flat plate can be described and they are illustrated in Figs. 3.1.2 and 3.1.3.

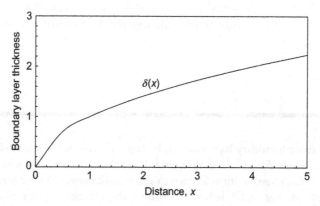

Fig. 3.1.2 Boundary layer thickness δ as a function of the horizontal distance x.

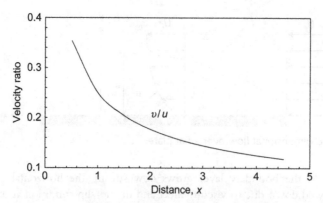

Fig. 3.1.3 Velocity ratio v/u at the edge of the boundary layer as a function of the horizontal distance x.

The latter figure further illustrates the decrease of the velocity ratio v/u further downstream of the fluid flow over the flat plate. At some downstream distance x, the change in the horizontal u velocity is appreciably small. Here, $\partial u/\partial x \to 0$, which also leads to $\partial v/\partial y \to 0$, hence satisfying the continuity Eq. (3.13).

Example 3.2

Consider the CFD case in Chapter 2 for the steady two-dimensional incompressible, laminar flow between two stationary parallel plates with the following dimensions: height $H = 0.1$ m and length $L = 0.5$ m (Fig. 3.2.1). Using CFD, plot the velocity vector along the channel length. Discuss the physical meaning of the continuity equation by plotting the velocity components u and v close to the bottom wall surface along the channel length with the working fluid taken as air and a uniformly distributed velocity profile of 0.01 m/s applied at the channel inlet (u_{in}).

Solution: The problem is described as follows:

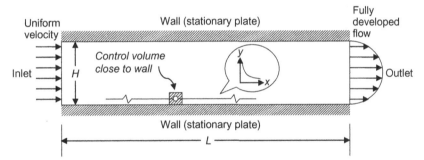

Fig. 3.2.1 Two-dimensional laminar flow between two stationary parallel plates.

From the CFD simulation, the flow field along the channel length is illustrated in Fig. 3.2.2. It is observed that the flow gradually changes from a uniform profile at the inlet surface to a parabolic profile as it travels downstream along the channel.

Fig. 3.2.2 Velocity profile distribution along the channel length.

The resultant velocity profiles of u and v close to the bottom wall surface along the channel length are given in Figs. 3.2.3 and 3.2.4.

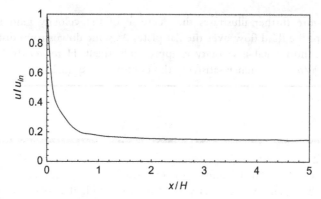

Fig. 3.2.3 Horizontal velocity u profile along the channel length.

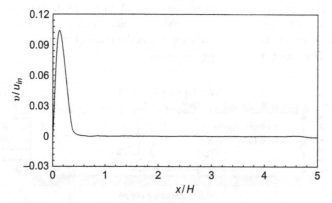

Fig. 3.2.4 Vertical velocity v profile along the channel length.

Discussion: It is observed that within the *hydrodynamic entrance region*, that is, $x < 3H$, the horizontal velocity u decreases along the channel length. This means that $\partial u/\partial x < 0$ close to the wall surface. The air flow is slowed along the x direction due to the no-slip boundary condition imposed near the wall as it flows over the wall surface. More fluid is therefore physically entering than leaving the control volume along the x direction. At the same time, there should be more fluid leaving than entering along the y direction, and the vertical velocity v increases, which implies that $\partial v/\partial y > 0$, in order to conserve the mass. On the other hand, when the flow is in the *fully developed region*, that is, $x \geq 3H$, it is observed that the horizontal velocity u does not appreciably change along the x direction, that is, $\partial u/\partial x = 0$. In accordance with the continuity equation, $\partial v/\partial y$ must also be zero within this flow region. This is clearly reflected by the constant vertical velocity v, as shown in Fig. 3.2.3 in the *fully developed region*. The physical meanings of the *hydrodynamic entrance region* and *fully developed region* are further discussed in Example 3.5.

Fig. 3.3 A pipe flow scenario.

3.2.3 Comments

In most Fluid Mechanics textbooks the principle of mass conservation is often explained by a fluid flowing in a pipe, see Fig. 3.3. The mass entering a pipe, denoted by the mass flow rate $\dot{m}_1$, is equivalent to the product of the density, cross-sectional average velocity at the inlet, and cross-sectional area, that is, $\rho\, u_1\, A_1$. This mass must equal the mass flow rate leaving the pipe, which is denoted by $\dot{m}_2$ given as $\rho\, u_2\, A_2$. If we take, for instance, that the cross-sectional areas are the same for both the inlet and outlet surfaces of the pipe, that is, $A_1 = A_2$, and based on the mass conservation, the outlet velocity u_2 must equal the inlet velocity u_1. If the cross-sectional areas are different at both ends of the pipe, take, for example, $A_1 = 2\, A_2$, then $u_2 = 2\, u_1$, which means that the flow is *accelerated*. On the other hand, if $A_1 = \tfrac{1}{2}\, A_2$, then $u_2 = \tfrac{1}{2}\, u_1$, which means that the flow is *decelerated*.

This principle of mass conservation as applied to the whole domain in one dimension is also applicable to any small control volume that is used in CFD to numerically solve the partial differential equations. In two dimensions the mass flow may not be conserved in one direction, but overall, it will be conserved throughout the control volume by either removing or adding the mass in the other direction. It is important to persevere with the physical meaning of the mathematical equations in mind when studying and analyzing CFD results, a philosophy that we urge the reader to embrace. Indeed, this philosophy is extrapolated to all mathematical equations and operations encompassing any physical problems. We have explored to some great length the physical meaning of the continuity Eq. (3.13), and in the following sections we will further explore the effects of various terms such as *advection* and *diffusion* in the other mathematical equations. Whatever they are, we strongly encourage the reader at all times to continuously adopt the philosophy of understanding the physical meaning of the terms in the equations that are being dealt with.

3.3 THE MOMENTUM EQUATION

3.3.1 Force Balance

In deriving this physical law we begin by considering the fluid element, as described in Fig. 3.2 for mass conservation. *Newton's second law of motion* states that the sum of forces that is acting on the fluid element, as illustrated in Fig. 3.4, equals the product between its mass and acceleration of the element. There are essentially three scalar relations along the

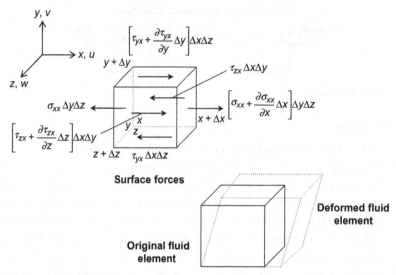

Fig. 3.4 Surface forces acting on the infinitesimal control volume for the velocity component *u*. Deformed fluid element due to the action of the surface forces.

x, *y*, and *z* directions of the Cartesian frame of which the fundamental law can be invoked. We begin by considering the *x* component of Newton's second law,

$$\sum F_x = ma_x \qquad (3.13)$$

where F_x and a_x are the force and acceleration along the *x* direction. The acceleration a_x at the right-hand side of Eq. (3.13) is simply the time rate change of *u*, which is given by the substantial derivative (see Appendix A for a more detailed description). Thus

$$a_x = \frac{Du}{Dt} \qquad (3.14)$$

Recalling that the mass of the fluid element *m* is $\rho\, \Delta x\, \Delta y\, \Delta z$, the rate of increase of *x*-momentum is

$$\rho \frac{Du}{Dt} \Delta x \Delta y \Delta z \qquad (3.15)$$

On the left-hand side of Eq. (3.13), there are two sources of this force that the moving fluid element experiences: *body forces* and *surface forces*. The type of body forces that may influence the rate of change of the fluid momentum are gravity, centrifugal, Coriolis, and electromagnetic forces. These effects are usually incorporated by introducing them into the momentum equations as additional source terms to the contribution of the surface forces. The surface forces for the velocity component *u*, as seen in Fig. 3.4, that deform the fluid element are due to the normal stress σ_{xx} and tangential stresses τ_{yx} and τ_{zx} acting on the surfaces of the fluid element. Combining the sum of these surface forces on the fluid element (see Appendix A for a more detailed derivation) and the time rate change of u from Eq. (3.15) into Eq. (3.13), the x-momentum equation becomes

$$\rho \frac{Du}{Dt} = \frac{\partial \sigma_{xx}}{\partial x} + \frac{\partial \tau_{yx}}{\partial y} + \frac{\partial \tau_{zx}}{\partial z} + \sum F_x^{body\,forces} \qquad (3.16)$$

In a similar fashion the y-momentum and z-momentum equations can be obtained as

$$\rho \frac{Dv}{Dt} = \frac{\partial \tau_{xy}}{\partial x} + \frac{\partial \sigma_{yy}}{\partial y} + \frac{\partial \tau_{zy}}{\partial z} + \sum F_y^{body\,forces} \qquad (3.17)$$

and

$$\rho \frac{Dw}{Dt} = \frac{\partial \tau_{xz}}{\partial x} + \frac{\partial \tau_{yz}}{\partial y} + \frac{\partial \sigma_{zz}}{\partial z} + \sum F_z^{body\,forces} \qquad (3.18)$$

The normal stresses σ_{xx}, σ_{yy}, and σ_{zz} in Eqs (3.16)–(3.18) are due to the combination of pressure p and normal viscous stress components τ_{xx}, τ_{yy}, and τ_{zz} acting perpendicular to the control volume. The remaining terms contain the tangential viscous stress components. In many fluid flows a suitable model for the viscous stresses is introduced. They are usually a function of the local deformation rate (or strain rate) that is expressed in terms of the velocity gradients. The formulation of the appropriate stress–strain relationships for a Newtonian fluid is found in Appendix A.

Based on the concept of *substantial derivative*, as illustrated in Appendix A, the *conservative* form of Eqs (3.16)–(3.18) can now be rewritten as

$$\frac{\partial(\rho u)}{\partial t} + \frac{\partial(\rho uu)}{\partial x} + \frac{\partial(\rho vu)}{\partial y} + \frac{\partial(\rho wu)}{\partial z} = \frac{\partial \sigma_{xx}}{\partial x} + \frac{\partial \tau_{yx}}{\partial y} + \frac{\partial \tau_{zx}}{\partial z} + \sum F_x^{body\,forces}$$
$$(3.19)$$

$$\frac{\partial(\rho v)}{\partial t} + \frac{\partial(\rho uv)}{\partial x} + \frac{\partial(\rho vv)}{\partial y} + \frac{\partial(\rho wv)}{\partial z} = \frac{\partial \tau_{xy}}{\partial x} + \frac{\partial \sigma_{yy}}{\partial y} + \frac{\partial \tau_{zy}}{\partial z} + \sum F_y^{body\,forces}$$
$$(3.20)$$

$$\frac{\partial(\rho w)}{\partial t} + \frac{\partial(\rho uw)}{\partial x} + \frac{\partial(\rho vw)}{\partial y} + \frac{\partial(\rho ww)}{\partial z} = \frac{\partial \tau_{xz}}{\partial x} + \frac{\partial \tau_{yz}}{\partial y} + \frac{\partial \sigma_{zz}}{\partial z} + \sum F_z^{body\,forces}$$
$$(3.21)$$

with the appropriate stress–strain relationships for a Newtonian fluid proposed in Appendix A.

For the two-dimensional case of the fluid flow between parallel plates (the flow being invariant along the z direction), the momentum equations with the inclusion of the stress–strain relationships for a compressible flow can be written as

$$\underbrace{\frac{\partial(\rho u)}{\partial t}}_{local\ accelration} + \underbrace{\frac{\partial(\rho uu)}{\partial x} + \frac{\partial(\rho vu)}{\partial y}}_{advection} = - \underbrace{\frac{\partial p}{\partial x}}_{pressure\ gradient} + \underbrace{\frac{\partial}{\partial x}\left(\mu\frac{\partial u}{\partial x}\right) + \frac{\partial}{\partial y}\left(\mu\frac{\partial u}{\partial y}\right)}_{diffusion}$$

$$+\underbrace{\frac{\partial}{\partial x}\left(\mu\frac{\partial u}{\partial x}\right) + \frac{\partial}{\partial y}\left(\mu\frac{\partial v}{\partial x}\right) + \frac{\partial}{\partial y}\left[\lambda\left(\frac{\partial u}{\partial x} + \frac{\partial v}{\partial y}\right)\right]}_{extra\ stress} + \underbrace{\sum F_x^{body\ forces}}_{body\ forces}$$

(3.22)

$$\underbrace{\frac{\partial(\rho v)}{\partial t}}_{local\ acclreation} + \underbrace{\frac{\partial(\rho uv)}{\partial x} + \frac{\partial(\rho vv)}{\partial y}}_{advection} = - \underbrace{\frac{\partial p}{\partial y}}_{pressure\ gradient} + \underbrace{\frac{\partial}{\partial x}\left(\mu\frac{\partial v}{\partial x}\right) + \frac{\partial}{\partial y}\left(\mu\frac{\partial v}{\partial y}\right)}_{diffusion}$$

$$+\underbrace{\frac{\partial}{\partial x}\left(\mu\frac{\partial u}{\partial y}\right) + \frac{\partial}{\partial y}\left(\mu\frac{\partial v}{\partial y}\right) + \frac{\partial}{\partial y}\left[\lambda\left(\frac{\partial u}{\partial x} + \frac{\partial v}{\partial y}\right)\right]}_{extra\ stress} + \underbrace{\sum F_y^{body\ forces}}_{body\ forces}$$

(3.23)

Alternatively, for the case of a *constant property* fluid flow, which implies that the density and viscosity are constants and in the absence of body forces particularly due to gravity (e.g. no density variation due to buoyancy), the momentum equations with the inclusion of the stress-strain relationships for an incompressible flow can be reduced to

$$\underbrace{\frac{Du}{Dt}}_{acceleration} = - \underbrace{\frac{1}{\rho}\frac{\partial p}{\partial x}}_{pressure\ gradient} + \underbrace{v\frac{\partial^2 u}{\partial x^2} + v\frac{\partial^2 u}{\partial y^2}}_{diffusion} + \frac{\partial}{\partial x}\left[\lambda\underbrace{\left(\overset{=0}{\frac{\partial u}{\partial x} + \frac{\partial v}{\partial y}}\right)}_{continuity\ equation}\right]$$

$$+v\frac{\partial}{\partial x}\underbrace{\left[\overset{=0}{\frac{\partial u}{\partial x} + \frac{\partial v}{\partial y}}\right]}_{continuity\ equation} + \underbrace{\overset{=0}{\frac{\partial v}{\partial x}\frac{\partial u}{\partial x} + \frac{\partial v}{\partial y}\frac{\partial v}{\partial x}}}_{constant\ property} + \underbrace{\overset{=0}{\sum F_x^{body\ forces}}}_{body\ forces}$$

$$\underbrace{\frac{Dv}{Dt}}_{acceleration} = - \underbrace{\frac{1}{\rho}\frac{\partial p}{\partial y}}_{pressure\ gradient} + \underbrace{v\frac{\partial^2 v}{\partial x^2} + v\frac{\partial^2 v}{\partial y^2}}_{diffusion} + \frac{\partial}{\partial y}\left[\lambda\underbrace{\left(\overset{=0}{\frac{\partial u}{\partial x} + \frac{\partial v}{\partial y}}\right)}_{continuity\ equation}\right]$$

$$+v\frac{\partial}{\partial y}\underbrace{\left[\overset{=0}{\frac{\partial u}{\partial x} + \frac{\partial v}{\partial y}}\right]}_{continuity\ equation} + \underbrace{\overset{=0}{\frac{\partial v}{\partial x}\frac{\partial u}{\partial y} + \frac{\partial v}{\partial y}\frac{\partial v}{\partial y}}}_{constant\ property} + \underbrace{\overset{=0}{\sum F_y^{body\ forces}}}_{body\ forces}$$

or

$$\underbrace{\frac{\partial u}{\partial t}}_{local\ acceleration} + \underbrace{u\frac{\partial u}{\partial x} + v\frac{\partial u}{\partial y}}_{advection} = - \underbrace{\frac{1}{\rho}\frac{\partial p}{\partial x}}_{pressure\ gradient} + \underbrace{v\frac{\partial^2 u}{\partial x^2} + v\frac{\partial^2 u}{\partial y^2}}_{diffusion}$$

(3.24)

$$\underbrace{\frac{\partial v}{\partial t}}_{\text{local acceleration}} + \underbrace{u\frac{\partial v}{\partial x} + v\frac{\partial v}{\partial y}}_{\text{advection}} = -\underbrace{\frac{1}{\rho}\frac{\partial p}{\partial y}}_{\text{pressure gradient}} + \underbrace{\nu\frac{\partial^2 v}{\partial x^2} + \nu\frac{\partial^2 v}{\partial y^2}}_{\text{diffusion}} \qquad (3.25)$$

where ν is the kinematic viscosity ($\nu = \mu/\rho$). Eqs (3.22)–(3.25) derived from *Newton's second law* describe the conservation of momentum in the fluid flow and are also known as the Navier-Stokes equations. For ease of describing the physical significance of each of the terms in the momentum equations, Eqs (3.24) and (3.25) are examined in the next section to provide suitable explanations for the terms within.

3.3.2 Physical Interpretation

Consider the motion of the fluid of a piston compressing the air within the enclosed cylinder. From the viewpoint of point A in Fig. 3.5, the fluid is locally accelerating due to the increase of the air velocity within the shrinking volume changing in time. In relation to the *x*-momentum Eq. (3.24), this is represented by the *acceleration* term $\partial u/\partial t$ for the horizontal velocity component u. Similarly, if the motion of fluid of the piston is inverted vertically, the local rate of increase of the air velocity is represented by the term $\partial v/\partial t$, which denotes the local *acceleration* of the vertical component v of Eq. (3.25).

The above example describes the motion of the fluid changing locally with time. We further explore the physical meaning of the fluid accelerating locally in space. Consider the motion of fluid through a venturi, as indicated in Fig. 3.6. As the fluid travels between the locations B and C along the *x* direction and assuming that the velocity in itself is not fluctuating with time, the horizontal velocity component u has a local *acceleration* in space where the velocity is accelerating between the incremental distance of locations B and C, that is, the velocity gradient in the term $u\partial u/\partial x$ of Eq. (3.24) is increasing. Similarly, if the

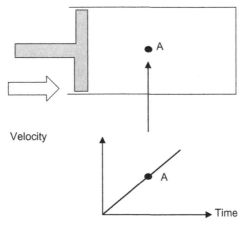

Fig. 3.5 The motion of fluid in a piston mechanism.

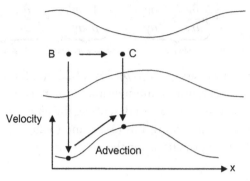

Fig. 3.6 The motion of fluid through a venturi.

venturi is vertically oriented, then the vertical velocity component v has a local *acceleration* gradient in space where the velocity gradient in the term $v\partial v/\partial y$ is increasing between the incremental distance of locations B and C. We usually refer to the fluid sweeping past point B and on its way to point C in the flow field as the *advection* term of the momentum equations.

We further investigate the physical interpretations of the pressure gradient and diffusion terms in the momentum Eqs (3.24) and (3.25) in the following examples below.

EXAMPLES

Example 3.3

Consider an incompressible, inviscid, laminar flow past a circular cylinder of diameter d in Fig. 3.3.1. The flow variation along the approaching stagnation streamline (A – B) can be expressed as

$$u(x) = U_\infty \left(1 - \frac{R^2}{x^2} \right)$$

With reference to Eqs (3.24) and (3.25), determine the total acceleration experienced by the fluid as it flows along the stagnation streamline. Also, determine the pressure distribution along the streamline by deriving Bernoulli's equation and the stagnation pressure at the stagnation point.

Solution: For an inviscid flow, the shear stresses are zero, that is, $\tau = 0$ for all shear stresses. The diffusion terms are zero as the viscosity is zero. Along the stagnation streamline, the vertical velocity component v is zero. Also, since the continuity equation applies, and considering the x-momentum Eq. (3.24), the following terms drop off from the equation:

$$\frac{\partial u}{\partial t} + u\frac{\partial u}{\partial x} + \overset{=0}{\cancel{v\frac{\partial u}{\partial y}}} = -\frac{1}{\rho}\frac{\partial p}{\partial x} + \overset{=0}{\cancel{v\frac{\partial^2 u}{\partial x^2}}} + \overset{=0}{\cancel{v\frac{\partial^2 u}{\partial y^2}}}$$

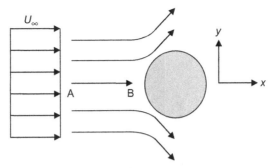

Fig. 3.3.1 Fluid motion over a circular cylinder.

The above equation reduces to

$$\frac{\partial u}{\partial t} + u\frac{\partial u}{\partial x} = -\frac{1}{\rho}\frac{\partial p}{\partial x}$$

The total acceleration of the fluid comprises the sum of the local *acceleration* and *advection* terms and is driven by the pressure gradient in the x direction. Assuming that the upstream velocity $U_\infty = 1$ m/s and a radius $R = 1$ m, the total acceleration is given by

$$a_x = \frac{\partial u}{\partial t} + u\frac{\partial u}{\partial x} = \left(1 - \frac{1}{x^2}\right)\left(\frac{2}{x^3}\right)$$

The equation derived represents another form of the momentum equation, that is, Euler's equation of the fluid flow. For a steady flow, Bernoulli's equation can be obtained by integrating the equation along a streamline. In other words,

$$u\frac{\partial u}{\partial x} = -\frac{1}{\rho}\frac{\partial p}{\partial x} \quad \Rightarrow \quad \frac{p(x)}{\rho} + \frac{u^2(x)}{2} = \frac{p_\infty}{\rho} + \frac{U_\infty^2}{2}$$

The above equation can be re-arranged and taking the upstream pressure p_∞ to be atmospheric, it becomes

$$p(x) - p_{atm} = \frac{\rho}{2}\left(U_\infty^2 - u^2(x)\right) = \frac{\rho}{2}\left[1 - \left(1 - \frac{1}{x^2}\right)^2\right]$$

Discussion: Along the streamline, the velocity profile $u(x)$ given above can be represented by the profile described in Fig. 3.3.2. The velocity drops very rapidly as the fluid approaches the cylinder. At the surface of the cylinder, the velocity is zero (stagnation point) and the surface pressure is a maximum.

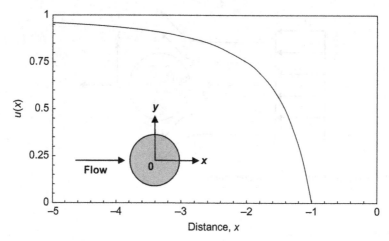

Fig. 3.3.2 Velocity profile $u(x)$ along the stagnation streamline.

The total acceleration profile depicted in Fig. 3.3.3 also shows the strong deceleration of the fluid as it approaches the cylinder. The maximum deceleration occurs at $x = -1.29$ m with a magnitude of -0.372 m/s^2.

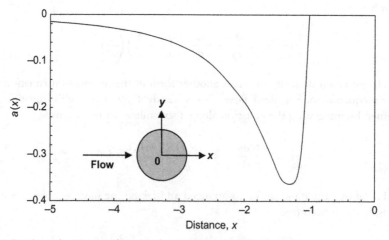

Fig. 3.3.3 Total acceleration profile $a(x)$ along the stagnation streamline.

The pressure difference $p(x) - p_{atm}$ derived above illustrated in Fig. 3.3.4 demonstrates that the pressure increases as the fluid approaches the stagnation point. With the density ρ taken to be unity, it reaches the maximum value of 0.5, that is, $p_{stag} - p_{atm} = (1/2)\rho U_\infty^2$ as $u(x) \to 0$ near the stagnation point.

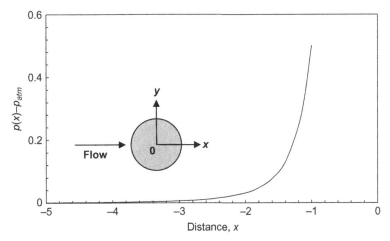

Fig. 3.3.4 Pressure difference $p(x) - p_{atm}$ along the stagnation streamline.

Example 3.4

Consider a steady incompressible laminar flow through the parallel-plate channel, as investigated in Example 3.2. For a constant property fluid with a fully developed flow, determine the velocity profile subject to the boundary condition where the vertical component v is zero everywhere.

Solution: The horizontal velocity component u depends only on x. The axial dependence on the horizontal velocity may be obtained by solving the appropriate form of the x-momentum Eq. (3.24). Since the vertical component v is zero everywhere, the continuity Eq. (3.13) reduces to

$$\frac{\partial u}{\partial x} + \underbrace{\frac{\partial v}{\partial y}}_{=0} = 0 \qquad \Rightarrow \qquad \frac{\partial u}{\partial x} = 0$$

This indicates that the velocity u is only a function of y. The x-momentum Eq. (3.24) becomes:

$$\underbrace{\frac{\partial u}{\partial t}}_{=0} + u\underbrace{\frac{\partial u}{\partial x}}_{=0} + v\underbrace{\frac{\partial u}{\partial y}}_{=0} = -\frac{1}{\rho}\frac{\partial p}{\partial x} + \nu\underbrace{\frac{\partial^2 u}{\partial x^2}}_{=0} + \nu\frac{\partial^2 u}{\partial y^2}$$

Since $\nu = \mu/\rho$, the above equation reduces to

$$\frac{\partial^2 u}{\partial y^2} = \frac{1}{\mu}\frac{\partial p}{\partial x}$$

Hence the momentum conservation requirement is just a simple balance between the shear and pressure forces in this case. Integrating once yields the velocity gradient $\partial u/\partial y$ with respect to y:

$$\frac{\partial u}{\partial y} = \frac{1}{\mu}\frac{\partial p}{\partial x}y + C_1$$

Integrating again yields the horizontal velocity u with respect to y:

$$u(y) = \frac{1}{2\mu}\left(\frac{\partial p}{\partial x}\right)y^2 + C_1 y + C_2$$

It is noted that the pressure gradient $\partial p/\partial x$ is treated as a constant as far as the integration is concerned since it is not a function of y (we can refer to the y-momentum equation later in this example).

The two boundary conditions required to determine the constants C_1 and C_2 are

$$u = 0 \quad \text{at} \quad y = \frac{H}{2} \quad \text{(no slip)}$$

$$\frac{\partial u}{\partial y} = 0 \quad \text{at} \quad y = 0 \quad \text{(symmetry)}$$

From the symmetry condition, based on the velocity gradient $\partial u/\partial y$ with respect to y, the constant C_1 is zero.

$$0 = \frac{1}{\mu}\left(\frac{\partial p}{\partial x}\right)\cdot 0 + C_1 \qquad \Rightarrow \qquad C_1 = 0$$

The constant C_2 can be determined by applying the no-slip condition above to the velocity profile $u(y)$ equation. Therefore

$$0 = \frac{1}{2\mu}\left(\frac{\partial p}{\partial x}\right)\left(\frac{H}{2}\right)^2 + C_2 \qquad \Rightarrow \qquad C_2 = -\frac{1}{2\mu}\left(\frac{\partial p}{\partial x}\right)\left(\frac{H}{2}\right)^2$$

The velocity profile $u(y)$ becomes

$$u(y) = \frac{1}{2\mu}\left(\frac{\partial p}{\partial x}\right)y^2 - \frac{1}{2\mu}\left(\frac{\partial p}{\partial x}\right)\left(\frac{H}{2}\right)^2 \Rightarrow$$

$$u(y) = \frac{3}{2}\frac{H^2}{12\mu}\left(-\frac{\partial p}{\partial x}\right)\left[1 - \frac{y^2}{(H/2)^2}\right] \Rightarrow \tag{3.4A}$$

$$u(y) = \frac{3}{2}U_m\left[1 - \frac{y^2}{(H/2)^2}\right]$$

In this equation the average velocity U_m is given as

$$U_m = \frac{H^2}{12\mu}\left(-\frac{\partial p}{\partial x}\right) \tag{3.4B}$$

The volume flow rate per unit width, q, passing between the plates can be obtained from the relationship

$$q = U_m H = -\frac{H^3}{12\mu}\left(\frac{\partial p}{\partial x}\right) \tag{3.4C}$$

The pressure gradient $\partial p/\partial x$ is negative, as the pressure decreases in the direction of the flow. If we let Δp represent the pressure drop between the inlet and outlet of the channel at a distance l apart, then

$$\frac{\Delta p}{l} = -\frac{\partial p}{\partial x}$$

and the volume flow rate can be expressed as

$$q = \frac{H^3 \Delta p}{12\mu l}$$

The flow is proportional to the pressure gradient, inversely proportional to the viscosity, and strongly dependent on the gap width ($\sim H^3$). In terms of the average velocity it becomes

$$U_m = \frac{H^2}{12\mu}\frac{\Delta p}{l}$$

The above equations provide convenient relationships for relating the pressure drop along the channel between the parallel plates and the rate of flow or average velocity. The maximum velocity U_{max} occurs midway ($y = 0$) between the two plates so that

$$U_{max} = \frac{3}{2}U_m$$

Based on the boundary condition where the vertical component v is zero everywhere and if the body force due to gravity is considered in the y-momentum Eq. (3.19), the equation becomes

$$\underbrace{\frac{\partial v}{\partial t}}_{=0} + u\underbrace{\frac{\partial v}{\partial x}}_{=0} + v\underbrace{\frac{\partial v}{\partial y}}_{=0} = -\frac{1}{\rho}\frac{\partial p}{\partial y} + v\underbrace{\frac{\partial^2 v}{\partial x^2}}_{=0} + v\underbrace{\frac{\partial^2 v}{\partial y^2}}_{=0} - g \quad \Rightarrow$$

Gravity

$$\frac{\partial p}{\partial y} = -\rho g$$

The gravitational force acts downward, which results in the negative body force being represented in the above equation. This equation can be integrated to yield

$$p = -\rho g y + \{C_3 = f(x)\}$$

which shows that the pressure varies hydrostatically in the y direction. The constant C_3 in the above equation can be expressed by a function $f(x)$, where it can be related to the varying pressure gradient $\partial p/\partial x$ along the x direction and some reference pressure p_o.

$$f(x) = \left(\frac{\partial p}{\partial x}\right)x + p_o$$

where p_o is at a location $x = y = 0$ and the pressure variation throughout the fluid can be obtained from

$$p = -\rho g y + \left(\frac{\partial p}{\partial x}\right)x + p_o \tag{3.4D}$$

Discussion: Details of a steady, laminar flow between infinite parallel plates are completely predicted by the solution formulated above from the Navier-Stokes equations. For instance, if the pressure gradient, viscosity, and plate spacing are specified, then the velocity profile can be determined, as well as those for the average velocity and flow rate – Eqs (3.4A), (3.4B), and (3.4C). Also, for a given fluid and reference pressure p_o, the pressure at any point can be predicted – Eq. (3.4D). The physical significance of the velocity distribution as seen in Eq. (3.4A) is that the profile is *parabolic*. The CFD simulation of Example 3.2 also predicts this particular velocity profile. The *parabolic* profile has been confirmed by many experiments where one of such by Nakayama et al. (1988) experimentally displayed a line of bubbles resembling a parabolic shape occurring at the downstream end of a channel. This fully developed flow distribution is known as the *Hagen-Poiseuille flow*, named after the first two investigators who reported this flow behavior (Hagen, 1839; Poiseuille, 1844). This relatively simple example of an exact solution illustrates the detail information about the flow field that can be obtained. The resultant velocity profile above being *parabolic* has physical implications as it is a typical distribution experienced for laminar flows in parallel channels.

Example 3.5

Consider the two-dimensional CFD case of an incompressible laminar flow between two stationary parallel plates to illustrate the physical meaning of the momentum equations. Here, the dimensions of the channel flow are given as: height $H = 0.1$ m and length $L = 1$ m (Fig. 3.5.1). Using CFD, discuss the development of the velocity profiles between the inlet and the *hydrodynamic entry length* (L_E) with air (density $\rho = 1.2$ kg/m^3) as the working fluid, for the following conditions:

(a) A fixed inlet velocity $u_{in} = 0.01$ m/s and dynamic viscosities $\mu_1 = 4 \times 10^{-5}$ kg/m · s and $\mu_2 = 10^{-5}$ kg/m · s.

(b) A fixed dynamic viscosity $\mu = 4 \times 10^{-5}$ kg/m · s and inlet velocities $u_{in1} = 0.01$ m/s and $u_{in2} = 0.04$ m/s.

Solution: The problem is described as follows:

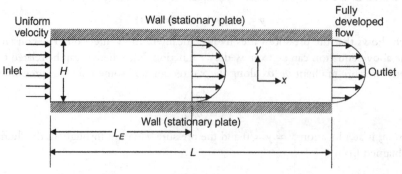

Fig. 3.5.1 Hydrodynamic entry length (L_E) location from the inlet surface in a two-dimensional laminar flow between two stationary parallel plates.

(a) By definition, the region from the inlet surface to the point at which the boundary layer merges at the centerline is called the *hydrodynamic entrance region* and the length of this region is called the *hydrodynamic entry length*. This entrance region is also referred to as the *hydrodynamically developing flow* since the velocity profile is still developing. Beyond this region, the velocity profile is fully developed and remains unchanged; it is called the *fully developed region*. The fluid flow at this instance is considered to be fully developed.

Based on the CFD simulation, the velocity profiles at different downstream locations x/H from the inlet surface for varying dynamic viscosities are given in Figs. 3.5.2 and 3.5.3. The former refers to the case where the flow has a higher dynamic viscosity than the latter.

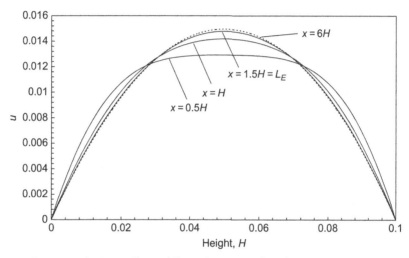

Fig. 3.5.2 Case A – Velocity profiles at different locations of x/H for $u_{in} = 0.01$ m/s and dynamic viscosity $\mu_1 = 4 \times 10^{-5}$ kg/m · s.

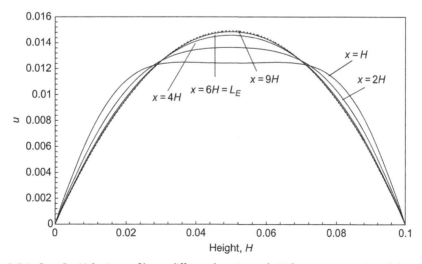

Fig. 3.5.3 Case B – Velocity profiles at different locations of x/H for $u_{in} = 0.01$ m/s and dynamic viscosity $\mu_2 = 10^{-5}$ kg/m · s.

(b) The velocity profiles at different downstream locations x/H from the inlet surface for varying inlet velocities are given in Figs. 3.5.4 and 3.5.5. The former refers to the case where the flow has a lower velocity than the latter. The case presented in Fig. 3.5.4 is obviously exactly the same as the case presented above in Fig. 3.5.2. It is repeated here for comparison against the higher velocity case.

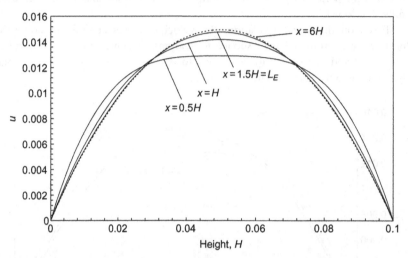

Fig. 3.5.4 Case C – Velocity profiles at different locations of x/H for $u_{in1} = 0.01$ m/s and dynamic viscosity $\mu = 4 \times 10^{-5}$ kg/m · s.

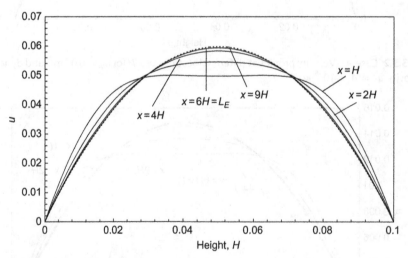

Fig. 3.5.5 Case D – Velocity profiles at different locations of x/H for $u_{in2} = 0.04$ m/s and dynamic viscosity $\mu = 4 \times 10^{-5}$ kg/m · s.

Discussion: We rewrite the steady two-dimensional x-momentum Eq. (3.24) as follows:

$$\underbrace{\rho u \frac{\partial u}{\partial x} + \rho v \frac{\partial u}{\partial y}}_{inertia} = -\underbrace{\frac{\partial p}{\partial x}}_{pressure} + \underbrace{\mu \left(\frac{\partial^2 u}{\partial x^2} + \frac{\partial^2 u}{\partial y^2} \right)}_{friction}$$

By comparing the results in Fig. 3.5.2 (Case A) with Fig. 3.5.3 (Case B), where the same *inertia* force is applied, it can be seen that the higher dynamic viscosity apparent in Case A produces a higher *friction* force which inhibits the fluid momentum. This leads to a quicker transition of the flow to a fully developed stage at a shorter distance. Case B has less frictional resistance, thus allowing a slower development of the hydrodynamic boundary layer, and reaches fully developed flow at a longer distance. The pressure gradients are the same for both of these cases. However, when the fluid has the same *friction* forces as demonstrated by the results in Fig. 3.5.4 (Case C) and Fig. 3.5.5 (Case D), the results for the higher velocity, that is, higher *inertia* force, are surprisingly exactly the same as those of the case of higher *friction* force in Fig. 3.5.3. The higher *inertia* force in Case C yields the effect of reaching the same *hydrodynamic entry length* as imposing a higher *friction* due to the higher wall shear stress resisting the fluid flow as in Case A. All the cases investigated demonstrated the relative physical contributions of the *inertia* and *friction* forces competing with each other to conserve the momentum transfer.

3.3.3 Comments

The principle of conservation of momentum has been explored with investigations of the various contributions from the *advection* and *diffusion* terms in the momentum equations that affect the fluid flow. The reader should be aware of the importance of keeping the physical meaning of the mathematical equations in mind and the roles they play in CFD analyses. In CFD the concept of *dynamic similarity* is frequently adopted. This involves normalizing the mathematical equations to yield the non-dimensional governing equations. For instance, we have observed from the previous CFD Example 3.5 that through the various combinations of different inlet velocities and dynamic viscosities, the same fluid flow effect is obtained in reference to the development of the flow in yielding the same *hydrodynamic entrance region*. We can now even change the air density from $\rho_1 = 1.2$ kg/m^3 to $\rho_2 = 4.8$ kg/m^3 while fixing the inlet velocity and dynamic viscosity at 0.01 m/s and $\mu = 4 \times 10^{-5}$ kg/m · s, respectively; the same results are obtained. This is because the increase of density contributes to the increase of the inertia force, which has the same effect as increasing the inlet velocity. There appears to be some homogeneity of the fluid flow behavior that results in the combination of these physical variables. One important non-dimensional parameter that describes the flow characteristics is the Reynolds number (*Re*), which is defined as the ratio of the *Inertia Force* over the *Friction Force*, that is,

$$\text{Re} = \frac{Inertia\ Force}{Friction\ Force} = \frac{\rho u_{in} H}{\mu} \tag{3.26}$$

It is observed that Eq. (3.25) encapsulates the three variables, density, dynamic viscosity, and inlet velocity. Through different combinations of these variables, the same *hydrodynamic entrance region* will be achieved if the resultant Reynolds number is the same. Another important use of the Reynolds number is to indicate whether the flow is laminar or turbulent. For a channel flow, the flow will remain laminar if the critical Reynolds number is below 1400. Further physical interpretation of this non-dimensional parameter will be demonstrated in Example 3.9.

3.4 THE ENERGY EQUATION

3.4.1 Energy Conservation

The equation for the conservation of energy is derived from the consideration of the *first law of thermodynamics*:

$$\text{Time rate of change of energy} = \text{Net rate of heat added}\left(\sum \dot{Q}\right)$$
$$+ \text{Net rate of work done}\left(\sum \dot{W}\right) \tag{3.27}$$

As discussed in Appendix A, the time rate of change of any arbitrary variable property ϕ is defined as *the product between the density and the substantial derivative of ϕ*. In keeping with our derivation of the Navier-Stokes (the momentum equations), we will refer again to the elemental volume in the Cartesian frame, as previously described in Fig. 3.2. The time rate of change of energy for the moving fluid element is just simply

$$\rho \frac{DE}{Dt} \Delta x \Delta y \Delta z \tag{3.28}$$

The two terms represented by $\sum \dot{Q}$ and $\sum \dot{W}$ describe the net rate of heat addition to the fluid within the control volume and the net rate of work done by surface forces on the fluid. We first consider the effects in the x direction, as illustrated in Fig. 3.7. The rate of work done and heat added in the y and z directions automatically follows from the analysis of the x direction.

The rate of work done on the control volume in the x direction is equivalent to the product between the surface forces (caused by the normal viscous stress σ_{xx} and tangential viscous stresses τ_{yx} and τ_{zx}) with the velocity component u. The formulae for the net rate work done in this direction as well as for the other coordinate directions due to the contributions of the normal and tangential surface forces are detailed in Appendix A. When we combine all the contributions of the surface forces in the x, y, and z directions, and substituting these expressions along with the time rate of change of

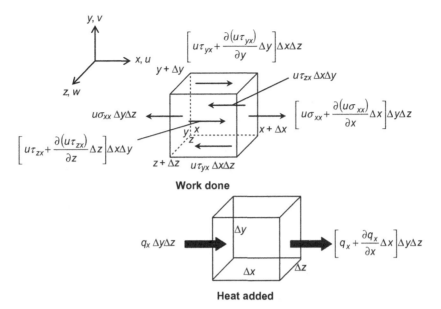

Fig. 3.7 Work done by surface forces on the fluid and heat added to the fluid within the infinitesimal control volume. Only the fluxes in the x direction are shown.

energy E, from Eq. (3.28) into Eq. (3.27), the *non-conservative* form of the equation for the conservation of energy is given as

$$\rho \frac{DE}{Dt} = \frac{\partial(u\sigma_{xx})}{\partial x} + \frac{\partial(v\sigma_{yy})}{\partial y} + \frac{\partial(w\sigma_{zz})}{\partial z}$$
$$+ \frac{\partial(u\tau_{yx})}{\partial y} + \frac{\partial(u\tau_{zx})}{\partial z} + \frac{\partial(v\tau_{xy})}{\partial x} + \frac{\partial(v\tau_{zy})}{\partial z} + \frac{\partial(w\tau_{xz})}{\partial x} + \frac{\partial(w\tau_{yz})}{\partial y} \qquad (3.29)$$
$$- \frac{\partial q_x}{\partial x} - \frac{\partial q_y}{\partial y} - \frac{\partial q_z}{\partial z}$$

The energy fluxes q_x, q_y, and q_z in Eq. (3.29) can be formulated by applying *Fourier's law of heat conduction* that relates the heat flux to the local temperature gradient:

$$q_x = -\lambda \frac{\partial T}{\partial x} \qquad q_y = -\lambda \frac{\partial T}{\partial y} \qquad q_z = -\lambda \frac{\partial T}{\partial z} \qquad (3.30)$$

where λ is the thermal conductivity. By substituting Eq. (3.30) into Eq. (3.29), and applying the normal stresses described in Appendix A, Eq. (3.29) becomes

$$\rho \frac{DE}{Dt} = \frac{\partial}{\partial x}\left[\lambda \frac{\partial T}{\partial x}\right] + \frac{\partial}{\partial y}\left[\lambda \frac{\partial T}{\partial y}\right] + \frac{\partial}{\partial z}\left[\lambda \frac{\partial T}{\partial z}\right]$$
$$- \frac{\partial(up)}{\partial x} - \frac{\partial(vp)}{\partial y} - \frac{\partial(wp)}{\partial z} + \Phi \qquad (3.31)$$

The effects due to the viscous stresses in Eq. (3.31) are described by the dissipation function Φ that can be shown to be

$$\Phi = \frac{\partial(u\tau_{xx})}{\partial x} + \frac{\partial(u\tau_{yx})}{\partial y} + \frac{\partial(u\tau_{zx})}{\partial z}$$
$$+ \frac{\partial(v\tau_{xy})}{\partial x} + \frac{\partial(v\tau_{yy})}{\partial y} + \frac{\partial(v\tau_{zy})}{\partial z} + \frac{\partial(w\tau_{xz})}{\partial x} + \frac{\partial(w\tau_{yz})}{\partial y} + \frac{\partial(w\tau_{zz})}{\partial z}$$

The dissipation function represents a source of energy due to deformation work done on the fluid. This work is extracted from the mechanical energy that causes fluid movement, which is converted into heat.

Based on the concept of *substantial derivative* as described in Appendix A, the *conservative* form of the energy equation can be rewritten as

$$\frac{\partial(\rho E)}{\partial t} + \frac{\partial(\rho u E)}{\partial x} + \frac{\partial(\rho v E)}{\partial y} + \frac{\partial(\rho w E)}{\partial z} = \frac{\partial}{\partial x}\left[\lambda\frac{\partial T}{\partial x}\right] + \frac{\partial}{\partial y}\left[\lambda\frac{\partial T}{\partial y}\right] + \frac{\partial}{\partial z}\left[\lambda\frac{\partial T}{\partial z}\right]$$
$$- \frac{\partial(up)}{\partial x} - \frac{\partial(vp)}{\partial y} - \frac{\partial(wp)}{\partial z} + \Phi$$

$$(3.32)$$

Thus far, we have not defined the specific energy E of a fluid. Often the energy of a fluid is defined as the sum of the *internal energy*, *kinetic energy*, and *gravitational potential energy*. We can regard the gravitational force as a body force and include the effects of potential energy changes as a source term. The energy equation for a *compressible flow* is often re-arranged to give an equation for the *enthalpy*, which can be written in the form of

$$\frac{\partial(\rho h)}{\partial t} + \frac{\partial(\rho u h)}{\partial x} + \frac{\partial(\rho v h)}{\partial y} + \frac{\partial(\rho w h)}{\partial z} = \frac{\partial p}{\partial t}$$
$$+ \frac{\partial}{\partial x}\left[\lambda\frac{\partial T}{\partial x}\right] + \frac{\partial}{\partial y}\left[\lambda\frac{\partial T}{\partial y}\right] + \frac{\partial}{\partial z}\lambda\left[k\frac{\partial T}{\partial z}\right] + \Phi$$

$$(3.33)$$

More detail discussion regarding the relationship between the specific energy E and enthalpy h can be found in Appendix A as well as in the textbook of Cengel (2003).

For a two-dimensional *compressible flow*, the equation for the conservation of energy can be reduced to

$$\underbrace{\frac{\partial(\rho h)}{\partial t}}_{\text{local acceleration}} + \underbrace{\frac{\partial(\rho u h)}{\partial x} + \frac{\partial(\rho v h)}{\partial y}}_{\text{advection}} = \underbrace{\frac{\partial p}{\partial t}}_{\text{local pressure time derivative}} + \underbrace{\frac{\partial}{\partial x}\left[\lambda\frac{\partial T}{\partial x}\right] + \frac{\partial}{\partial y}\left[\lambda\frac{\partial T}{\partial y}\right]}_{\text{diffusion}} + \Phi$$

$$(3.34)$$

Let us also consider for the special case where the fluid is incompressible. For such flows in most practical fluid engineering problems, the local pressure time derivative $\partial p/\partial t$ and the dissipation function Φ can be neglected. Based on the definition of enthalpy, h can be reduced to $C_p T$, where C_p is the specific heat and is assumed to be constant. With the temperature being invariant along the z direction and the thermal conductivity λ being constant, the equation for the conservation of energy in two dimensions for an *incompressible flow* can be expressed as

$$\rho C_p \underbrace{\frac{DT}{Dt}}_{acceleration} = \underbrace{\lambda \frac{\partial^2 T}{\partial x^2} + \lambda \frac{\partial^2 T}{\partial y^2}}_{diffusion}$$

or

$$\underbrace{\frac{\partial T}{\partial t}}_{local\ acceleration} + \underbrace{u \frac{\partial T}{\partial x} + v \frac{\partial T}{\partial y}}_{advection} = \underbrace{\frac{\lambda}{\rho C_p} \frac{\partial^2 T}{\partial x^2} + \frac{\lambda}{\rho C_p} \frac{\partial^2 T}{\partial y^2}}_{diffusion} \tag{3.35}$$

3.4.2 Physical Interpretation

Physically, Eq. (3.35) defines the temperature of a differential fluid control volume as it travels past a point, taking into consideration the local acceleration derivative (where the temperature in itself may be fluctuating with time at a given point) and also the advection derivative (where the temperature changes spatially from one point to another). To reinforce the physical meaning of these derivatives, imagine you are sitting on a high seat, close to the ceiling, in a sauna room, where the buoyant heat flow causes the air to be hottest. You decide to move to a lower seat in the sauna and make your way down to the floor. As you descend, the air is a little cooler and you experience a temperature decrease – this is analogous to the *advection* derivative in Eq. (3.35). Additionally as you sit on the lower seat, the sauna door opens as someone enters and you feel a sudden rush of cold air. The temperature around you immediately drops for that moment – this is analogous to the local *acceleration* derivative in Eq. (3.35). The net temperature change you experience is therefore a combination of both the act of descending from the highest seat to the floor and also the rush of cool air as the sauna door opens.

The remaining term in Eq. (3.35) represents the temperature flow due to heat conduction (the *diffusion* derivative), where λ is the thermal conductivity of the fluid. We examine the physical interaction of this term with the previous two derivatives. To further reinforce the physical meaning, imagine the problem concerning the fluid flowing between parallel plates in Fig. 3.2, of which the plates are now heated. If the surrounding fluid velocity is very low, the surrounding fluid temperature within the channel increases due to the heat flowing from the flat plate into the bulk fluid – this is analogous to the heat conduction through the heat *diffusion* derivative dominating over the local *acceleration* and *advection* derivatives of the fluid. However, if the surrounding fluid velocity is high, the heat of the fluid is carried away by the relatively cooler fluid. The high temperatures are only found near the hot surface of the flat plates – this is analogous to the local *acceleration* and *advection* derivatives dominating over the heat *diffusion* derivative.

EXAMPLES

Example 3.6

To illustrate the application of the energy Eq. (3.35), consider the steady heat conduction across an infinite long solid slab with a finite thickness, as illustrated in Fig. 3.6.1. Determine the analytical expressions based on the boundary conditions:

(a) $x = 0$, $T = T_0$ and $x = L$, $T = T_L$
(b) $x = 0$, $T = T_0$ and $x = L$, $\dot{q}_L = -k\frac{\partial T(L)}{\partial x}$.

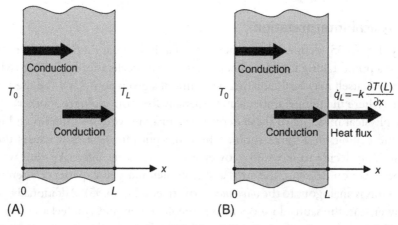

Fig. 3.6.1 Heat conduction across the solid slab subjected to various temperature boundary conditions.

Solution: Since the heat transfer is steady and it is a solid problem, the acceleration and advection terms in Eq. (3.34) vanish.

$$\underbrace{\frac{\partial T}{\partial t}}_{=0} + \underbrace{u\frac{\partial T}{\partial x}}_{=0} + \underbrace{v\frac{\partial T}{\partial y}}_{=0} = \frac{\lambda}{\rho\,C_p}\frac{\partial^2 T}{\partial x^2} + \frac{\lambda}{\rho\,C_p}\frac{\partial^2 T}{\partial y^2}$$

The above equation reduces to

$$\frac{\partial^2 T}{\partial x^2} + \frac{\partial^2 T}{\partial y^2} = 0$$

For an infinite long slab, we can restrict ourselves to a one-dimensional analysis; the equation can be further simplified to

$$\frac{\partial^2 T}{\partial x^2} + \cancelto{=0}{\frac{\partial^2 T}{\partial y^2}} = 0 \qquad \Rightarrow \qquad \frac{\partial^2 T}{\partial x^2} = 0$$

Integrating the differential equation once with respect to x yields

$$\frac{\partial T}{\partial x} = C_1$$

where C_1 is an arbitrary constant. Subsequent integration of the above equation yields

$$T(x) = C_1 x + C_2$$

which is the analytical expression of the differential equation. The general solution contains two unknown constants, C_1 and C_2. We need two equations that can be determined through the boundary conditions imposed on the left and right surfaces of the solid slab.

Analysis:

(a) For this case, the left side boundary condition is applied to the general solution by replacing the x's with zero and $T(x)$ with T_0. In other words

$$T_0 = C_1 \times 0 + C_2 \qquad \Rightarrow \qquad C_2 = T_0$$

The right side boundary condition is applied by replacing the x's with L and $T(x)$ with T_L, which gives

$$T_L = C_1 L + C_2 \quad \Rightarrow \quad T_L = C_1 L + T_0 \quad \Rightarrow \quad C_1 = \frac{T_L - T_0}{L}$$

Substituting the C_1 and C_2 expressions into the general solution, we obtain

$$T(x) = \frac{T_L - T_0}{L} x + T_0$$

(b) For case b, the left side boundary condition is the same as in part (a). Thus

$$C_2 = T_0$$

On the right side boundary condition, a heat flux $\dot{q}_L$ is specified rather than a specified temperature T_L. Noting that

$$\frac{\partial T}{\partial x} = C_1$$

the application of the boundary condition yields

$$-\lambda \frac{\partial T(L)}{\partial x} = \dot{q}_L \quad \Rightarrow \quad -\lambda C_1 = \dot{q}_L \quad \Rightarrow \quad C_1 = -\frac{\dot{q}_L}{\lambda}$$

Substituting the above equations into the general solution, we obtain

$$T(x) = -\frac{\dot{q}_L}{\lambda} x + T_0$$

Discussion: The above equations are the analytical solutions for the temperature distribution across the finite thickness L in the solid slab. They satisfy not only the one-dimensional partial differential form of the energy equation but also the two specified boundary conditions. Both general solutions exhibit a linear distribution whose slopes are $T_L - T_0/L$ and $-\dot{q}_L/k$, respectively. During the integration process in establishing the general one-dimensional partial differential form of the energy equation, the thermal diffusivity $\alpha = \lambda/\rho\, C_p$ disappears. This may imply to the reader that the heat conduction across the slab is not influenced by any thermophysical properties such as the thermal conductivity k, density ρ, and specific heat C_p of the solid material. Nevertheless, in part (b) the final expression of the general solution demonstrates the influence of the material thermal conductivity that is obtained from the physical boundary condition based on Fourier's law. A better understanding of the types of physical boundary conditions for the temperature will be addressed later at the end of this chapter.

Example 3.7

Consider a two-dimensional CFD case of the incompressible laminar flow between two stationary parallel plates to illustrate the physical meaning of the energy equation. Using CFD and the dimensions of height $H = 0.1$ m and length $L = 1$ m (Fig. 3.7.1), with air as the working fluid (density $\rho = 1.2$ kg/m^3), a specified uniform temperature of 330 K at the inlet, and a wall temperature maintained at 300 K, discuss the development of the temperature profiles between the inlet and the *thermal entry length* (L_T) for the following cases:

(a) A fixed inlet velocity $u_{in} = 0.01$ m/s and thermal conductivities $\lambda_1 = 0.04$ W/m · K and $\lambda_2 = 0.01$ W/m · K.

(b) A fixed thermal conductivity $\lambda = 0.04$ W/m · K and inlet velocities $u_{in1} = 0.01$ m/s and $u_{in2} = 0.1$ m/s.

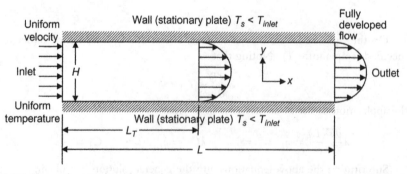

Fig. 3.7.1 Thermal entry length (L_T) location from the inlet surface in a two-dimensional laminar flow between two stationary parallel plates.

Solution: The problem is described as follows:

(a) By definition, the region from the inlet surface to the point at which the thermal boundary layer merges at the centerline is called the *thermal entrance region* and the length of this region is called the *thermal entry length*. Flow in this entrance region is also called the *thermally developing flow* since the dimensionless temperature profile, $(T_s - T) / (T_s - T_m)$, where T_m denotes the mean temperature, is still developing. Beyond this region, the dimensionless temperature profile is fully developed and remains unchanged. Similar to the fluid flow, the region in which the flow is thermally developed and thus both the velocity and dimensionless temperature profiles remain unchanged is called *fully developed flow*.

Based on the CFD simulation, the dimensionless temperature profiles at different downstream locations x/H from the inlet surface for varying thermal conductivities are given in Figs. 3.7.2 and 3.7.3. The former refers to the case where the flow has a higher thermal conductivity than the latter.

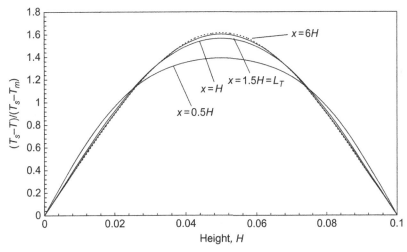

Fig. 3.7.2 Dimensionless temperature profiles at different locations of x/H for $u_{in} = 0.01$ m/s and thermal conductivity $\lambda_1 = 0.04$ W/m · K.

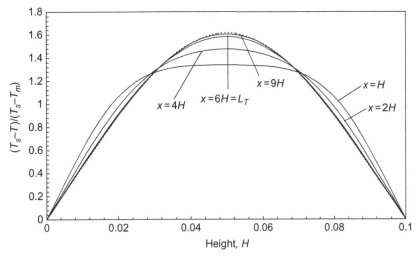

Fig. 3.7.3 Dimensionless temperature profiles at different locations of x/H for $u_{in} = 0.01$ m/s and thermal conductivity $\lambda_2 = 0.01$ W/m · K.

(b) The temperature profiles at different downstream locations x/H from the inlet surface for varying inlet velocities are given in Figs. 3.7.4 and 3.7.5. The former refers to the case where the flow has a lower velocity than the latter. For comparison purposes, the case presented in Fig. 3.7.4 is of course exactly the same as the case presented above in Fig. 3.7.2 and is repeated here against the higher-velocity case.

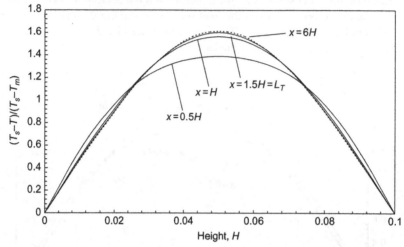

Fig. 3.7.4 Dimensionless temperature profiles at different locations of x/H for $u_{in1} = 0.01$ m/s and thermal conductivity $\lambda = 0.04$ W/m · K.

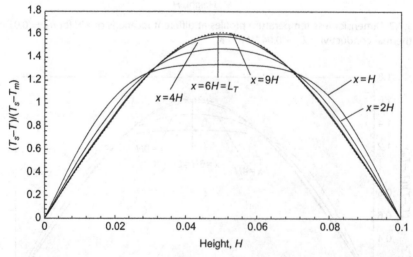

Fig. 3.7.5 Dimensionless temperature profiles at different locations of x/H for $u_{in2} = 0.1$ m/s and thermal conductivity $k = 0.04$ W/m · K.

Discussion: As in the previous Example 3.5 for the momentum equations, we can rewrite the steady two-dimensional energy Eq. (3.34) as follows:

$$\underbrace{u\frac{\partial T}{\partial x} + v\frac{\partial T}{\partial y}}_{advection} = \underbrace{\frac{\lambda}{\rho C_p}\left(\frac{\partial^2 T}{\partial x^2} + \frac{\partial^2 T}{\partial y^2}\right)}_{diffusion}$$

The competition between the *advection* and *diffusion* in the energy equation is highlighted from the above results through the variation of the thermal conductivities and inlet velocities. For the case of a higher thermal conductivity, the *diffusion* term dominates. This leads to a shorter thermal entry length (i.e. the fully developed stage is reached at a shorter distance) due to a quicker development of the thermal boundary layer. This appears to have the same effect as that for the case of a higher *friction* force that inhibits the fluid momentum. However, a lower thermal conductivity induces less resistance to the heat flow, thus allowing a slower development of the thermal boundary layer and the flow reaches the fully developed stage at a longer distance. With a higher inlet velocity, this culminates in a higher *advection* overcoming the *diffusion* effect of the heat flow resulting in a longer development of the thermal entry length. It is noted that the thermal entry lengths for all the cases in the current example correspond exactly to the same hydrodynamic entry lengths obtained in Example 3.5. More explanation of the relationship between these hydrodynamic and thermal behaviors is given in the next section. All the cases investigated demonstrated the relative physical contributions of the *advection* and *diffusion* effects competing with each other to conserve the energy transfer.

3.4.3 Comments

Here, the principle of conservation of energy, like in the conservation of momentum, has been explored with investigations of the various contributions of the *advection* and *diffusion* terms in the energy equation. Through Example 3.7, there also appears to be some homogeneity for the thermal behavior that results in the combination of the various flow variables. The dimensionless parameter previously introduced in Section 3.3.3, the Reynolds number (*Re*), depicts the contribution between the *Inertia Force* and the *Friction Force*, which describes the flow characteristics. Here, we introduce another important dimensionless parameter, the Prandtl number (*Pr*), which denotes the ratio of the *Molecular diffusivity of momentum* over the *Molecular diffusivity of heat*, that is,

$$\text{Pr} = \frac{Molecular\ diffusivity\ of\ momentum}{Molecular\ diffusivity\ of\ heat} = \frac{\nu}{\alpha} = \frac{\mu C_p}{\lambda} \tag{3.36}$$

During laminar flow, the magnitude of the dimensionless Prandtl number is a measure of the relative growth of the velocity and thermal boundary layers. For fluids with $Pr \approx 1$, such as gases, the two boundary layers essentially coincide with each other. From Example 3.7, if the specific heat of air is taken to be 1000 J/kg, various combinations of the dynamic viscosity and thermal conductivity of air can enable the Prandtl number to be unity. Here, the two boundary layers coincide with each other and the same hydrodynamic and thermal entry lengths are obtained. For fluids with $Pr \gg 1$, such as water or oils, the velocity boundary layer outgrows the thermal boundary layer, while the opposite is true for fluids with $Pr \ll 1$ such as liquid metals. In thermal flow problems we can therefore infer that the heat transfer characteristics are controlled by the combination of the Prandtl number and Reynolds number (representing the *advection* term). Further interpretation of the Prandtl number will be demonstrated in Example 3.9.

3.5 THE ADDITIONAL EQUATIONS FOR TURBULENT FLOW

3.5.1 What Is Turbulence

Many if not most flows of engineering significance are turbulent in nature. The turbulent flow regime is therefore not just of theoretical interest among academics but a problematic source for engineers who need to capture the effects of turbulence in solving everyday problems.

Flows in the laminar regime are completely described by the continuity and momentum equations as aforementioned. In simple cases they can be solved analytically (Examples 3.1, 3.3, and 3.4). More complex flows may have to be tackled numerically with CFD techniques. It is well known that small disturbances associated with disturbances in the fluid streamlines of a laminar flow can eventually lead to a chaotic and random state of motion: a turbulent condition. These disturbances may originate from the free stream of the fluid motion, or induced by the surface roughness, where they may be amplified in the direction of the flow, in which case turbulence will occur. The onset of turbulence depends on the ratio of the inertia force to viscous force, which is indicated by the Reynolds number, Eq. (3.20). At a low Reynolds number, inertia forces are smaller than the viscous forces. The naturally occurring disturbances are dissipated away and the flow remains laminar. At a high Reynolds number, the inertia forces are sufficiently large to amplify the disturbances, and a transition to turbulence occurs. Here, the motion becomes intrinsically unstable even with constant imposed boundary conditions. The velocity and all other flow properties are varying in a random and chaotic way.

Turbulence is associated with the existence of *random fluctuations* in the fluid. This behavior can be exemplified by a typical point velocity measurement as a function of time at some location in the turbulent flow shown in Fig. 3.8. The random nature of flow precludes computations based on the equations that describe the fluid motion. Although the conservation equations remain applicable, the dependent variable such as the transient

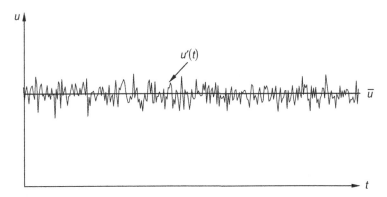

Fig. 3.8 Velocity fluctuating with time at some point in a turbulent flow.

velocity distribution in Fig. 3.6 must be interpreted as an instantaneous velocity, a phenomenon that is impossible to predict as the fluctuating velocity occurs randomly with time. Instead, the velocity can be decomposed into a steady mean value $\bar{u}$ with a fluctuating component $u'(t)$ superimposed on it: $u(t) = \bar{u} + u'(t)$. In general, it is most attractive to characterize a turbulent flow by the mean values of flow properties ($\bar{u}, \bar{v}, \bar{w}, \bar{p}$ etc.) with its corresponding statistical fluctuating property (u', v', w', p' etc.).

Turbulent fluctuations always have a *three-dimensional* spatial character. Visualizations of turbulent flows have revealed rotational flow structures, so-called turbulent eddies, with a wide range of length and velocity scales called turbulent scales. The largest eddies have a characteristic velocity and a characteristic length of the same order as the velocity and length scale of the mean flow. This suggests that for turbulent flows where $ul \gg \nu$ (i.e. inertia effects dominate viscous effects), the largest eddies whose scales are comparable with the mean flow are dominated by inertia effects rather than the viscous effects. The large eddies are therefore effectively inviscid. Transport of eddies is attained by the extraction of energy from the mean flow by a process called vortex stretching. The presence of velocity gradients in the mean flow causes deformation of the fluid such as shear and linear strain and rotation that "stretch" eddies that are appropriately aligned by forcing one end of the eddies to move faster than the other. During vortex stretching, the angular momentum is conserved and the stretching work done by the mean flow on the large eddies provides the energy that maintains the turbulence. These larger eddies then breed new instabilities creating smaller eddies that are transported mainly by vortex stretching from the larger eddy rather than from the mean flow. Thus the energy is handed down from the larger eddy to the smaller eddy. This process continues until the eddies become so small that viscous effects become important (the eddy length scales $ul \ll \nu$). Work is performed against the action of the viscous stresses so that the energy associated with the eddy motions is dissipated and converted into thermal internal energy. The continual transfer of energy from the larger eddy to smaller and smaller eddies is termed *energy cascade*. Larger eddies are flow dependent as they are generated from mean flow characteristics; thus their turbulent scales are large compared with viscosity, causing the structure of the eddy to be highly *anisotropic* (i.e. varying in all directions). Small eddies have much smaller turbulent scales (with scales up to the order of 10^{-4}) compared with viscosity causing the flow to be *isotropic* since the diffusive effects of viscosity dominate and smear out the directionality of the flow structure.

3.5.2 *k-ε* Two-Equation Turbulence Model

There is a crucial difference when modeling the physical phenomena between laminar and turbulent flow. For the latter, the appearance of turbulence eddies occurs over a wide range of length scales. A typical flow domain having a cross-sectional area of 0.1 m by 0.1 m with a high Reynolds number turbulent flow might contain eddies down to 10 or 100 μm in size. In order to describe the flow processes at all length scales, we would require computing meshes of 10^9 to 10^{12} grid points. Also, the fastest events can take place with a frequency of the order 10 kHz, which we would need to discretize time into steps of about 100 μs.

With the present-day computing power, the computing requirements for a direct numerical solution (DNS) of the time-dependent Navier-Stokes equations of fully turbulent flows at high Reynolds numbers are still truly phenomenal. Meanwhile, engineers require computational procedures that can supply adequate information about the turbulent processes but wish to avoid the need to predict all the effects associated with each and every eddy in the flow. This category of CFD users is almost always satisfied with information about the time-averaged properties of the flow (e.g. mean velocities, mean pressures, mean stresses etc.). Since engineers are content to focus their attention on mean quantities, by adopting a suitable time-averaging operation on the momentum equations, we are able to discard all details concerning the state of the flow contained in the instantaneous fluctuations. This process, of obtaining mean quantities, can be applied to the compressible, two-dimensional equations of continuity and the *conservative* forms of momentum and energy that produce the time-averaged governing equations or more popularly known as the Reynolds-averaged Navier-Stokes (RANS) equations, given by

$$\frac{\partial \rho}{\partial t} + \frac{\partial(\rho \bar{u})}{\partial x} + \frac{\partial(\rho \bar{v})}{\partial y} = 0 \tag{3.37}$$

$$\frac{\partial(\rho \bar{u})}{\partial t} + \frac{\partial(\rho \bar{u}\bar{u})}{\partial x} + \frac{\partial(\rho \bar{v}\bar{u})}{\partial y} = -\frac{\partial \bar{p}}{\partial x} + \frac{\partial}{\partial x}\left(\mu \frac{\partial \bar{u}}{\partial x}\right) + \frac{\partial}{\partial y}\left(\mu \frac{\partial \bar{u}}{\partial y}\right)$$

$$+ \frac{\partial}{\partial x}\left[\mu \frac{\partial \bar{u}}{\partial x}\right] + \frac{\partial}{\partial y}\left[\mu \frac{\partial \bar{v}}{\partial x}\right] - \left[\frac{\partial(\rho \overline{u'u'})}{\partial x} + \frac{\partial(\rho \overline{u'v'})}{\partial y}\right] \tag{3.38}$$

$$\frac{\partial(\rho v)}{\partial t} + \frac{\partial(\rho \bar{u}\bar{v})}{\partial x} + \frac{\partial(\rho \bar{v}\bar{v})}{\partial y} = -\frac{\partial \bar{p}}{\partial y} + \frac{\partial}{\partial x}\left(\mu \frac{\partial \bar{v}}{\partial x}\right) + \frac{\partial}{\partial y}\left(\mu \frac{\partial \bar{v}}{\partial y}\right)$$
$$+ \frac{\partial}{\partial x}\left[\mu \frac{\partial \bar{u}}{\partial y}\right] + \frac{\partial}{\partial y}\left[\mu \frac{\partial \bar{v}}{\partial y}\right] - \left[\frac{\partial(\rho \overline{u'v'})}{\partial x} + \frac{\partial(\rho \overline{v'v'})}{\partial y}\right] \tag{3.39}$$

$$\frac{\partial(\rho \overline{T})}{\partial t} + \frac{\partial(\rho \bar{u}\overline{T})}{\partial x} + \frac{\partial(\rho \bar{v}\overline{T})}{\partial y} = \frac{\partial}{\partial x}\left(\frac{k}{C_p}\frac{\partial \overline{T}}{\partial x}\right) + \frac{\partial}{\partial y}\left(\frac{k}{C_p}\frac{\partial \overline{T}}{\partial y}\right)$$
$$- \left[\frac{\partial(\rho \overline{u'T'})}{\partial x} + \frac{\partial(\rho \overline{v'T'})}{\partial y}\right] \tag{3.40}$$

where $\bar{u}$, $\bar{v}$, $\bar{p}$, and $\overline{T}$ are mean values and u', v', p', and T' are turbulent fluctuations. The equations above are similar to those formulated for laminar flows, except for the presence of additional terms of the form $\overline{a'b'}$. As a result, we have three additional unknowns (in three dimensions, we will have nine additional unknowns), known as the Reynolds stresses, in the time-averaged momentum equations. Similarly, the time-averaged temperature equation shows extra terms $\overline{u'T'}$ and $\overline{v'T'}$ (in three dimensions, we have an extra term $\overline{w'T'}$). For incompressible, two-dimensional equations of continuity and the *non-conservative* forms of momentum and energy, the time-averaged governing equations can be expressed as

$$\frac{\partial \bar{u}}{\partial x} + \frac{\partial \bar{v}}{\partial y} = 0 \tag{3.41}$$

$$\frac{\partial \bar{u}}{\partial t} + \frac{\partial(\overline{uu})}{\partial x} + \frac{\partial(\overline{vu})}{\partial y} = -\frac{1}{\rho}\frac{\partial \bar{p}}{\partial x} + \frac{\partial}{\partial x}\left(\nu \frac{\partial \bar{u}}{\partial x}\right) + \frac{\partial}{\partial y}\left(\nu \frac{\partial \bar{u}}{\partial y}\right) + \frac{\partial}{\partial x}\left[\nu \frac{\partial \bar{u}}{\partial x}\right]$$
$$+ \frac{\partial}{\partial y}\left[\nu \frac{\partial \bar{v}}{\partial x}\right] - \left[\frac{\partial(\overline{u'u'})}{\partial x} + \frac{\partial(\overline{u'v'})}{\partial y}\right] \tag{3.42}$$

$$\frac{\partial v}{\partial t} + \frac{\partial(\overline{uv})}{\partial x} + \frac{\partial(\overline{vv})}{\partial y} = -\frac{1}{\rho}\frac{\partial \bar{p}}{\partial y} + \frac{\partial}{\partial x}\left(\nu \frac{\partial \bar{v}}{\partial x}\right) + \frac{\partial}{\partial y}\left(\nu \frac{\partial \bar{v}}{\partial y}\right) + \frac{\partial}{\partial x}\left[\nu \frac{\partial \bar{u}}{\partial y}\right]$$
$$+ \frac{\partial}{\partial y}\left[\nu \frac{\partial \bar{v}}{\partial y}\right] - \left[\frac{\partial(\overline{u'v'})}{\partial x} + \frac{\partial(\overline{v'v'})}{\partial y}\right] \tag{3.43}$$

$$\frac{\partial \overline{T}}{\partial t} + \frac{\partial(\bar{u}\overline{T})}{\partial x} + \frac{\partial(\bar{v}\overline{T})}{\partial y} = \frac{\partial}{\partial x}\left(\frac{k}{\rho C_p}\frac{\partial \overline{T}}{\partial x}\right) + \frac{\partial}{\partial y}\left(\frac{k}{\rho C_p}\frac{\partial \overline{T}}{\partial y}\right) - \left[\frac{\partial(\overline{u'T'})}{\partial x} + \frac{\partial(\overline{v'T'})}{\partial y}\right] \tag{3.44}$$

where the term $k/\rho\, C_p$ in Eq. (3.44) is the thermal diffusivity α of the fluid.

The time-averaged equations can be solved if the Reynolds stresses and extra temperature transport terms can be related to the mean flow and heat quantities. It was proposed by Boussinesq (1868) that the Reynolds stresses could be linked to the mean rates of deformation. We obtain

$$-\rho\overline{u'u'} = 2\mu_T \frac{\partial \overline{u}}{\partial x} - \frac{2}{3}\rho k \qquad -\rho\overline{v'v'} = 2\mu_T \frac{\partial \overline{v}}{\partial y} - \frac{2}{3}\rho k$$

$$-\rho\overline{u'v'} = \mu_T \left(\frac{\partial \overline{v}}{\partial x} + \frac{\partial \overline{u}}{\partial y} \right)$$

(3.45)

The right-hand side is analogous to *Newton's law of viscosity*, except for the appearance of the turbulent or eddy viscosity μ_T and turbulent kinetic energy k.

In Eq. (3.45) the turbulent momentum transport is assumed to be proportional to the mean gradients of velocity. Similarly the turbulent transport of temperature is taken to be proportional to the gradient of the mean value of the transported quantity. In order words

$$-\rho\overline{u'T'} = \Gamma_T \frac{\partial \overline{T}}{\partial x} \qquad -\rho\overline{v'T'} = \Gamma_T \frac{\partial \overline{T}}{\partial y}$$

(3.46)

where Γ_T is the turbulent diffusivity. Since the turbulent transport of momentum and heat is due to the same mechanisms, namely, eddy mixing, the value of the turbulent viscosity can be taken to be close to that of turbulent viscosity μ_T. Based on the definition of the turbulent Prandtl number Pr_T, we obtain

$$Pr_T = \frac{\mu_T}{\Gamma_T}$$

Experiments have established that this ratio is often nearly constant. Most CFD procedures assume this to be the case and use values of Pr_T around unity.

Since the complexity of turbulence in most engineering flow problems precludes the use of any simple formulae, it is possible to develop similar transport equations to accommodate the turbulent quantity k and other turbulent quantities, one of which is the rate of dissipation of turbulent energy ε. Here we indicate the form of a typical two-equation turbulence model that is commonly used in handling many turbulent fluid engineering problems, the *standard k-ε model* by Launder and Spalding (1974).

Some preliminary definitions are required first. The turbulent kinetic energy k and rate of dissipation of turbulent energy ε can be defined and expressed in Cartesian tensor notation as

$$k = \frac{1}{2}\overline{u'_i u'_i} \text{ and } \varepsilon = \nu_T \overline{\left(\frac{\partial u'_i}{\partial x_j} \right)\left(\frac{\partial u'_i}{\partial x_j} \right)} \text{ where } i,j = 1,2,3$$

From the local values of k and ε, a local turbulent viscosity μ_T can be evaluated as

$$\mu_T = \rho C_\mu \frac{k^2}{\varepsilon}$$

(3.47)

and the kinematic turbulent or eddy viscosity is denoted by $\nu_T = \mu_T / \rho$.

By substituting the Reynolds stress expressions in Eqs (3.38), (3.39), (3.42), and (3.43) and the extra temperature transport terms in Eqs (3.40) and (3.44), removing the overbar that is by default indicating the average quantities, we obtain the compressible form of governing equations as

$$\frac{\partial \rho}{\partial t} + \frac{\partial(\rho u)}{\partial x} + \frac{\partial(\rho v)}{\partial y} = 0 \tag{3.48}$$

$$\frac{\partial(\rho u)}{\partial t} + \frac{\partial(\rho u u)}{\partial x} + \frac{\partial(\rho v u)}{\partial y} = -\frac{\partial p}{\partial x} + \frac{\partial}{\partial x}\left[(\mu + \mu_T)\frac{\partial u}{\partial x}\right] + \frac{\partial}{\partial y}\left[(\mu + \mu_T)\frac{\partial u}{\partial y}\right]$$
$$+ \frac{\partial}{\partial x}\left[(\mu + \mu_T)\frac{\partial u}{\partial x}\right] + \frac{\partial}{\partial y}\left[(\mu + \mu_T)\frac{\partial v}{\partial x}\right] \tag{3.49}$$

$$\frac{\partial(\rho v)}{\partial t} + \frac{\partial(\rho u v)}{\partial x} + \frac{\partial(\rho v v)}{\partial y} = -\frac{\partial p}{\partial y} + \frac{\partial}{\partial x}\left[(\mu + \mu_T)\frac{\partial v}{\partial x}\right] + \frac{\partial}{\partial y}\left[(\mu + \mu_T)\frac{\partial v}{\partial y}\right]$$
$$+ \frac{\partial}{\partial x}\left[(\mu + \mu_T)\frac{\partial u}{\partial y}\right] + \frac{\partial}{\partial y}\left[(\mu + \mu_T)\frac{\partial v}{\partial y}\right] \tag{3.50}$$

$$\frac{\partial(\rho T)}{\partial t} + \frac{\partial(\rho u T)}{\partial x} + \frac{\partial(\rho v T)}{\partial y} = \frac{\partial}{\partial x}\left[\left(\frac{\mu}{\mathrm{Pr}} + \frac{\mu_T}{\mathrm{Pr}_T}\right)\frac{\partial T}{\partial x}\right] + \frac{\partial}{\partial y}\left[\left(\frac{\mu}{\mathrm{Pr}} + \frac{\mu_T}{\mathrm{Pr}_T}\right)\frac{\partial T}{\partial y}\right] \tag{3.51}$$

and the incompressible form of governing equations as

$$\frac{\partial u}{\partial x} + \frac{\partial v}{\partial y} = 0 \tag{3.52}$$

$$\frac{\partial u}{\partial t} + \frac{\partial(uu)}{\partial x} + \frac{\partial(vu)}{\partial y} = -\frac{1}{\rho}\frac{\partial p}{\partial x} + \frac{\partial}{\partial x}\left[(\nu + \nu_T)\frac{\partial u}{\partial x}\right] + \frac{\partial}{\partial y}\left[(\nu + \nu_T)\frac{\partial u}{\partial y}\right]$$
$$+ \frac{\partial}{\partial x}\left[(\nu + \nu_T)\frac{\partial u}{\partial x}\right] + \frac{\partial}{\partial y}\left[(\nu + \nu_T)\frac{\partial v}{\partial x}\right] \tag{3.53}$$

$$\frac{\partial v}{\partial t} + \frac{\partial(uv)}{\partial x} + \frac{\partial(vv)}{\partial y} = -\frac{1}{\rho}\frac{\partial p}{\partial y} + \frac{\partial}{\partial x}\left[(\nu + \nu_T)\frac{\partial v}{\partial x}\right] + \frac{\partial}{\partial y}\left[(\nu + \nu_T)\frac{\partial v}{\partial y}\right]$$
$$+ \frac{\partial}{\partial x}\left[(\nu + \nu_T)\frac{\partial u}{\partial y}\right] + \frac{\partial}{\partial y}\left[(\nu + \nu_T)\frac{\partial v}{\partial y}\right] \tag{3.54}$$

$$\frac{\partial T}{\partial t} + \frac{\partial(uT)}{\partial x} + \frac{\partial(vT)}{\partial y} = \frac{\partial}{\partial x}\left[\left(\frac{\nu}{\mathrm{Pr}} + \frac{\nu_T}{\mathrm{Pr}_T}\right)\frac{\partial T}{\partial x}\right] + \frac{\partial}{\partial y}\left[\left(\frac{\nu}{\mathrm{Pr}} + \frac{\nu_T}{\mathrm{Pr}_T}\right)\frac{\partial T}{\partial y}\right] \tag{3.55}$$

The term ν/Pr appearing in the temperature Eq. (3.55) is obtained from the definition of the laminar Prandtl number that is already defined in Eq. (3.36) as $Pr = \nu/\alpha$, where $\alpha = k/\rho\, C_p$. Interestingly, the time-averaged equations above have the same form as those developed for the laminar equations except for the additional turbulent viscosity found in the diffusion and non-pressure gradient terms for the momentum equations and also found in the diffusion term for the energy equation. Hence the solution to turbulent flow in engineering problems entails greater diffusion that is imposed by the turbulent nature of the fluid flow.

Additional differential transport equations that are required for the standard k-ε model, for the case of a variable fluid property in *conservative* form, are given by

$$\frac{\partial(\rho k)}{\partial t} + \frac{\partial(\rho u k)}{\partial x} + \frac{\partial(\rho v k)}{\partial y} = \frac{\partial}{\partial x}\left(\frac{\mu_T}{\sigma_k}\frac{\partial k}{\partial x}\right) + \frac{\partial}{\partial y}\left(\frac{\mu_T}{\sigma_k}\frac{\partial k}{\partial y}\right) + P - D \qquad (3.56)$$

$$\frac{\partial(\rho \varepsilon)}{\partial t} + \frac{\partial(\rho u \varepsilon)}{\partial x} + \frac{\partial(\rho v \varepsilon)}{\partial y} = \frac{\partial}{\partial x}\left(\frac{\mu_T}{\sigma_\varepsilon}\frac{\partial \varepsilon}{\partial x}\right) + \frac{\partial}{\partial y}\left(\frac{\mu_T}{\sigma_\varepsilon}\frac{\partial \varepsilon}{\partial y}\right) + \frac{\varepsilon}{k}(C_{\varepsilon 1}P - C_{\varepsilon 2}D)$$

$$(3.57)$$

with the destruction term D given by $\rho\varepsilon$ and the production term P formulated as

$$P = 2\mu_T\left[\left(\frac{\partial u}{\partial x}\right)^2 + \left(\frac{\partial v}{\partial y}\right)^2\right] + \mu_T\left(\frac{\partial u}{\partial y} + \frac{\partial v}{\partial x}\right)^2 - \frac{2}{3}\mu_T\left(\frac{\partial u}{\partial x} + \frac{\partial v}{\partial y}\right)^2$$
$$- \frac{2}{3}\rho\mu_T k\left(\frac{\partial u}{\partial x} + \frac{\partial v}{\partial y}\right)$$

For the case of a constant fluid property, the differential transport equations in *non-conservative* form are expressed as

$$\frac{\partial k}{\partial t} + u\frac{\partial k}{\partial x} + v\frac{\partial k}{\partial y} = \frac{\partial}{\partial x}\left(\frac{\nu_T}{\sigma_k}\frac{\partial k}{\partial x}\right) + \frac{\partial}{\partial y}\left(\frac{\nu_T}{\sigma_k}\frac{\partial k}{\partial y}\right) + P - D \qquad (3.58)$$

$$\frac{\partial \varepsilon}{\partial t} + u\frac{\partial \varepsilon}{\partial x} + v\frac{\partial \varepsilon}{\partial y} = \frac{\partial}{\partial x}\left(\frac{\nu_T}{\sigma_\varepsilon}\frac{\partial \varepsilon}{\partial x}\right) + \frac{\partial}{\partial y}\left(\frac{\nu_T}{\sigma_\varepsilon}\frac{\partial \varepsilon}{\partial y}\right) + \frac{\varepsilon}{k}(C_{\varepsilon 1}P - C_{\varepsilon 2}D) \qquad (3.59)$$

where the destruction term D is given by ε and the production term P is formulated as

$$P = 2\nu_T\left[\left(\frac{\partial u}{\partial x}\right)^2 + \left(\frac{\partial v}{\partial y}\right)^2\right] + \nu_T\left(\frac{\partial u}{\partial y} + \frac{\partial v}{\partial x}\right)^2$$

The physical significance of the above equations is: *the rate of change and the advection transport of k or ε equals the diffusion transport combined with the rate of production and destruction of k or ε.* The equations contain five adjustable constants: C_μ, σ_k, σ_ε, $C_{\varepsilon 1}$, and $C_{\varepsilon 2}$. These constants have been arrived at by comprehensive data fitting for a wide range of turbulent flows (Launder and Spalding, 1974):

$$C_\mu = 0.09, \qquad \sigma_k = 1.0, \qquad \sigma_\varepsilon = 1.3, \qquad C_{\varepsilon 1} = 1.44, \qquad C_{\varepsilon 2} = 1.92$$

The production and destruction of turbulent kinetic energy are always closely linked in the k-equations (3.56) and (3.58). The dissipation rate ε is large where the production of k is large. The model Eqs (3.56) and (3.58) assume that the production and destruction terms are proportional to the production and destruction terms of the k-equation. Adoption of such terms ensures that ε increases rapidly if k increases rapidly and that it decreases sufficiently fast to avoid non-physical (negative) values of turbulent kinetic energy if k decreases. The factor ε/k in the production and destruction terms makes these terms dimensionally correct in the ε-equations (3.57) and (3.59).

EXAMPLE

Example 3.8

Consider the two-dimensional CFD case of the flow between two stationary parallel plates to demonstrate the laminar and turbulent nature of the fluid flow. Using CFD with the dimensions of height $H = 0.1$ m and length $L = 10$ m of the channel (Fig. 3.8.1), observe the velocity and viscosity profiles in the fully developed region with the working fluid taken as air (density $\rho = 1.2$ kg/m^3 and dynamic viscosity $\mu = 2 \times 10^{-5}$ kg/m · s) for inlet velocities of $u_{in1} = 0.02$ m/s and $u_{in2} = 1$ m/s.

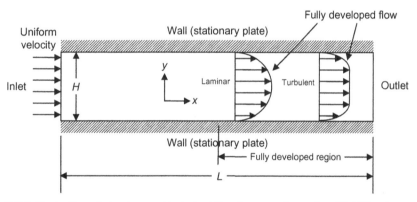

Fig. 3.8.1 Illustration of a laminar and turbulent flow in a two-dimensional fluid flow between two stationary parallel plates.

Solution: The problem is described as follows:

For an inlet velocity of $u_{in1} = 0.02$ m/s, the Reynolds number determined from Eq. (3.26) is given as 120. The flow is laminar. However, for an inlet velocity of $u_{in2} = 1$ m/s, the Reynolds number is calculated to be 6000. This Reynolds number is well above the critical Reynolds number of 1400; hence the flow is turbulent.

Based on the CFD simulation, the laminar and turbulent velocity profiles in the fully developed region are given in Fig. 3.8.2.

The laminar (dynamic) and turbulent viscosities corresponding to the same location of the velocity profiles are shown in Fig. 3.8.3.

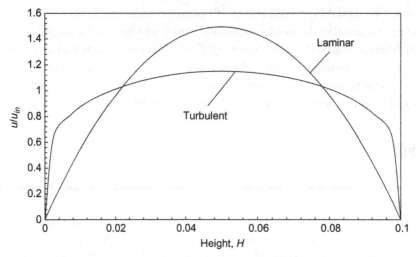

Fig. 3.8.2 Laminar and turbulent velocity profiles in the fully developed region.

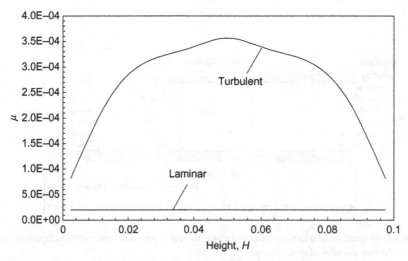

Fig. 3.8.3 Laminar and turbulent viscosities in the fully developed region.

Discussion: It is observed that there is a significant difference in the flow structure for the laminar and turbulent flow regimes. In the fully developed region the velocity profile for laminar flow is parabolic but the velocity profile for turbulent flow is rather blunt primarily at the center of the channel with a steep velocity gradient near the wall surface. The turbulent velocity profile being blunt or flat at the center is due to the effect of the high turbulent viscosity diffusing the flow in the momentum equations (this is exemplified in Fig. 3.8.3, where the turbulent viscosity is significantly higher by orders of magnitude than the laminar [dynamic] viscosity at the center of the channel). To further illustrate the physical characteristics of the turbulent viscosity that is derived in Eq. (3.36), profiles of the turbulent kinetic energy and dissipation along the height of the channel at the same location of the velocity profiles are plotted, which are respectively depicted in Figs. 3.8.4 and 3.8.5. The results show that the turbulent kinetic energy k and dissipation rate ε peaking near the walls. Away from the wall and toward the center of the channel, the kinetic energy decreases but at a slower rate than the dissipation rate. Even though the kinetic energy is high near the wall, the dissipation rate is much higher, thereby resulting in a lower turbulent viscosity. At the center, the kinetic energy is higher than the dissipation rate, which indicates a greater generation of turbulence, thereby resulting in a higher turbulent viscosity.

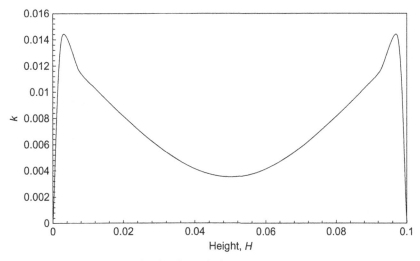

Fig. 3.8.4 Kinetic energy k profile for the turbulent flow in the fully developed region.

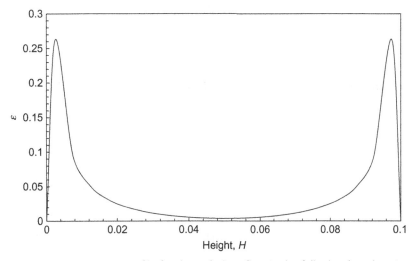

Fig. 3.8.5 Dissipation rate ε profile for the turbulent flow in the fully developed region.

3.5.3 Comments

The two-equation k-ε model is the most widely used and validated turbulence model. The model's performance has been assessed against a number of practical flows. It has achieved notable successes in predicting thin shear layers, boundary layers, and duct flows without the need for case-by-case adjustment of the model constants. It has also been shown to perform extremely well where the Reynolds shear stresses are important in confined flows. This accommodates a wide range of flows with industrial engineering applications, which explains the model's popularity among CFD users. Extension of the model to incorporate the buoyancy effects has led to the application of the models to study environmental flows such as pollutant dispersion in the atmosphere and in lakes and the modeling of enclosure fires.

Despite the many successful applications in handling industrial problems, the standard k-ε model demonstrates only moderate agreement when predicting unconfined flows. The weakness of the model is particularly amplified for weak shear layers – far-wake and mixing-layer unconfined separated flows. Also, for the case of the axisymmetric jets in stagnant surroundings, the spreading rate is severely overpredicted, where in major parts of these flows, the rate of production of the turbulent kinetic energy is much less than the rate of dissipation. The difficulties can, however, be overcome by making *ad hoc* adjustments to the model constants, thereby reducing the model's generality and robustness.

Numerous problems are also experienced with the model in predicting swirling flows and flows with large, rapid, extra strains (e.g. highly curved boundary layers and diverging passages) since the model is unable to fully describe the subtle effects of the streamline curvature on turbulence. One major weakness of the standard k-ε model is the assumption of an *isotropic* eddy viscosity. Owing to the deficiencies of the treatment of the normal stresses, secondary flows that exist in long non-circular ducts, which are driven by anisotropic normal Reynolds stresses, could not be predicted. Finally, the model is oblivious to body forces due to rotation of the frame of reference.

3.6 GENERIC FORM OF THE GOVERNING EQUATIONS FOR CFD

From the governing equations derived above for either the laminar or turbulent conditions, there are significant commonalities between these various equations. Here, we will present the three-dimensional form of the governing equations for the conservation of mass, momentum, energy, and the turbulent quantities. If we introduce a general variable ϕ, and expressing all the fluid flow equations in the conservative *incompressible* form, the equation can be written as

$$\frac{\partial \phi}{\partial t} + \frac{\partial (u\phi)}{\partial x} + \frac{\partial (v\phi)}{\partial y} + \frac{\partial (w\phi)}{\partial z} = \frac{\partial}{\partial x}\left[\Gamma \frac{\partial \phi}{\partial x}\right] + \frac{\partial}{\partial y}\left[\Gamma \frac{\partial \phi}{\partial y}\right] + \frac{\partial}{\partial z}\left[\Gamma \frac{\partial \phi}{\partial z}\right] + S_\phi$$

$$(3.60)$$

while in the conservative *compressible* form, the equation is given by

$$\frac{\partial(\rho\phi)}{\partial t} + \frac{\partial(\rho u\phi)}{\partial x} + \frac{\partial(\rho v\phi)}{\partial y} + \frac{\partial(\rho w\phi)}{\partial z} = \frac{\partial}{\partial x}\left[\Gamma\frac{\partial\phi}{\partial x}\right] + \frac{\partial}{\partial y}\left[\Gamma\frac{\partial\phi}{\partial y}\right] + \frac{\partial}{\partial z}\left[\Gamma\frac{\partial\phi}{\partial z}\right] + S_\phi$$

(3.61)

Eqs (3.60) and (3.61) are the so-called transport equations for the property ϕ. Each of them illustrates the various physical transport processes occurring in the fluid flow: the *local acceleration* and *advection* terms on the left-hand side are respectively equivalent to the *diffusion* term (Γ = diffusion coefficient) and the *source* term (S_ϕ) on the right-hand side. Tables 3.1 and 3.2 present the governing equations for the incompressible and compressible flows in the Cartesian framework. In order to bring forth the common features, we have, of course, combined the terms that are not shared between the equations inside the source terms. It is noted that the additional source terms in the momentum equations S'_u, S'_v, and S'_w comprise the pressure and non-pressure gradient terms and other possible sources such as gravity that influence the fluid motion, while the additional source term S_T in the energy equation may contain heat sources or sinks within the flow domain. It is also noted that the production P in the incompressible form of the turbulence equations can be obtained from its compressible counterpart by invoking the incompressible form of the continuity equation and division by constant density.

For compressible flow, the density and temperature can be evaluated through the equations of state, which provide the linkage between the energy equation and those of the mass and momentum equations. For a perfect gas, the following equations of state are: $p = \rho R T$, where R is the gas constant, and $e = C_v T$, where C_v is the specific heat of constant volume. For the dynamics viscosity and thermal conductivity, the variables can usually be determined via a linear or polynomial dependence on temperature.

This equation is usually used as the starting point for computational procedures in either the finite difference or the finite volume method. Algebraic expressions of this equation for the various transport properties are formulated and hereafter solved. For incompressible flow, by setting the transport property ϕ equal to 1, u, v, w, T, k or ε, and selecting appropriate values for the diffusion coefficient Γ and source terms S_ϕ, we obtain the special forms presented in Table 3.3 for each of the partial differential equations for the conservation of mass, momentum, energy, and the turbulent quantities. In Table 3.4, by setting the transport property ϕ equal to 1, u, v, w, h, k or ε, and selecting appropriate values for Γ and S_ϕ, we nonetheless obtain the special forms presented in Table 3.4 for the compressible form of the partial differential equations for the conservation of mass, momentum, energy, and the turbulent quantities.

Although we have systematically walked through the derivation of the complete set of governing equations in detail from basic conservation principles, the final general form pertaining to the fluid motion, heat transfer, etc., conforms simply to the generic form of

Table 3.1 Governing Equations for Incompressible Flow in Cartesian Coordinates.

Mass conservation

$$(m) \quad \frac{\partial u}{\partial x} + \frac{\partial v}{\partial y} + \frac{\partial w}{\partial z} = 0$$

Momentum equations

$$(M_x) \quad \frac{\partial u}{\partial t} + \frac{\partial(uu)}{\partial x} + \frac{\partial(vu)}{\partial y} + \frac{\partial(wu)}{\partial z} = \frac{\partial}{\partial x}\left[(\nu + \nu_T)\frac{\partial u}{\partial x}\right] + \frac{\partial}{\partial y}\left[(\nu + \nu_T)\frac{\partial u}{\partial y}\right] + \frac{\partial}{\partial z}\left[(\nu + \nu_T)\frac{\partial u}{\partial z}\right] + \left(S_u = -\frac{1}{\rho}\frac{\partial p}{\partial x} + S_u'\right)$$

$$(M_y) \quad \frac{\partial v}{\partial t} + \frac{\partial(uv)}{\partial x} + \frac{\partial(vv)}{\partial y} + \frac{\partial(wv)}{\partial z} = \frac{\partial}{\partial x}\left[(\nu + \nu_T)\frac{\partial v}{\partial x}\right] + \frac{\partial}{\partial y}\left[(\nu + \nu_T)\frac{\partial v}{\partial y}\right] + \frac{\partial}{\partial z}\left[(\nu + \nu_T)\frac{\partial v}{\partial z}\right] + \left(S_v = -\frac{1}{\rho}\frac{\partial p}{\partial y} + S_v'\right)$$

$$(M_z) \quad \frac{\partial w}{\partial t} + \frac{\partial(uw)}{\partial x} + \frac{\partial(vw)}{\partial y} + \frac{\partial(ww)}{\partial z} = \frac{\partial}{\partial x}\left[(\nu + \nu_T)\frac{\partial w}{\partial x}\right] + \frac{\partial}{\partial y}\left[(\nu + \nu_T)\frac{\partial w}{\partial y}\right] + \frac{\partial}{\partial z}\left[(\nu + \nu_T)\frac{\partial w}{\partial z}\right] + \left(S_w = -\frac{1}{\rho}\frac{\partial p}{\partial z} + S_w'\right)$$

Energy equation

$$(E) \quad \frac{\partial T}{\partial t} + \frac{\partial(uT)}{\partial x} + \frac{\partial(vT)}{\partial y} + \frac{\partial(wT)}{\partial z} = \frac{\partial}{\partial x}\left[\left(\frac{\nu}{Pr} + \frac{\nu_T}{Pr_T}\right)\frac{\partial T}{\partial x}\right] + \frac{\partial}{\partial y}\left[\left(\frac{\nu}{Pr} + \frac{\nu_T}{Pr_T}\right)\frac{\partial T}{\partial y}\right] + \frac{\partial}{\partial z}\left[\left(\frac{\nu}{Pr} + \frac{\nu_T}{Pr_T}\right)\frac{\partial T}{\partial z}\right] + S_T$$

Turbulence equations

$$(k) \quad \frac{\partial k}{\partial t} + \frac{\partial(uk)}{\partial x} + \frac{\partial(vk)}{\partial y} + \frac{\partial(wk)}{\partial z} = \frac{\partial}{\partial x}\left[\frac{\nu_T}{\sigma_k}\frac{\partial k}{\partial x}\right] + \frac{\partial}{\partial y}\left[\frac{\nu_T}{\sigma_k}\frac{\partial k}{\partial y}\right] + \frac{\partial}{\partial z}\left[\frac{\nu_T}{\sigma_k}\frac{\partial k}{\partial z}\right] + (S_k = P - D)$$

$$(\varepsilon) \quad \frac{\partial \varepsilon}{\partial t} + \frac{\partial(u\varepsilon)}{\partial x} + \frac{\partial(v\varepsilon)}{\partial y} + \frac{\partial(w\varepsilon)}{\partial z} = \frac{\partial}{\partial x}\left[\frac{\nu_T}{\sigma_\varepsilon}\frac{\partial \varepsilon}{\partial x}\right] + \frac{\partial}{\partial y}\left[\frac{\nu_T}{\sigma_\varepsilon}\frac{\partial \varepsilon}{\partial y}\right] + \frac{\partial}{\partial z}\left[\frac{\nu_T}{\sigma_\varepsilon}\frac{\partial \varepsilon}{\partial z}\right] + \left(S_\varepsilon = \frac{\varepsilon}{k}(C_{\varepsilon 1}P - C_{\varepsilon 2}D)\right)$$

where $P = 2\nu_T\left[\left(\frac{\partial u}{\partial x}\right)^2 + \left(\frac{\partial v}{\partial y}\right)^2 + \left(\frac{\partial w}{\partial z}\right)^2\right] + \nu_T\left[\left(\frac{\partial u}{\partial y} + \frac{\partial v}{\partial x}\right)^2 + \left(\frac{\partial v}{\partial z} + \frac{\partial w}{\partial y}\right)^2 + \left(\frac{\partial w}{\partial x} + \frac{\partial u}{\partial z}\right)^2\right]$ and $D = \varepsilon$

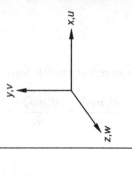

Table 3.2 Governing Equations for Compressible Flow in Cartesian Coordinates.

Mass conservation

(m) $\dfrac{\partial \rho}{\partial t} + \dfrac{\partial(\rho u)}{\partial x} + \dfrac{\partial(\rho v)}{\partial y} + \dfrac{\partial(\rho w)}{\partial z} = 0$

Momentum equations

(M_x) $\dfrac{\partial(\rho u)}{\partial t} + \dfrac{\partial(\rho u u)}{\partial x} + \dfrac{\partial(\rho v u)}{\partial y} + \dfrac{\partial(\rho w u)}{\partial z} = \dfrac{\partial}{\partial x}\left[(\mu+\mu_T)\dfrac{\partial u}{\partial x}\right] + \dfrac{\partial}{\partial y}\left[(\mu+\mu_T)\dfrac{\partial u}{\partial y}\right] + \dfrac{\partial}{\partial z}\left[(\mu+\mu_T)\dfrac{\partial u}{\partial z}\right] + \left(S_u = -\dfrac{\partial p}{\partial x} + S'_u\right)$

(M_y) $\dfrac{\partial(\rho v)}{\partial t} + \dfrac{\partial(\rho u v)}{\partial x} + \dfrac{\partial(\rho v v)}{\partial y} + \dfrac{\partial(\rho w v)}{\partial z} = \dfrac{\partial}{\partial x}\left[(\mu+\mu_T)\dfrac{\partial v}{\partial x}\right] + \dfrac{\partial}{\partial y}\left[(\mu+\mu_T)\dfrac{\partial v}{\partial y}\right] + \dfrac{\partial}{\partial z}\left[(\mu+\mu_T)\dfrac{\partial v}{\partial z}\right] + \left(S_v = -\dfrac{\partial p}{\partial y} + S'_v\right)$

(M_z) $\dfrac{\partial(\rho w)}{\partial t} + \dfrac{\partial(\rho u w)}{\partial x} + \dfrac{\partial(\rho v w)}{\partial y} + \dfrac{\partial(\rho w w)}{\partial z} = \dfrac{\partial}{\partial x}\left[(\mu+\mu_T)\dfrac{\partial w}{\partial x}\right] + \dfrac{\partial}{\partial y}\left[(\mu+\mu_T)\dfrac{\partial w}{\partial y}\right] + \dfrac{\partial}{\partial z}\left[(\mu+\mu_T)\dfrac{\partial w}{\partial z}\right] + \left(S_w = -\dfrac{\partial p}{\partial z} + S'_w\right)$

Energy equation

(E) $\dfrac{\partial(\rho h)}{\partial t} + \dfrac{\partial(\rho u h)}{\partial x} + \dfrac{\partial(\rho v h)}{\partial y} + \dfrac{\partial(\rho w h)}{\partial z} = \dfrac{\partial}{\partial x}\left[\lambda\dfrac{\partial T}{\partial x}\right] + \dfrac{\partial}{\partial y}\left[\lambda\dfrac{\partial T}{\partial y}\right] + \dfrac{\partial}{\partial z}\left[\lambda\dfrac{\partial T}{\partial z}\right] + \dfrac{\partial}{\partial x}\left[\dfrac{\mu_T}{Pr_T}\dfrac{\partial h}{\partial x}\right] + \dfrac{\partial}{\partial y}\left[\dfrac{\mu_T}{Pr_T}\dfrac{\partial h}{\partial y}\right] + \dfrac{\partial}{\partial z}\left[\dfrac{\mu_T}{Pr_T}\dfrac{\partial h}{\partial z}\right] + \dfrac{\partial p}{\partial t} + \Phi + S_T$

Turbulence equations

(k) $\dfrac{\partial(\rho k)}{\partial t} + \dfrac{\partial(\rho u k)}{\partial x} + \dfrac{\partial(\rho v k)}{\partial y} + \dfrac{\partial(\rho w k)}{\partial z} = \dfrac{\partial}{\partial x}\left[\dfrac{\mu_T}{\sigma_k}\dfrac{\partial k}{\partial x}\right] + \dfrac{\partial}{\partial y}\left[\dfrac{\mu_T}{\sigma_k}\dfrac{\partial k}{\partial y}\right] + \dfrac{\partial}{\partial z}\left[\dfrac{\mu_T}{\sigma_k}\dfrac{\partial k}{\partial z}\right] + \left(S_k = \rho(P-D)\right)$

(ε) $\dfrac{\partial(\rho \varepsilon)}{\partial t} + \dfrac{\partial(\rho u \varepsilon)}{\partial x} + \dfrac{\partial(\rho v \varepsilon)}{\partial y} + \dfrac{\partial(\rho w \varepsilon)}{\partial z} = \dfrac{\partial}{\partial x}\left[\dfrac{\mu_T}{\sigma_\varepsilon}\dfrac{\partial \varepsilon}{\partial x}\right] + \dfrac{\partial}{\partial y}\left[\dfrac{\mu_T}{\sigma_\varepsilon}\dfrac{\partial \varepsilon}{\partial y}\right] + \dfrac{\partial}{\partial z}\left[\dfrac{\mu_T}{\sigma_\varepsilon}\dfrac{\partial \varepsilon}{\partial z}\right] + \left(S_\varepsilon = \rho\dfrac{\varepsilon}{k}(C_{\varepsilon 1}P - C_{\varepsilon 2}D)\right)$

where

$$P = 2\mu_T\left[\left(\frac{\partial u}{\partial x}\right)^2 + \left(\frac{\partial v}{\partial y}\right)^2 + \left(\frac{\partial w}{\partial z}\right)^2\right] + \mu_T\left[\left(\frac{\partial u}{\partial y}+\frac{\partial v}{\partial x}\right)^2 + \left(\frac{\partial v}{\partial z}+\frac{\partial w}{\partial y}\right)^2 + \left(\frac{\partial u}{\partial z}+\frac{\partial w}{\partial x}\right)^2\right] - \frac{2}{3}\mu_T\left(\frac{\partial u}{\partial x}+\frac{\partial v}{\partial y}+\frac{\partial w}{\partial z}\right) - \frac{2}{3}\rho\mu_T k\left(\frac{\partial u}{\partial x}+\frac{\partial v}{\partial y}+\frac{\partial w}{\partial z}\right)$$

and $D = \varepsilon$

Table 3.3 General Form of Governing Equations for Incompressible Flow in Cartesian Coordinates.

Φ	Γ_Φ	S_Φ
1	0	0
u	$\nu + \nu_T$	$-\dfrac{1}{\rho}\dfrac{\partial p}{\partial x} + S'_u$
v	$\nu + \nu_T$	$-\dfrac{1}{\rho}\dfrac{\partial p}{\partial y} + S'_v$
w	$\nu + \nu_T$	$-\dfrac{1}{\rho}\dfrac{\partial p}{\partial z} + S'_w$
T	$\dfrac{\nu}{\text{Pr}} + \dfrac{\nu_T}{\text{Pr}_T}$	S_T
k	$\dfrac{\nu_T}{\sigma_k}$	$P - D$
ε	$\dfrac{\nu_T}{\sigma_\varepsilon}$	$\dfrac{\varepsilon}{k}\left(C_{\varepsilon 1}P - C_{\varepsilon 2}D\right)$

Table 3.4 General Form of Governing Equations for Compressible Flow in Cartesian Coordinates.

Φ	Γ_Φ	S_Φ
1	0	0
u	$\mu + \mu_T$	$-\dfrac{\partial p}{\partial x} + S'_u$
v	$\mu + \mu_T$	$-\dfrac{\partial p}{\partial y} + S'_v$
w	$\mu + \mu_T$	$-\dfrac{\partial p}{\partial z} + S'_w$
h	$\dfrac{\mu_T}{\text{Pr}_T}$	$\dfrac{\partial}{\partial x}\left[\lambda\dfrac{\partial T}{\partial x}\right] + \dfrac{\partial}{\partial y}\left[\lambda\dfrac{\partial T}{\partial y}\right] + \dfrac{\partial}{\partial z}\left[\lambda\dfrac{\partial T}{\partial z}\right] + \dfrac{\partial p}{\partial t} + \Phi + S_T$
k	$\dfrac{\mu_T}{\sigma_k}$	$P - D$
ε	$\dfrac{\mu_T}{\sigma_\varepsilon}$	$\dfrac{\varepsilon}{k}\left(C_{\varepsilon 1}P - C_{\varepsilon 2}D\right)$

Eq. (3.60) for incompressible flow and Eq. (3.61) for compressible flow. These equations are important generic transport equations as they can accommodate increasing complexity within the CFD model for solving more complicated problems generally found in engineering applications.

Let us focus on some typical complex engineering flow problems that are of significant interest such as multi-step combustion processes of swirling turbulent reactive flows in combustors and multi-phase flows involving interactions between gas bubbles and liquids in bubble columns. Solutions to these processes can easily be obtained by modeling them through additional transport equations expressed in the simple generic form of Eq. (3.60) for incompressible flow and Eq. (3.61) for compressible flow. For reactive flows, the transport of the various *chemical species* can be handled by the additional scalar quantities representing each of the reactive species and appropriately formulating the reaction rates in the source terms to account for the chemical reaction processes that are occurring. For bubbly flows, additional transport equations of the *number density* can be formulated and solved for the various gas bubble sizes that migrate alongside the liquid in the bubble columns.

Understanding CFD is not meant to be an arduous process. On the contrary, Eqs (3.60) and (3.61), originally formulated from first principles, reinforce the adherent *simplicity* that is embraced for any transport property that may be required to be solved within the CFD framework.

In Sections 3.3.3 and 3.4.3 we introduced the dimensionless parameters such as the Reynolds number (*Re*) and Prandtl number (*Pr*) that may be useful to describe some similar physical phenomena of the flow and heat transfer processes. It will be demonstrated in the next example that the governing equations of mass, momentum, and energy can be non-dimensionalized to reduce the number of parameters that appear in the equations. Other similar characteristics of the fluid flow are also discussed.

EXAMPLE

Example 3.9

Consider the *dynamic similarity* of the partial differential equations that govern a two-dimensional CFD case for a steady incompressible laminar flow between two stationary parallel plates. The channel dimensions are: height $H = 0.1$ m and length $L = 1$ m. This is the same model geometry as that previously investigated in Examples 3.7 and 3.8.

(a) Non-dimensionalize the continuity, momentum, and energy equations given by Eqs (3.12), (3.24), (3.25), and (3.35), respectively.

(b) Using CFD, determine the flow field for both air and water with the same Reynolds number while adjusting the inlet velocity. Discuss the velocity profiles in the fully developed region.

(c) Also determine the temperature field for both air and water with the same Reynolds number as in part (b). Discuss the non-dimensional temperature profiles in the fully developed region with a prescribed uniform temperature of 330 K at the inlet and a wall temperature specified at 300 K.

Solution:

(a) The non-dimensional form of the governing equations (and boundary conditions) can be achieved by dividing all the dependent and independent flow variables by relevant and meaningful constant quantities. For lengths, the variable can be divided by a characteristic length H (which is the width of the channel), all velocities by a reference velocity u_{in} (which is the inlet velocity), pressure by ρu_{in}^2 (which is twice the dynamic pressure for the channel), and temperature by a suitable temperature difference (which is $T_\infty - T_s$ for the channel). We therefore obtain

$$x^* = \frac{x}{H}, \qquad y^* = \frac{y}{H}, \qquad u^* = \frac{u}{u_{in}}, \qquad v^* = \frac{v}{u_{in}},$$

$$p^* = \frac{p}{\rho u_{in}^2} \qquad \text{and} \qquad T^* = \frac{T - T_s}{T_\infty - T_s}$$

where the asterisks denote the non-dimensional variables. Introducing these variables into the governing equations of mass, momentum, and energy produces

$$\frac{\partial u^*}{\partial x^*} + \frac{\partial v^*}{\partial y^*} = 0 - continuity$$

$$u^* \frac{\partial u^*}{\partial x^*} + v^* \frac{\partial u^*}{\partial y^*} = \frac{1}{Re} \left(\frac{\partial^2 u^*}{\partial x^{*2}} + \frac{\partial^2 u^*}{\partial y^{*2}} \right) - \frac{\partial p^*}{\partial x^*} - x - momentum$$

$$u^* \frac{\partial v^*}{\partial x^*} + v^* \frac{\partial v^*}{\partial y^*} = \frac{1}{Re} \left(\frac{\partial^2 v^*}{\partial x^{*2}} + \frac{\partial^2 v^*}{\partial y^{*2}} \right) - \frac{\partial p^*}{\partial y^*} - y - momentum$$

$$u^* \frac{\partial T^*}{\partial x^*} + v^* \frac{\partial T^*}{\partial y^*} = \frac{1}{RePr} \left(\frac{\partial^2 T^*}{\partial x^{*2}} + \frac{\partial^2 T^*}{\partial y^{*2}} \right) - energy$$

Discussion: A major advantage of non-dimensionalizing the governing equations is the significant reduction of parameters to be considered. By grouping the dimensional parameters, originally 8 (H, u_{in}, T_∞, T_s, k, ρ, μ, and C_p), the non-dimensionalized problem now involves only just two parameters (Re and Pr). Hence, for a given geometry, problems having the same values for the similarity parameters will have identical solutions. Another advantage of the use of similarity parameters is that results from a large number of experiments can be grouped and reported conveniently in terms of such parameters.

(b) The physical significance of the Reynolds number (Re) is investigated herein. As previously defined in Eq. (3.20), this dimensionless number requires the values of density (ρ) and dynamic viscosity (μ). For air, the values are $\rho = 1.2$ kg/m^3 and $\mu = 2 \times 10^{-5}$ kg/m · s, while for water, they are $\rho = 1000$ kg/m^3 and $\mu = 10^{-3}$ kg/m · s. If a laminar flow, for example, $Re = 120$, is assumed for both fluids, the inlet velocities for air and water are 0.02 m/s and 0.0012 m/s, respectively.

Based on CFD simulations, the axial dimensional and non-dimensional velocity profiles for air and water in the fully developed region are given in Figs. 3.9.1 and 3.9.2.

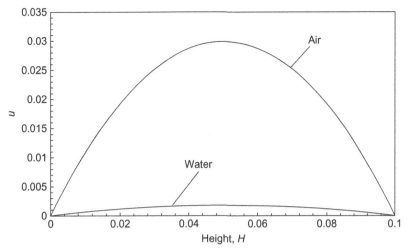

Fig. 3.9.1 Dimensional axial velocity profiles in the fully developed region.

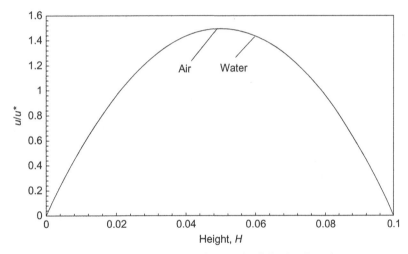

Fig. 3.9.2 Non-dimensional axial velocity profiles in the fully developed region.

Discussion: Although the two flows have different fluid properties, one being air while the other is water, we obtain the same flow behavior. This is because they have the same Reynolds numbers, and the non-dimensional governing equations of the *x-momentum* and *y-momentum* are identical, which leads to the same numerical results. If we consider another fluid mechanics example that is the flow of air or water over a flat plate (Fig. 3.9.3) having different lengths but with the same inlet velocities at the same Reynolds numbers, these two geometrically similar bodies have the same physical phenomena since they have the same friction coefficients (C_f).

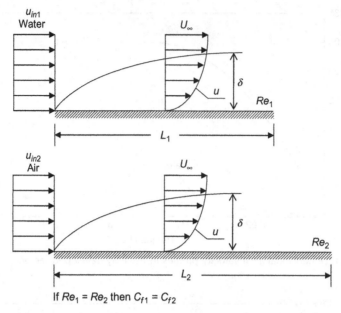

Fig. 3.9.3 Two geometrically similar bodies having the same friction coefficients at the same Reynolds numbers.

(c) The physical significance of the Prandtl number (Pr) is subsequently investigated herein. As previously defined in Eq. (3.35), it requires the values of the kinematic viscosity (ν) and thermal diffusivity (α). For air, the values are $\nu = 1.667 \times 10^{-5}$ m^2/s and $\alpha = 1.667 \times 10^{-5}$ m^2/s, while for water, they are $\nu = 1 \times 10^{-6}$ m^2/s and $\alpha = 1.435 \times 10^{-7}$ m^2/s. This yields $Pr = 1$ for air and $Pr \approx 7$ for water.

Based on the CFD simulations, the non-dimensional temperature profiles for air and water in the fully developed region are given in Fig. 3.9.4.

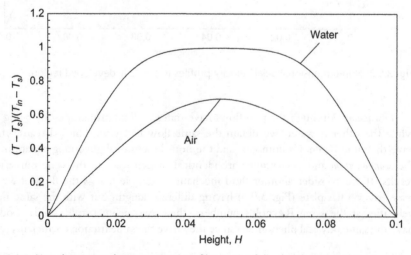

Fig. 3.9.4 Non-dimensional temperature profiles in the fully developed region.

Discussion: With reference to the energy *diffusion* term, $1/(RePr)$ in the non-dimensional energy equation, the Reynolds number of air and water are the same; however, the Prandtl number of air is found to be much less than water, leading to heat flow being diffused more in air than in water. Therefore, with less *diffusion*, more *advection* is encouraged in water than in air, which leads to a higher temperature in the fully developed region. Nevertheless, we have the same momentum *diffusion* for the velocity fields since the Reynolds numbers are the same for both fluids in the *x-momentum* and *y-momentum* equations; both fluids therefore have the same velocity profiles.

3.7 PHYSICAL BOUNDARY CONDITIONS OF THE GOVERNING EQUATIONS

In Chapter 2 we have introduced various types of boundary conditions that are required in order to perform the numerical calculations of fluid flow in a channel. Here, we explore the physical meanings of the boundary conditions that were applied and other important boundary conditions that may be required to close the fluid flow system.

The continuity, momentum, energy, and turbulent quantities equations above govern the flow and heat transfer of a fluid. They are the same equations, whether the flow is over high-rise buildings or bridges, through subsonic wind tunnels, or within various narrow channels of intricate electronic components in a computer box. However, the flow fields are quite *different* for each of these cases, although the governing equations are the *same*. The reason for the difference lies in the specification of the *boundary conditions*. The boundary conditions, and sometimes the *initial conditions,* strongly dictate the particular solutions to be obtained from the governing equations. For example, when all the geometrical shapes of the high-rise buildings are treated, when certain physical boundary conditions are applied on the surfaces and when appropriate boundary conditions are imposed on the far free-stream, the resulting solution of the governing partial differential equations listed in Table 3.1 will yield the complex flow field over the high-rise buildings. In contrast, the solution of the flow field within heated channels in a computer box demands other physical boundary conditions, for example, the specific geometrical shape and locations of air entrainment and exhaust ports within the computer box.

We have formulated the governing equations that have been developed and described in the previous sections. However, the real driver for any particular solution is the boundary conditions. This creates particular significance especially in CFD, as any numerical solution of the governing equations must result in a strong and compelling numerical representation of the proper boundary conditions.

Let us now review the physical boundary conditions for a subsonic viscous flow. We focus initially on the so-called *no-slip* condition. Here, the boundary condition on a solid surface assumes zero relative velocity between the surface and the fluid immediately at the surface. If the surface is stationary, with the flow moving past it, then all the velocity components can be taken to be zero. In other words

$$u = v = w = 0 \quad \text{at the surface} \tag{3.62}$$

In Chapter 2 and Examples 3.2, 3.5, 3.7, 3.8, and 3.9, we were introduced to the inflow boundary condition for a channel flow. The solution of the governing equations for any of the transport property ϕ for most flows requires at least one velocity component to be given at the inflow boundary. For the channel flow, this is provided by the *Dirichlet* boundary condition on the velocity in the x direction:

$$u = f \quad \text{and} \quad v = w = 0 \quad \text{at the inflow boundary} \tag{3.63}$$

where f can either be specified as a constant value or a velocity profile at the surface. Computationally, *Dirichlet* conditions can be applied accurately as long as f is continuous. We were also introduced in Chapter 2 as well as in Examples 3.2, 3.5, 3.7, 3.8, and 3.9, the use of an outflow boundary condition. Commonly, outflow boundaries are positioned at locations where the flow is approximately unidirectional and where surface stresses take known values. In a fully developed flow exiting from the channel there is no change in the velocity component in the direction across the boundary. To satisfy stress continuity, the shear forces along the surface are taken to be zero; this gives the outflow condition

$$\frac{\partial u}{\partial n} = \frac{\partial v}{\partial n} = \frac{\partial w}{\partial n} = 0 \quad \text{at the outflow boundary} \tag{3.64}$$

where n is the direction normal to the surface, which for the channel flow problem is the x direction. This condition is usually known as the *Neumann* boundary condition. Physically, in reference to the continuity Eq. (3.12), it is clear that the appropriate boundary conditions (3.62), (3.63), and (3.64) that are imposed at any location on the surface of the channel walls in Chapter 2 close the system mathematically and satisfy local and overall mass conservation.

There is an analogous *no-slip* condition associated with the temperature at the surface. If the material temperature of the surface is denoted by T_w, then the temperature fluid layer immediately in contact with the surface is also T_w. In a given problem where the wall temperature is known, the *Dirichlet* boundary condition applies and the fluid temperature is

$$T = T_w \quad \text{at the wall} \tag{3.65}$$

The application of this boundary condition was illustrated in Example 3.6, part (a), and Examples 3.8 and 3.9. However, if the wall temperature is not known (e.g. if the temperature is changing as a function of time due to the heat transfer to or from the surface), then Fourier's law of heat condition can be applied to provide the necessary boundary condition at the surface. If we denote the instantaneous wall heat flux as q_w, then according to Fourier's law

$$q_w = -\left(k\frac{\partial T}{\partial n}\right)_w \quad \text{at the wall} \tag{3.66}$$

The application of this boundary condition was also illustrated by the same Example 3.6, but for part (b). Here, the changing surface temperature T_w is responding to the thermal response of the wall material through the heat transfer to the wall q_w. This type of boundary condition, as far as the flow is concerned, is a boundary condition of the temperature *gradient* at the wall. For the case where there is no heat transfer to the surface, this wall temperature, by definition, is called an adiabatic wall temperature T_{adia}. The proper boundary condition comes from Eq. (3.66) with $q_w = 0$, hence

$$\left(\frac{\partial T}{\partial n}\right)_w = 0 \quad \text{at the wall} \tag{3.67}$$

This condition falls in line with the *Neumann* boundary condition for the velocity at the outflow boundaries. On the inflow and outflow boundaries of the flow domain, it is common to have the temperature specified at the inflow boundary and the adiabatic condition adopted at the outflow boundary.

Other commonly used boundary conditions include the open boundary condition. If we refer back to the description of the boundary conditions for the flow over high-rise buildings at the beginning of this section, the far free-stream boundary requires the application of an open boundary, which simply states that the normal gradient of any of the transport property ϕ is zero, that is, $\partial\phi/\partial n = 0$. Gresho and Sani (1990) reviewed the intricacies of open boundary conditions in an incompressible flow and stated that there are some *theoretical concerns* regarding this boundary condition. However, its success in CFD practice left them to recommend it as the simplest and cheapest method when compared with theoretically more satisfying selections. Furthermore, symmetric and cyclic boundary conditions can be employed to take advantage of special geometrical features of the solution region. For the symmetry boundary condition, the normal velocity at the surface is zero while the normal gradients of the other velocity components are zero, where the latter condition also applies for any scalar quantity. For the cyclic boundary condition, the transport property of the one surface ϕ_1 is equivalent to the transport property of the second surface ϕ_2, that is, $\phi_1 = \phi_2$ depending on which two surfaces of the flow domain experience periodicity. It is worth noting that the cyclic boundary condition usually refers to a type of periodic boundary condition with no pressure drop across the periodic planes. If a periodic pressure drop is to occur across translationally periodic boundaries, resulting in "fully developed" or "streamwise-periodic" flow, special measures have to be taken to define the periodic change of the pressure.

For the turbulent quantities, *Dirichlet* and *Neumann* boundary conditions for the turbulent kinetic energy k and its dissipation ε are usually applied at the inflow and outflow boundaries. However, special treatment of the boundary conditions for k and ε is required to properly account for the laminar sub-layer that exists near the wall surface in turbulent flow. We will further discuss in detail the wall treatment for the turbulent quantities in Chapter 6.

For a supersonic flow, *Dirichlet* and *Neumann* boundary conditions can be applied at the inflow boundary but not at the outflow boundary. All variables and properties at the outflow boundary are determined from the upstream flow. If the flow is assumed to be inviscid, the flow velocity is usually taken to be a finite, non-zero value at the solid surface. This means that the wall shear stress is zero. Also, there cannot be any mass flow into and out of the wall. The component velocity perpendicular to the wall is thus zero and the flow at the surface is tangent to the solid surface.

3.8 SUMMARY

In this chapter we have formulated the mathematical basis for a comprehensive general-purpose model of fluid flow and heat transfer from the basic principles of conservation of mass, momentum, and energy. The governing equations are derived by the consideration of an infinitesimal small control volume to conserve mass and energy, and that the net force acting on the control volume is equivalent to the time rate of change of linear momentum. Although we have employed the Newtonian model of viscous stresses to close the system of equations, the accommodation of fluids having non-Newtonian characteristics can be easily incorporated within the framework of these equations. The appearance of viscosity and other thermophysical properties such as the density, thermal conductivity, and specific heat in the governing equations may vary with local conditions and to account for these effects is a straightforward process.

The reader was also introduced to some aspects of turbulence modeling, which resulted in the derivation of the widely applied two-equation k-ε model. This turbulence model still comes highly recommended for general-purpose CFD computations and is the default model used in many commercial codes.

Whether the fluid flow is laminar or turbulent, there are significant commonalities between these conservation equations. This leads to the formulation of the generic form of the governing equations accompanied by a discussion of the physical boundary conditions commonly employed within the CFD framework to close the fluid flow system. A road map to encapsulate all these aforementioned aspects, beginning from the conception of the fluid flow and ending with the system of equations to be solved, is shown in Fig. 3.9 below.

The means of obtaining a solution to these governing equations are discussed in the next chapter. We will present some of the basic computational techniques that can be employed to solve such partial differential equations. The equations presented in this chapter are usually regarded as the starting point for the application of the many *discretization* procedures predicting the fluid motion and heat transfer processes, which will be further elaborated on in the next chapter.

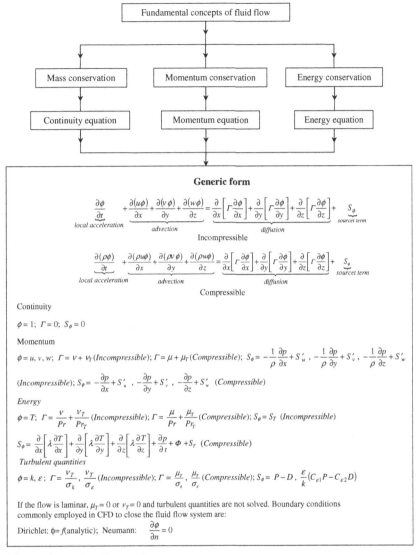

Fig. 3.9 The road map for Chapter 3.

REVIEW QUESTIONS

3.1 Simplify the general continuity equation: $\dfrac{\partial \rho}{\partial t} + \dfrac{\partial(\rho u)}{\partial x} + \dfrac{\partial(\rho v)}{\partial y} + \dfrac{\partial(\rho w)}{\partial z} = 0$, for a two-dimensional constant density case.

3.2 In a converging nozzle the flow accelerates due to the narrowing geometry. Discuss the changes in the velocity gradients $\dfrac{\partial u}{\partial x}$ and $\dfrac{\partial v}{\partial y}$ during the flow (assume a constant density).

3.3 What is Newton's second law of motion?

3.4 Write a force balance equation for all the forces acting on a differential control volume.

3.5 For the momentum of a fluid property in the x direction, discuss how the local acceleration $\dfrac{\partial u}{\partial t}$ and the advection terms $u\dfrac{\partial u}{\partial x} + v\dfrac{\partial u}{\partial y}$ contribute to the overall transport of the fluid.

3.6 A simplified one-dimensional inviscid, incompressible, laminar flow is defined by the following momentum equation in the x direction: $\dfrac{\partial u}{\partial t} + u\dfrac{\partial u}{\partial x} = -\dfrac{1}{\rho}\dfrac{\partial p}{\partial x}$. Name each term and discuss their contribution to the flow.

3.7 The momentum of a fluid in the y direction is given by the following equation: $\dfrac{\partial v}{\partial t} + u\dfrac{\partial v}{\partial x} + v\dfrac{\partial v}{\partial y} = -\dfrac{1}{\rho}\dfrac{\partial p}{\partial y} + \nu\dfrac{\partial^2 v}{\partial x^2} + \nu\dfrac{\partial^2 v}{\partial y^2} - g$. Discuss the forces that act to transport the fluid.

3.8 What are the differences between the momentum equation in question 3.7 and the following momentum equation: $\rho u\dfrac{\partial u}{\partial x} + \rho v\dfrac{\partial u}{\partial y} = -\dfrac{\partial p}{\partial x} + \mu\left(\dfrac{\partial^2 u}{\partial x^2} + \dfrac{\partial^2 u}{\partial y^2}\right)$.

3.9 The hydrodynamic length for a channel flow shown below is equal to L_e when air is used ($u_{in} = 0.03$ m/s, $\mu_1 = 1.65 \times 10^{-5}$ kg/m $\cdot$ s, and $\rho = 1.2$ kg/m^3). To obtain the same hydrodynamic length for water, what inlet velocity is required ($\mu_1 = 1.003 \times 10^{-3}$ kg/m $\cdot$ s, $\rho = 1000$ kg/m^3)?

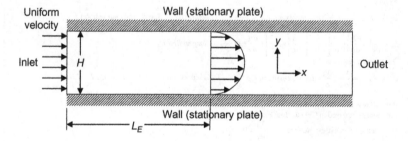

3.10 The Reynolds number is a ratio of two fluid properties. What are they?

3.11 If the Reynolds is very high ($Re \gg 10{,}000$), what does this suggest? If it is very low ($Re \ll 100$), what does this imply?

3.12 Which fluid property changes from a constant to a variable when the fluid is considered a compressible flow?

3.13 Write down the continuity equation for a compressible and incompressible flow. What differences do you notice in the equations?

3.14 For an incompressible flow, what two fluid properties affect density variation? What general equation relates these two properties to the density?

3.15 Explain the first law of thermodynamics.

3.16 Name the sources of energy that contribute to the energy equation.

3.17 Write an equation for the energy balance using the sources of energy defined in question 3.11 for the first law of thermodynamics.

3.18 Apply *Fourier's law of heat conduction* to obtain the heat flux in the x direction.

3.19 Write the equation that defines the substantial derivative for the transport of temperature in terms of the local acceleration derivative and the advection derivative of temperature, and why?

3.20 If a car travels across a warmer environment, the car body will experience a sudden rise in temperature. Is this an example of the local *acceleration* derivative or *advection* derivative at work?

3.21 In what type of situation can you simplify the general 2D energy equation, $\frac{\partial T}{\partial t} + u\frac{\partial T}{\partial x} + v\frac{\partial T}{\partial y} = \frac{k}{\rho C_p}\frac{\partial^2 T}{\partial x^2} + \frac{k}{\rho C_p}\frac{\partial^2 T}{\partial y^2}$, to reach the well-known Laplace's equation: $\frac{\partial^2 T}{\partial x^2} + \frac{\partial^2 T}{\partial y^2} = 0$.

3.22 Obtain the general analytical solution for Laplace's equation for a one-dimensional case.

3.23 The Prandtl number is a ratio of two fluid properties. What are they?

3.24 Fluids such as oils have a high Prandtl number ($Pr \gg 1$). What does this suggest?

3.25 What is the significance of the Prandtl number equaling 1 in terms of entry lengths?

3.26 What is the *energy cascade* process in turbulence?

3.27 Why do large eddies tend to be anisotropic in nature? Why are small-scaled eddies isotropic in nature?

3.28 The use of direct numerical simulation (DNS) remains a problem for engineering applications. Why?

3.29 Which profile as indicated below is the laminar or turbulent velocity profile?

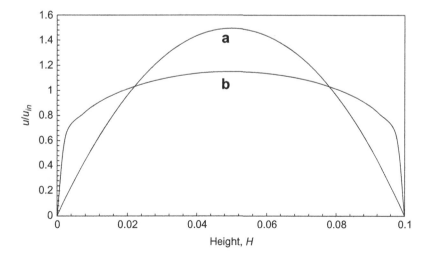

3.30 From question 3.26, describe the shape of the turbulent velocity profile? Explain why the profile is so different from the laminar profile?

3.31 For incompressible flow, the non–dimensional transport equation can be simplified as $u\dfrac{\partial u}{\partial x} = -\dfrac{\partial p}{\partial x} + \dfrac{1}{Re}\dfrac{\partial^2 u}{\partial x^2}$ Provide an explanation for why for a laminar flow, the diffusion term (the right-hand side of the equation) will be more dominant than the convective term (the left-hand side of the equation). At a highly turbulent flow, the situation is opposite; why?

3.32 Indicate the terms in the following general equation: $u\dfrac{\partial \phi}{\partial x} = \Gamma\dfrac{\partial^2 \phi}{\partial x^2} + S_\phi.$

3.33 The equation in question 3.29 represents a transport process for the property ϕ. Under what circumstances can this equation be applied?

3.34 Spot the errors in the equation below. (Hint: It is a 3D x-momentum equation.)

$$\frac{\partial u}{\partial t} + \frac{\partial(uu)}{\partial x} + \frac{\partial(vu)}{\partial x} + \frac{\partial(uu)}{\partial z} = \frac{\partial}{\partial x}\left[(\nu + \nu_T)\frac{\partial u}{\partial x}\right] + \frac{\partial}{\partial y}\left[(\nu + \nu_T)\frac{\partial u}{\partial x}\right]$$

$$+ \frac{\partial}{\partial z}\left[(\nu + \nu_T)\frac{\partial u}{\partial x}\right] + \left(S_u = -\frac{1}{\rho}\frac{\partial p}{\partial x} + S'_u\right)$$

3.35 Referring to the diagram below, answer the following questions.

(a) At the wall, what is the value of u and v?

(b) At the symmetry plane, what is the value of u, v, and p?

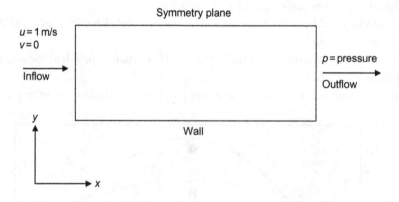

CHAPTER 4

CFD Mesh Generation – A Practical Guideline

4.1 INTRODUCTION

Mesh or grid generation represents an important consideration in attaining the numerical solutions to the governing partial differential equations of the CFD problem. We would like to stress that the term grid or mesh is generally used interchangeably with identical meaning throughout this chapter. In Chapter 2, grid generation has been viewed as one of the key important considerations during the pre-process stage following the definition of a domain geometry. The aspect of grid generation is an important numerical issue where the specific mesh for a given flow problem can determine the *success* or *failure* in obtaining a computational solution. Principally, it must be fine enough to provide an adequate resolution in capturing the important flow features and geometrical structures. For flows with bounded walls, it is recommended that recirculation vortices or steep flow gradients within the viscous boundary layers are properly resolved through locally refining or clustering the mesh in the vicinity of wall boundaries. Mesh concentration may also be required for fluid flows having high shear and/or high temperature gradients. Furthermore, the quality of the mesh has significant implications on the convergence and stability of the numerical simulation and accuracy of the computational result. These grid generation issues will be examined in this chapter.

What is a mesh in the context of CFD? By definition, a mesh in itself consists of an arrangement of a discrete number of points overlaying the whole domain geometry. Through the sub-division of this domain, a number of smaller mesh or grid cells are thereby generated. It is therefore generally expected that the discretized domain is required to adequately resolve the important physics and to capture all the geometrical details of the domain within the flow region. Designing a suitable mesh is certainly by no means trivial. The quest to yield a well-constructed mesh deserves as much attention as prescribing the necessary physics to the flow problem. Because of this, grid generation, as it is commonly known within the CFD community, has become a separate entity by itself and remains a very active area of research and development.

In retrospect, grid generation represents only a means to an end. It forms a necessary tool for the computational simulation of physical flow phenomena and processes. At the same time, grid generation can also be regarded as an art as well as a scientific requirement from a technological standpoint. Focusing on the discipline of mathematics, the role of

Computational Fluid Dynamics
https://doi.org/10.1016/B978-0-323-93938-6.00007-5

mathematics provides a pathway for the essential foundation to move the grid generation process from a user-intensive craft or skill to an automated system. There is both art and science in the design of mathematics but unfortunately not for grid generation since there are no inherent conservation laws such can be utilized for the derivation of CFD governing equations described in Chapter 3. Therefore grid generation is essentially not unique, but rather it must be designed. In this chapter some practical guidelines through proper illustrations will be provided for the design of suitable mesh systems that would be appropriate for CFD calculations.

Historically, early computing machines such as digital computers, being limited in resources, have focused mainly on CFD problems having regular domains. The use of rectangular Cartesian meshes has therefore been the preferred choice due to the adoption of a *discretization* approach such as the finite difference method. In essence, such a rectangular lattice structure provides easy identification of neighboring grid points. The derivatives of the partial differential equations of the CFD problem can be directly approximated via the use of these neighboring grid points. Another form of a regular lattice that also promotes easy identification of neighboring grid points especially in the ability of handling irregular domains is the curvilinear coordinate systems. Generally speaking, a rectangular Cartesian system represents a special case of an orthogonal curvilinear system. As a matter of fact, the curvilinear coordinate system can be considered to be logically rectangular, and it is conceptually no different from the Cartesian coordinate system. For the consideration of domains that are nonetheless highly irregular, the nature of filling the domain by cells of general shapes has brought about the application of unstructured meshes. A network of such cells is made to purposefully fill any arbitrarily shaped domains, and each of these cells represents an entity unto itself. With the advancement of digital computers, another *discretization* approach such as the finite element method has been the nature of its construction of these general-shaped cells. With regard to commonly encountered CFD problems, the finite volume method is nonetheless the preferred *discretization* approach. The use of the finite volume method in the majority of currently developed commercial CFD codes stems basically from the conceptually identical representation of the control volume approach that has been considered for the derivation of the governing equations.

The grid generation process has certainly matured to the point where the use of commercial codes is generally recommended over the construction of computer codes by end users performing computational simulations. We would like to draw the reader to many existing commercial CFD codes in the market having their own in-built powerful mesh generators and there is also a choice of a number of independent grid generation packages such as GRIDGEN by Pointwise (ww.pointwise.com) and GRIDPRO by Program Development Company (www.gridpro.com) for the reader's perusal. Although these independent grid generation packages as well as the different in-built mesh generators in the commercial CFD codes have been designed to be very user-friendly and easy to utilize, the prerequisite to proficiently managing these software packages still relies on the reader's aptitude to operate them properly. In spite of this particular feature being available in commercial CFD codes or independent software packages, some understanding of the process of grid generation especially for the underlying principles, mathematics, and technology remains, however, relevant and important. It is from this standpoint that this chapter is geared toward a basic understanding of different types of mesh systems and how the reader can best adopt these different types of mesh systems to attain appropriate solutions for specific CFD problems being solved. At the very least, this chapter will serve to provide the reader with sufficient background knowledge and philosophy of grid generation in the prospect of handling more sophisticated flow configurations.

4.2 TYPES OF MESHES

4.2.1 Structured Mesh

Of all the techniques that currently exist for grid generation, the simplest is the creation of a structured mesh. For the purpose of illustration and a demonstration of how a mesh can be generated, let us consider a fluid flowing within a rectangular conduit. In the first instance a uniformly distributed Cartesian mesh can be conveniently generated where the spacing of the grid points along the x direction is uniform and the spacing of the points along the y direction is also uniform, as illustrated in Fig. 4.1. An exploded view of a

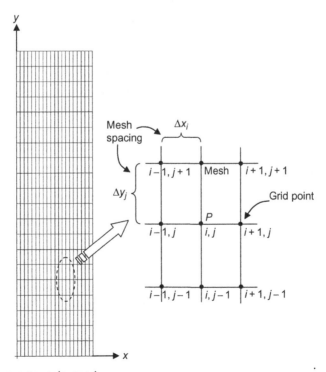

Fig. 4.1 A uniform rectangular mesh.

section of the discrete mesh in the x-y plane is also drawn to further demonstrate the arrangement of the discrete points within the domain. We would like the reader to pay attention to the regular-shape four-nodal grid points of the rectangular element as described in Fig. 4.1.

In a two-dimensional structured mesh the grid points are normally addressed by the indices (i, j), where the index i represents points that run in the x direction while the index j represents points that run in the y direction. If (i, j) are the indices for point P, the neighboring points immediately to the right, to the left, directly above, and directly below are defined by increasing or reducing one of the indices by unity. By allocating appropriate discrete values for Δx_i and Δy_j, the coordinates in the x direction and y direction inside the physical space can henceforth be incrementally determined, resulting in a rectangular mesh covering the whole domain.

For uniformly distributed grid points, the spacing of Δx_i or Δy_j is essentially a single representative value in the x direction or y direction. For non-uniformly distributed grid points, the spacing of Δx_i or Δy_j can, however, effectively take a number of discrete values; hence we could have easily generated a mesh with totally unequal spacing in the x direction or y direction. For example, a finely spaced mesh in the x direction can be generated to adequately resolve the viscous boundary layer of the flow in the vicinity of the wall geometry while a uniformly spaced mesh is retained in the y direction, as shown in Fig. 4.2. This particular arrangement is usually regarded as a "stretched" or "concentrated" mesh where the grid points are considered of being *biased* toward the wall boundaries.

In three dimensions any grid point in space can now be addressed by the indices (i, j, k), where the introduction of index k represents points that run in the z direction. The consideration of an additional dimension requires now the knowledge of the spacing of Δz_k in addition to the spacing of Δx_i and the spacing of Δy_j to construct the appropriate mesh covering the three-dimensional geometry.

4.2.2 Body-Fitted Mesh

Consider the following problem of a flow inside a 90° bend, as illustrated in Fig. 4.3. In order to apply an orthogonal mesh such as a structured mesh to the geometry, compromises are required to be made especially on the curved section through characterizing the boundaries through staircase-like steps. Nevertheless, such an approach raises two kinds of problems. Firstly, such an approximate boundary description is tedious and rather time consuming to set up. Secondly, the steps at the boundary may introduce errors in computations of the wall stresses, heat fluxes, boundary layer effects, etc. The treatment of the boundary conditions at stepwise walls generally requires a fine Cartesian mesh to cover the wall regions but the requirement of the highly regular structure of grid lines may cause further wastage of computer storage due to unnecessary refinement in interior regions,

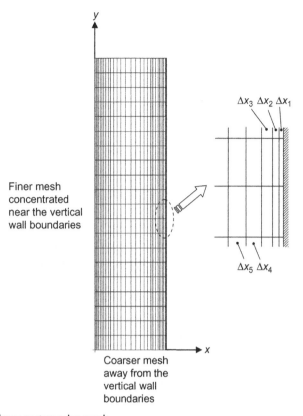

Fig. 4.2 A non-uniform rectangular mesh.

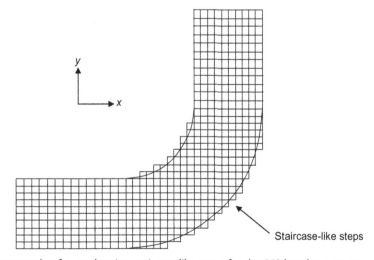

Fig. 4.3 An example of a mesh using staircase-like steps for the 90° bend geometry.

which are of minimal interest. This example clearly shows that mesh systems that are based on Cartesian coordinate systems have severe limitations in irregular geometries. It would therefore be more advantageous to work with meshes that can handle curvature and geometric complexity more naturally.

When a body-fitted mesh is applied to the 90° bend geometry, let us alternatively consider that the walls coincide with lines of constant η (see Fig. 4.4). Location along the geometry, say from A to B or D to C, subsequently corresponds to specific values of ξ in the computational domain. Corresponding points on AB and CD connected by a particular η line will have the same value of ξ_i but different η values. At a particular point (i, j) along this η line, $\xi = \xi_i$ and $\eta = \eta_j$. A corresponding point $x = x(\xi_i, \eta_j)$ and $y = y(\xi_i, \eta_j)$ in the computational domain exists in the physical domain.

In Fig. 4.4 the transformation of the 90° bend geometry should be defined such that there is a one-to-one correspondence between the rectangular mesh in the computational domain and the curvilinear mesh in the physical domain. The algebraic forms of the governing equations for the flow problems are carried out in the computational

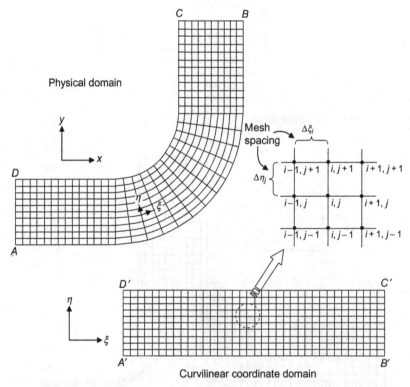

Fig. 4.4 An example of a body-fitted or curvilinear mesh for the 90° bend geometry and corresponding computational geometry.

domain, which has a uniform spacing of $\Delta\xi$ and uniform spacing of $\Delta\eta$. Computed information is then directly fed back to the physical domain via a one-to-one correspondence of grid points. Because of the need to solve the equations in the computational domain, the governing equations are required to be expressed in terms of the curvilinear coordinates rather than the Cartesian coordinates, which means that they must be transformed from (x, y) to (ξ, η) as the new independent variables.

The mesh construction of the internal region of the 90° bend geometry can normally be achieved via two approaches. On the one hand, the Cartesian coordinates may be algebraically determined through interpolation from the boundary values. This methodology requires no iterative procedure and it is computationally inexpensive. On the other hand, a system of partial differential equations of the respective Cartesian coordinates may be solved numerically with the set of boundary values as boundary conditions in order to yield a highly smooth mesh in the physical domain. The former is commonly known as the transfinite interpolation method and the latter is typically the elliptic grid generation method (Smith, 1982; Thompson et al., 1982).

The method of transfinite interpolation according to Gordon and Thiel (1982) consists of generating the interior mesh from the boundary grid data using appropriate interpolation functions or "blending" functions. In multiple dimensions the method can be easily developed by merely extending the one-dimensional interpolation function. Defining the position vector $\mathbf{r} \equiv (x, y)$, one-dimensional interpolation between two boundaries on a varying index i can be expressed as

$$\mathbf{r}_{i,j} = f_i \mathbf{r}_{1,j} + (1 - f_i)\mathbf{r}_{I,j} \tag{4.1}$$

where f_i varies monotonically from $f_1 = 0$ to $f_I = 1$ for $i = 1, 2, ..., I$. Analogous expressions involving function g_j apply for the interpolation in the j direction for $j = 1, 2, ..., J$. The interpolation operation given by Eq. (4.1) can be defined as the "projector" P^i, that is,

$$\mathbf{r}_{i,j} = P^i = f_i \mathbf{r}_{1,j} + (1 - f_i)\mathbf{r}_{I,j} \tag{4.2}$$

Based on Eq. (4.2), two-dimensional transfinite interpolation can be effectively provided by the following relationship:

$$\mathbf{r}_{i,j} = P^i + P^j - P^i P^j \tag{4.3}$$

where

$$P^i P^j = f_i g_j \mathbf{r}_{I,J} + f_i \left(1 - g_j\right)\mathbf{r}_{I,1} + (1 - f_i)g_j \mathbf{r}_{1,J} + (1 - f_i)\left(1 - g_j\right)\mathbf{r}_{1,1} \tag{4.4}$$

For simplicity, the blending functions f_i and g_j may be taken to be linear $- f_i = (i - 1)/(I - 1)$ and $g_j = (j - 1)/(J - 1)$. They may also be formulated with added complexity to concentrate more grid nodal points near the boundaries of the physical domain or to include the specification of derivative boundary conditions to force the grid lines to intersect the boundaries orthogonally of the physical domain. On the latter aspect, the method of transfinite interpolation based on Hermite interpolation by Shih et al. (1991), which allows the specification of derivatives at the end points of curves, has been shown to generate high-quality meshes where the orthogonality of the grid lines intersecting the boundaries can be appropriately realized.

For the elliptic grid generation method, the interior mesh can be constructed via the numerical evaluation of the Poisson equation, which can be written in the form

$$\frac{\partial^2 \xi}{\partial x^2} + \frac{\partial^2 \xi}{\partial y^2} = P(\xi, \eta) \tag{4.5}$$

$$\frac{\partial^2 \eta}{\partial x^2} + \frac{\partial^2 \eta}{\partial y^2} = Q(\xi, \eta) \tag{4.6}$$

where P and Q are functions that are generally unknown at the beginning of the grid generation procedure. They can either be specified externally or determined from the boundary-point distributions to control the spacing and orientation of the grid lines. If the functions of P and Q are set to zero, Eqs (4.5) and (4.6) reduce to the homogeneous Laplace equation. It can be readily shown by calculus of variations that a mesh generated as the solution of the Laplace equation is the smoothest possible mesh that can be attained. The actual solution of Eqs (4.5) and (4.6) needs, however, to be performed in the computational domain in which the roles of x and y are required to be switched as the dependent variables while ξ and η as the independent variables. Hence the partial differential equations are transformed in this domain to

$$\alpha \frac{\partial^2 x}{\partial \xi^2} - 2\beta \frac{\partial^2 x}{\partial \xi \partial \eta} + \gamma \frac{\partial^2 x}{\partial \eta^2} + \alpha P(\xi, \eta) \frac{\partial x}{\partial \xi} + \gamma Q(\xi, \eta) \frac{\partial x}{\partial \eta} = 0 \tag{4.7}$$

$$\alpha \frac{\partial^2 y}{\partial \xi^2} - 2\beta \frac{\partial^2 y}{\partial \xi \partial \eta} + \gamma \frac{\partial^2 y}{\partial \eta^2} + \alpha P(\xi, \eta) \frac{\partial y}{\partial \xi} + \gamma Q(\xi, \eta) \frac{\partial y}{\partial \eta} = 0 \tag{4.8}$$

where

$$\alpha = \left(\frac{\partial x}{\partial \eta}\right)^2 + \left(\frac{\partial y}{\partial \eta}\right)^2, \quad \beta = \left(\frac{\partial x}{\partial \xi}\right)\left(\frac{\partial x}{\partial \eta}\right) + \left(\frac{\partial y}{\partial \xi}\right)\left(\frac{\partial y}{\partial \eta}\right), \quad \gamma = \left(\frac{\partial x}{\partial \xi}\right)^2 + \left(\frac{\partial y}{\partial \xi}\right)^2$$

Eqs (4.3) and (4.4) are merely elliptic partial differential equations with ξ and η as independent variables and x and y as dependent variables (Thompson et al., 1982). Bear in mind that Eqs (4.3), (4.7), and (4.8) have *nothing* to do with the *physics* of the flow field. They are just simply algebraic and elliptic partial differential equations in which ξ and η have been chosen to relate to x and y. This constitutes a transformation from the physical to the computational domain, a one–to–one correspondence of grid nodal points.

4.2.3 Unstructured Mesh

In addition to the use of a body-fitted mesh, an unstructured mesh could be constructed to fill the interior region of the 90° bend geometry (see Fig. 4.5). In the figure shown there appears to be no regularity to the arrangement of the cells in the overlay mesh. Herein, the cells are totally unstructured oriented and there are no coordinate lines that correspond to the curvilinear directions ξ and η such as in a body-fitted mesh. Maximum flexibility is thus allowed in matching the cells especially with highly curved boundaries and required cells can be purposefully inserted to resolve the flow regions where they matter most such as in areas of high gradients. Triangle and tetrahedral meshing are by far the most common forms of unstructured grid generation.

As depicted in Fig. 4.5, the initial set of boundary nodes of the geometry can be effectively triangulated according to the Delaunay triangulation criterion. Here, the most important property of a Delaunay triangulation is that it has the *empty circumcircle (circumscribing circle) property* (Shewchuk, 2002). By definition, the circumcircle of a triangle is the unique circle that passes through its three vertices. The Delaunay triangulation of a set of vertices can therefore be regarded as the triangulation (usually, but not always, unique) in which every triangle has an empty circumcircle, meaning that the circle encloses no vertex of the triangulation. It can be shown that the circumcircle of every Delaunay triangle of the mesh generated for the 90° bend geometry is empty. All algorithms for computing Delaunay triangulations rely on the fast operations for detecting when a grid point is within a triangle's circumcircle and an efficient data structure for storing triangles and edges. The most straightforward way of computing the Delaunay triangulation is to repeatedly add one vertex at a time, then re-triangulating the affected

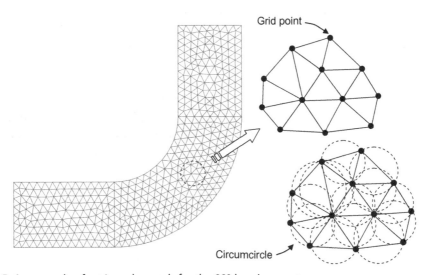

Fig. 4.5 An example of a triangular mesh for the 90° bend geometry.

parts thereafter. When a vertex is added, a search is done for all triangles' circumcircles containing the vertex. Then, those triangles are removed whose circumcircles contain the newly inserted point. All new triangulation is then formed by joining the new point to all boundary vertices of the cavity created by the previous removal of intersected triangles. Delaunay triangulation techniques based on point insertion extend naturally to three dimensions by considering the *circumsphere* (*circumscribing sphere*) associated with a tetrahedron. The Delaunay method is normally more efficient than the advancing-front method owing to the simplicity of the algorithm. The reader is strongly encouraged to refer to an excellent review paper by Mavriplis (1997) and a book by De Berg et al. (2000) for a more thorough understanding of the basic concepts of Delaunay triangulation and meshing.

Relying on Delaunay triangulation alone does not, however, resolve the many extensive problems concerning the application of generating an unstructured mesh. In particular, the Delaunay method tends to maximize the minimum angle of the triangle, but the angle may be still small and might not conform to the domain boundaries. The obvious solution is to add more vertices, but the question remains, "Where should they be inserted?" This has brought about much concerted developments of a number of Delaunay refinement algorithms. For the particular refinement algorithm proposed by Chew (1989), the order of insertion that is based on the minimum angle of any triangle proceeds until the minimum angle is greater than the predefined minimum ($\approx 30°$). Such an approach has been shown to yield high-quality meshes, albeit without any guarantees of grading (ratio of the size between subsequent cells) or size optimality. The problem was rectified by Ruppert (1993), where an algorithm was introduced to produce a mesh with good grading as well as size optimality.

Other meshing algorithms in unstructured grid generation have also been introduced including the advancing front method (Lo, 1985; Gumbert et al., 1989; Marcum and Weatherill, 1995) and the quadtree/octree method (Yerry and Shepard, 1984; Shepard and Georges, 1991).

The main idea of the advancing front method is to generate an unstructured mesh through the adding of individual elements one at a time to an existing front of generated elements. Once the boundary nodes are generated, these edges form the initial front that is to be advanced out into the field. Triangular cells are formed on each line segment, through which, in turn, these cells create more line segments on the front. The front thus constitutes a stack, and edges are continuously added to or removed from the stack. The process terminates when the stack is empty, which is when all fronts have merged upon each other and the domain is entirely covered. For three-dimensional grid generation, a surface grid is first constructed by generating a two-dimensional triangular mesh on the surface boundaries of the domain. This mesh forms the initial front, which is then advanced into the physical space by placing new points ahead of the front and forming tetrahedral elements. The required intersection checking now involves triangular front faces rather than edges as in the two-dimensional case.

Alternatively, the quadtree/octree method involves the generation of an unstructured mesh through a recursive sub-division of the physical space down to a prescribed (spatially varying) resolution. The vertices of the resulting quadtree or octree structure are used as grid points, and the tree quadrants or octants are divided up into triangular or tetrahedral elements, in two or three dimensions, respectively. It should, however, be mentioned that the quadtree/octree cells intersecting the boundary surfaces and the vertices at boundaries must somehow be required to be displaced or wrapped in order to coincide with the boundary. The method is relatively simple and inexpensive and produces good quality meshed in interior regions of the domain. One drawback of the method is that it has a tendency of generating an irregular cell distribution near boundaries.

Another unstructured mesh system that is gaining significant interest in CFD calculations is the employment of a mesh containing *polyhedral* cells. A polyhedral mesh can be created by combining tetrahedral cells into polyhedral ones. Considering the tetrahedral mesh that has been generated for the 90° bend geometry in Fig. 4.5, a polyhedral mesh such as shown in Fig. 4.6 can be created via cell agglomeration, which results in a considerable reduction of the overall cell count. More importantly, cell agglomeration has the capacity of improving the original mesh by converting particular regions with highly skewed tetrahedral cells to polyhedral, thereby improving mesh quality. The use of a polyhedral mesh also leads to quicker convergence of the numerical solution. A clear potential benefit of applying polyhedral mesh is that it allows the flexibility of an unstructured mesh to be applied to a complex geometry without the computational overheads associated with a large tetrahedral mesh. The application of polyhedral meshing is gaining significant traction in the computational fluid dynamics community. Polyhedral meshing has been shown thus far to have considerable advantages over tetrahedral meshing with regard to the attained accuracy and efficiency of numerical computations.

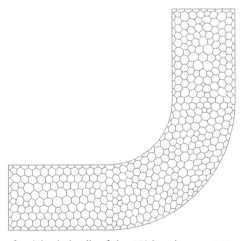

Fig. 4.6 A grid consisting of polyhedral cells of the 90° bend geometry.

4.3 COMMENTS ON MESH TOPOLOGY

Application of a *structured* mesh in any aspect of grid generation has certain advantages and disadvantages. The advantage of such a mesh is that the points of an elemental cell can be easily addressed by a double of indices (i, j) in two dimensions or a triple of indices (i, j, k) in three dimensions. The connectivity is straightforward because cells adjacent to a given elemental face are identified by the indices and the cell edges form continuous mesh lines that begin and end on opposite elemental faces, as illustrated in Fig. 4.7. In two dimensions the central cell is connected by four neighboring cells. In three dimensions the central cell is connected by six neighboring cells. It also allows easy data management and connectivity occurs in a regular fashion, which makes programming easy. Nevertheless, the reader should note that the disadvantage of adopting such a mesh particularly for more

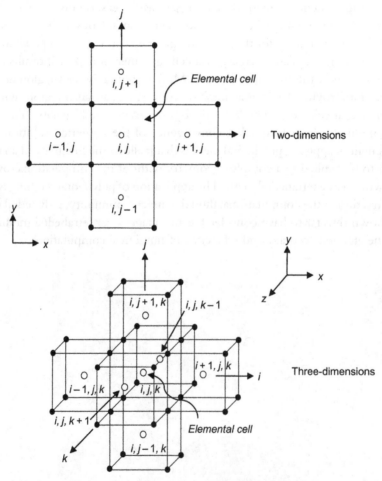

Fig. 4.7 Nodal indexing of elemental cells in two and three dimensions for a structured mesh.

complex geometries is the increase in grid non-orthogonality or skewness that can cause unphysical solutions due to the transformation of the governing equations. The transformed equations that accommodate the non-orthogonality act as the link between the structured coordinate system (such as Cartesian coordinates) and the body-fitted coordinate system and contain additional terms, thereby augmenting the cost of numerical calculations and difficulties in programming. Because of this, such a mesh may also affect the accuracy and efficiency of the numerical algorithm that is being applied.

One special case of the clever use of a *structured* mesh is the adoption of block structured or multi-block meshes. Through this approach, the mesh of the domain in question is now assembled from a number of structured blocks being attached to one another. Here, the attachments of each face of adjacent blocks may be regular (i.e. having matching cell faces) or arbitrary (i.e. having non-matching cell interfaces), as exemplified in Fig. 4.8. Generation of grids especially with non-matching cell interfaces is certainly much simpler than the creation of a single-block mesh fitted to the whole domain and to circumvent the increase in grid non-orthogonality or skewness of the grid cells. Such an approach offers great flexibility in determining the best grid topology for each of the sub-divided blocks. The reader may select an appropriate grid topology based on structured H-, O-, or C-grid, to fill each block. Fig. 4.9 presents an example of an H-grid designed for the flow calculation in a symmetry segment of a staggered tube bank.

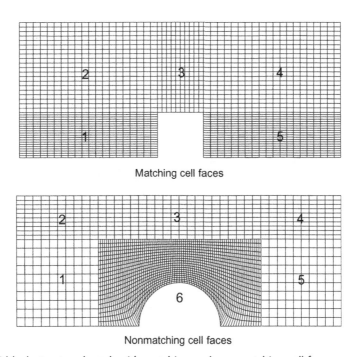

Matching cell faces

Nonmatching cell faces

Fig. 4.8 Multi-block structured mesh with matching and non-matching cell faces.

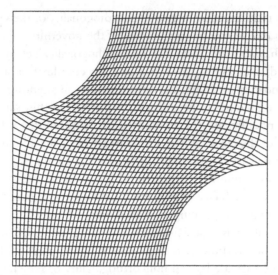

Fig. 4.9 The generation of a structured H-grid for the flow calculation in a symmetry segment of a staggered tube bank.

It is also possible to remove the highly skewed cells for a circular cylinder through the use of an O-grid, as exemplified in Fig. 4.10.

On the other hand, the use of an *unstructured* mesh has become more prevalent and widespread in many CFD applications. The majority of commercial codes nowadays are based on the use of *unstructured* meshing, whereby the cells are allowed to be assembled

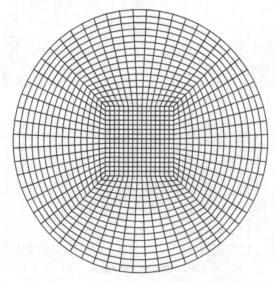

Fig. 4.10 The generation of a structured O-grid for a circular cylinder.

freely within the computational domain. The connectivity information for each face thus requires appropriate storage in some form of a table. The most typical shape of an *unstructured* element is a triangle in two dimensions or a tetrahedron in three dimensions. Nevertheless, any other elemental shape including quadrilateral or hexahedral cells is also possible. *Unstructured* meshes are certainly well suited for handling arbitrary shape geometries especially for domains having high curvature boundaries. Fig. 4.11 illustrates the interior of a circular cylinder that has been filled by the overlaying of *structured* and *unstructured* meshes. For such a geometrical feature, the *structured* body-fitted non–orthogonal grid has a tendency to generate highly skewed cells at the four vertices, as indicated in Fig. 4.11, since the interior of the domain must be built to satisfy the geometrical constraints imposed by the domain boundary. This type of mesh generally leads to numerical instabilities and deterioration of the computational results. It may have been more preferable to re-mesh the geometry with an *unstructured* triangular mesh (see Fig. 4.11).

In spite of the favorable features of *unstructured* meshing, the reader should be aware of a number of disadvantages that exist with the employment of such a mesh for CFD simulations. In comparison to *structured* mesh, the points of an elemental cell for an *unstructured* mesh generally cannot be simply treated or addressed by a double of indices (i,j) in two dimensions or a triple of indices (i,j,k) in three dimensions. An elemental cell may have an arbitrary number of neighboring cells attaching to it, making the data treatment and connection arduously complicated. More importantly, triangular (two-dimensional) or tetrahedral (three-dimensional) cells in comparison to quadrilateral (two-dimensional) or hexahedral (three-dimensional) cells are usually ineffective to resolve wall boundary layers. In most cases the grid yields very long, thin triangular or tetrahedral cells adjacent to the wall boundaries, thereby creating major problems in the approximation of the diffusive fluxes. Another disadvantage in connection with data treatment and connectivity

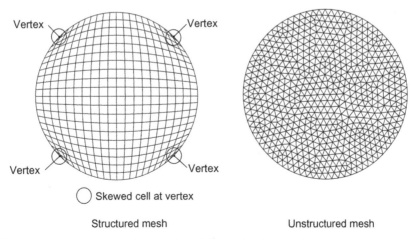

Fig. 4.11 A structured and an unstructured mesh for a circular cylinder.

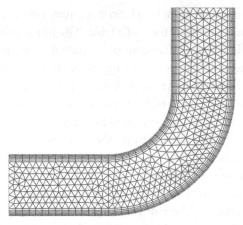

Fig. 4.12 A grid consisting of structured quadrilateral elements near the walls and unstructured triangular elements in the remaining part of the 90° bend geometry.

of elemental cells is the requirement of more complex solution algorithms to solve the flow field variables. This may result in increased computational times to obtain a solution and erode the gains in computational efficiency.

The use of hybrid grids that combine different element types such as triangular and quadrilateral in two dimensions or tetrahedral, hexahedra, prisms, and pyramids in three dimensions can provide maximum flexibility in matching mesh cells with the boundary surfaces, and allocating cells of various element types in other parts of the complex flow regions. As a common practice, grid quality is usually enhanced through the placement of quadrilateral or hexahedral elements in resolving boundary layers near solid walls, while triangular or tetrahedral elements or polyhedral cells, as exemplified in Fig. 4.6, are generated for the rest of the flow domain. This generally leads to both accurate solutions and better convergence for the numerical solution methods. Fig. 4.12 illustrates an example of a mesh consisting of quadrilateral elements near the walls and triangular elements for the rest of the flow domain. Note again that the idea of a "stretched" mesh as previously considered for the rectangular conduit in Fig. 4.2 has been constructed in the vicinity of the wall boundaries for the 90° bend geometry.

4.4 LOCAL MESH REFINEMENT

One local mesh refinement technique that is widely used in many CFD applications is the concept of a stretched grid in the near vicinity of domain walls. For a viscous flow bounded with solid boundaries, the need to cluster a large number of small cells within the physical boundary layer is more than just attempting to minimize the truncation error with the closely spaced grid points. Rather, it is a matter of utter importance that the actual flow physics is appropriately encapsulated.

Let us revisit the case for the fluid flowing between two stationary parallel plates, as investigated in previous chapters. In a real physical flow the existence of a developing boundary layer will grow in thickness as the fluid enters the left boundary and migrates downstream along the bottom wall of the domain, as illustrated in Fig. 4.13. By denoting the local thickness of the boundary layer as δ, where $\delta = \delta(x)$, it is evidently clear that the use of a coarse uniform mesh misses the physical boundary layer. The predicted *viscous-like* velocity profile shown at some point downstream in the fully developed region is simply due to the application of the no-slip boundary condition at the bottom wall. In contrast, the coarse stretched grid at the very least captures some of the essential features of the actual physical boundary layer. It is therefore not surprising that the accuracy of the computational solution is greatly influenced by the grid distribution inside the

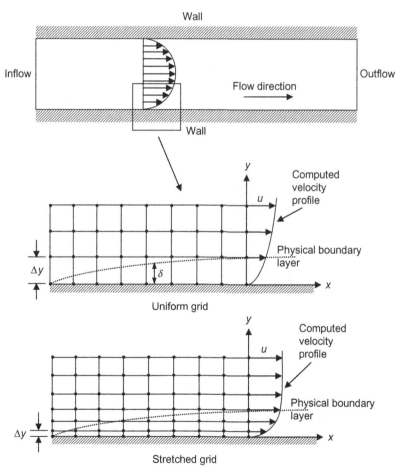

Fig. 4.13 Two schematic illustrations demonstrating the need for local refinement in the near vicinity of the bottom wall to resolve the physical boundary layer.

boundary layer region. We can further investigate this by imposing the concept of *consistency* to be discussed in Chapter 6 for the incompressible mass conservation in two dimensions where the truncation error yields the following:

$$\frac{\Delta x^2}{6}\frac{\partial^3 u}{\partial x^3} + \frac{\Delta y^3}{6}\frac{\partial^3 v}{\partial y^3} \tag{4.9}$$

If the solution error is expected to follow the truncation error, it is imperative that the grid needs to be appropriately refined in order to sufficiently resolve the steep gradient of the velocity profile that exists within this region. This will help to minimize the solution error associated with the truncation error.

Local mesh refinement is also important to better resolve specific fluid dynamics problems such as upward stagnation flow and backward-facing step geometry. The latter is one of the basic geometries used commonly in many engineering applications to better understand the phenomena of flow separation, flow reattachment, and free shear jet. The placement of stretched grids along both the horizontal and vertical directions to encapsulate the essential feature of the recirculation vortex is illustrated in Fig. 4.14.

While applying the stretched grid described above, one useful consideration is to exercise care in avoiding sudden changes of the grid size away from the domain boundary. The mesh spacing should be continuous and grid size discontinuities should be removed as much as possible in regions of large flow changes, particularly when dealing

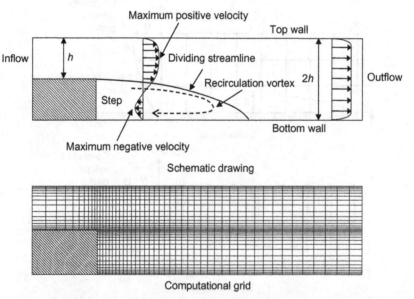

Fig. 4.14 A schematic drawing for the backward-facing step geometry and the computational grid to capture the essential feature of the recirculation vortex.

with multi-block meshing of arbitrary mesh coupling, non-matching cell faces, or extended changes of element types. Discontinuity in the grid size destabilizes the numerical procedure due to the accumulation of truncation errors in the critical flow regions. These errors usually contain the diffusive terms (second derivatives) where the discretization imposed on these derivatives requires very smooth grid changes. Making sure that the grid changes slowly and smoothly away from the domain boundary as well as within the domain interior will assist in overcoming the divergence tendencies of the numerical calculations. It is also worthwhile noting that most in-built mesh generators in commercial codes and independent grid generation packages have the means of prescribing suitable mesh stretching or expansion ratios (rates of change of cell size for adjacent cells). The specification of these ratios should always be negotiated within the codes' requirements while generating the appropriate stretched grid.

By the mere construction of a stretched grid, the local refinement technique provides the possibilities of allocating additional grid nodal points to resolve the important fluid flow action and reducing or removing the grid nodal points from other regions where there is little or no action. Nevertheless, it should be noted that a stretched grid is an algebraically generated grid prescribed *prior* to the solution of the flow field being calculated. The question that needs to be carefully addressed is whether the generated stretched grid sufficiently captures the major fluid flow action or whether the real flow action is far away from the intended significant flow activity to be resolved by the generated stretched grid region, which is not known *a priori*.

In the event where finer details of the local flow physics are required to be predicted, the use of embedded meshes can also be adopted. Here such meshes do not necessarily have to be constrained to have matching cell faces. Fig. 4.15 illustrates an example of embedded Cartesian mesh without matching cell faces near the bottom boundary for the rectangular-type geometry. We can view this type of mesh as a special case of a local mesh refinement strategy. It is noted that the locally refined region in Fig. 4.15 could also

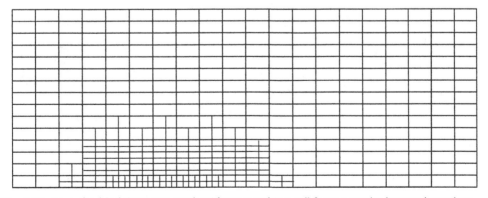

Fig. 4.15 An embedded Cartesian mesh without matching cell faces near the bottom boundary.

have been achieved through the use of other types of elements such as triangular or poly-hedral elements or a combination of different types of elements.

4.5 OVERLAPPING MESH TECHNIQUES

Instead of block structured mesh where the attachment of a number of adjacent blocks is realized at block boundaries, the use of overlapping grids to cover the irregular flow domains further presents another grid generation approach in handling complex geom-etries. Here, rectangular, cylindrical, spherical, or non-orthogonal grids can be com-bined with the parent Cartesian grids in the solution domain. An example of an overlapping grid for a cylinder in a channel with inlet-outlet mappings is shown in Fig. 4.16. This approach is attractive because the structured mesh blocks can be placed freely in the domain to fit any geometrical boundary while satisfying the essential res-olution requirements. Information between the different grids is achieved through the interpolation process. Block structured grids with overlapping blocks are sometimes referred to as *Chimera* grids. The advantages of employing such grids are that complex domains are treated with ease and they can especially be employed to follow moving bodies in stagnant surroundings. Some examples can be found in Tu and Fuchs (1992) and Hubbard and Chen (1994, 1995). The disadvantages of these grids are that conser-vation is usually not maintained or enforced at block boundaries, and the interpolation process may introduce errors or convergence problems if the solution exhibits strong variation near the interface.

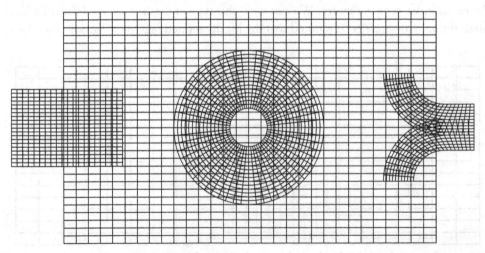

Fig. 4.16 A structured overlapping grid for a cylinder in a channel with inlet-outlet mappings.

4.6 ADAPTIVE MESH WITH SOLUTION

Solution adaptation through the use of an adaptive mesh is a grid network that automatically or dynamically clusters the grid points in regions where large gradients exist in the flow field. It therefore employs the solution of the flow properties to locate the grid points in the physical flow domain. During the course of the solution, the grid points in the physical flow domain *migrate* in such a manner as to *adapt* to the evolution of the large flow gradients or discontinuities. Hence the actual grid points are constantly in motion during the solution of the flow field and become stationary when the flow solution approaches some quasi-steady-state condition. An adaptive mesh is therefore intimately linked to the flow field solution and alters as the flow field develops, unlike the stretched grid described in the previous section, where the grid generation is completely separate from the flow field solution. For this purpose, unstructured meshes are well suited especially in automating the generation of elements such as triangular or tetrahedral meshes of various sizes to solve the critical flow regions. If a solution method can be applied to an unstructured mesh with cells of varying topology, the adaptive mesh is thus subjected to fewer constraints.

Adaptive gridding techniques can generally be categorized into two broad classes: adaptive mesh redistribution and adaptive mesh refinement. Techniques for adaptive mesh redistribution continuously reposition a fixed number of cells so that they improve the resolution in particular locations of the fluid flow domain. Techniques for adaptive mesh refinement add new cells as required and delete other cells that are no longer required. A sample illustration of solution adaptation via adaptive mesh refinement for the fluid flowing over two cylinders investigated in Chapter 2 is given in Fig. 4.17.

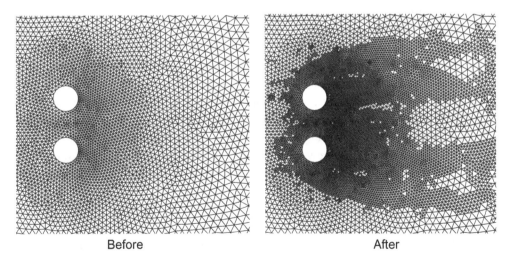

Before   After

Fig. 4.17 A demonstration of solution adaptation through the use of triangular meshes for the fluid flowing over two cylinders.

For this particular flow problem, the wake region has been further resolved to capture the essential formation and shedding of vortices behind the two cylinders using triangular elements. Adaptive mesh refinement by further sub-dividing the cells into smaller ones can also be made possible, whereby a non-refined neighboring cell, although it retains its original shape, becomes a *polyhedron* since its cell face is now replaced by a set of sub-faces. An adaptive mesh with *polyhedral* elements certainly offers many challenges and potential prospects in CFD. Most commercial codes nowadays offer some form of automated techniques for solution adaptation.

4.7 MOVING MESHES

The solution adaption via adaptive mesh redistribution is described herein. In the context of finite volumes and fixed number of grid points, this particular solution adaption is alternatively known as moving meshes. A requirement of this solution adaption approach is to generate a suitable mapping from a regular domain in computational space whereby the discretized equations are solved to predict the physical processes occurring in an irregularly shaped domain in physical space. Three key ingredients are described in the following.

The first ingredient is the use of mesh equations. Principally, they determine a one-to-one mapping from a regular domain in the computational space to an irregularly shaped domain in the physical space. By connecting points in the physical space corresponding to discrete points in the computational space, the physical domain can be overlaid with a suitable computational mesh for the solution of the discretized equations via the finite volume method. Selecting suitable mesh equations and solving them efficiently are crucial for the effectiveness of solution adaption via moving meshes.

The second ingredient is the consideration of a mesh density function, which is also known as monitor function, to guide the mesh redistribution. This monitor function is usually restricted both to redistribute the point movement and to find a mesh relaxation in the search of its redistributed state. The choice of a monitor function may depend on the arc length of the solution in one-dimensional problems, in the solution curvature, and in errors calculated *a posteriori*. Most importantly, the monitor function must be based on both physical and geometrical considerations that account for the distance or the area between the interfaces, and the new mesh being adapted is as close as possible to the reference (old) mesh. Once the monitor function is determined, the mesh is redistributed in some way and should be verified. The redistribution problem by itself is an algebraic nonlinear problem. In practice, some smoothing or spatial and time relaxation is necessary to improve the mesh quality, thereby decreasing the distortion of the elements.

The third ingredient is the requirement to introduce interpolations. If the mesh equations are time dependent and are solved simultaneously with the given differential governing equations, then interpolation of dependent variables from the old mesh to the new mesh is unnecessary. Such interpolation-free moving meshes allow problems with large solution gradients or discontinuities for steep wave fronts and moving boundaries to be handled easily, unlike static meshes, where some kind of interpolation is required to pass the solution information on the newly generated mesh from the old mesh. They can also assist in reducing time variation to allow larger time steps to be employed.

An example of the application of moving meshes for the movement of a limb swinging backward and forward is illustrated in Fig. 4.18. During the re-meshing process, constraints on the cell and face skewness are imposed to control the size distribution as the limb undergoes its swinging motion. Adjacent cells and faces around the limb are also allowed to move with the limb rather than being deformed in order to properly resolve the boundary layer fluid flow.

Despite the ability to resolve large solution variations occurring within the physical domain, the limitation of moving meshes is manifested by the difficulty of proper definition of an adequate time interval due to the fact that the grid points vary their position with time, increasing the stiffness of the system which requires implicit time integration to work efficiently. The applicability of moving meshes is also limited to the fixed number of degrees of freedom and to a constant connectivity to the mesh polygons. More importantly, extra attentions have to be paid to guarantee that the important properties of the physical solutions will not be lost after moving the grid points. One of the main

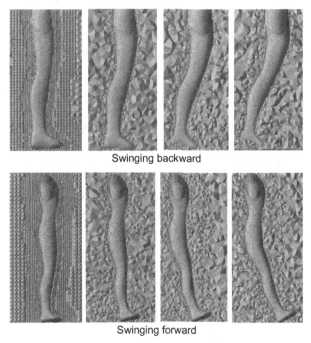

Swinging backward

Swinging forward

Fig. 4.18 A demonstration of moving meshes for a swinging limb.

difficulties in developing a scheme for moving meshes is how mass conservation of the flow field can be properly maintained throughout the course of the numerical calculations. The reader should bear in mind that grid quality (skewness and/or aspect ratio) might suffer or not improve as a consequence of adaptive gridding techniques. Adaptive meshes are still progressively being enhanced in CFD. The concept of solution adaptation described in this section as well as in the previous section is by no means exhaustive. We strongly encourage the reader to consult the literature for a more in-depth understanding of this extensive subject.

4.8 GUIDELINES FOR MESH QUALITY AND MESH DESIGN

As aforementioned, grid generation is by no means a trivial exercise. The mesh for each class of problem is usually unique to the specific CFD geometry, thereby demanding a more measured consideration during the grid generation process.

Firstly, adopting the mindset of applying a coarse mesh topology for a particular CFD problem at the initial step is one of the recommended strategies as part of grid design, as it offers the opportunity to evaluate the specific computer code's storage and running time. More importantly, the utilization of a suitable coarse mesh allows a number of "test-runs" to be carried out with the aim of better assessing the convergence or divergence behavior of the numerical calculations. When a solution is found to be converging, mesh refinement in the flow domain can then be undertaken to achieve the eventual CFD solution. For a diverging solution, the user needs to negotiate and investigate the problems and causes arising during the numerical calculations. We note that "test-runs" performed on such a mesh also provide the means of rectifying possible sources of solution errors such as *physical modeling* and *human* errors that may be present during the course of the numerical simulations.

Evidently, "test-runs" are not recommended on a fine mesh since this could take hours or days to examine the numerical solutions. Care should also be taken in hastily applying a fine mesh at the first instance because of the plausible diverging tendency that may occur during the iterative procedure. Let us elucidate this point by examining the convective (first-order gradients) and diffusive (second-order gradients) terms along the Cartesian x and y directions in the governing transport equation, which can be approximated according to the following discretizations:

$$u\frac{\partial\phi}{\partial x}\approx u\frac{\Delta\phi}{\Delta x}, v\frac{\partial\phi}{\partial y}\approx v\frac{\Delta\phi}{\Delta y} \quad \text{and} \quad \frac{\partial^2\phi}{\partial x^2}\approx\frac{\Delta^2\phi}{\Delta x^2}, \frac{\partial^2\phi}{\partial y^2}\approx\frac{\Delta^2\phi}{\Delta y^2} \tag{4.9}$$

On a fine mesh, the mesh spacing of Δx and Δy may be very small, resulting in

$$\frac{\Delta^2\phi}{\Delta x^2}, \frac{\Delta^2\phi}{\Delta y^2} \gg u\frac{\Delta\phi}{\Delta x}, v\frac{\Delta\phi}{\Delta y} \tag{4.10}$$

When the above Eq. (4.10) is coupled with poor initial conditions for the transport variable ϕ during the first iteration, we observe that the values of the second order gradients can become extremely large because of the step change of ϕ being small. This can cause the flow solver to misbehave and the iterative procedure to subsequently diverge in due process. One practical way to overcome the poor initial guesses or unresolved steep gradients in the flow field is through the use of *under-relaxation factors*, as will be discussed in Section 6.4.3, to curtail the iterative advancement of the numerical calculations. The other strategy that is worthwhile noting and is available in many commercial codes is to initially solve the problem in a coarse mesh. The solution of this mesh is later interpolated onto a fine mesh in the subsequent calculations to aid convergence and promote numerical stability and efficiency.

Secondly, we address the grid quality of a generated mesh that depends on the consideration of the cell shape: aspect ratio, skewness, warp angle, or included angle of adjacent faces.

Fig. 4.19 illustrates a quadrilateral cell having mesh spacing of Δx and Δy and an angle of θ between the grid lines of the cell. Accordingly, we can define the grid *aspect ratio* of the cell as $AR = \Delta y/\Delta x$. One pertinent guideline to bear in mind during the course of grid generation is that large aspect ratios should always be avoided in important regions inside the interior flow domain as they can degrade the solution accuracy and may result in possible poor iterative convergence (or divergence) depending on the flow solver during the numerical computations. Whenever possible, it is recommended that AR is maintained within the range of $0.2 < AR < 5$ within the interior region. For near wall boundaries, the condition for AR can, however, be relaxed. If the fluid flow is in the y direction, the need to appropriately choose small Δx mesh spacing in the x direction will generally yield $AR > 5$. In such a case the approximated first- and second-order gradients are now only biased in the y direction, mimicking more of a one-dimensional flow behavior along this direction since

$$\frac{\Delta^2\phi}{\Delta x^2} << \frac{\Delta^2\phi}{\Delta y^2} \quad \text{and} \quad \frac{\Delta\phi}{\Delta x} << \frac{\Delta\phi}{\Delta y} \tag{4.11}$$

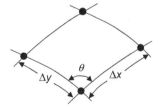

Fig. 4.19 A quadrilateral cell having mesh spacing of Δx and Δy and an angle of θ between the grid lines of the cell.

This behavior is also exemplified where $AR < 0.2$ if the fluid flow is in the x direction. Eq. (4.11) can assist in possibly alleviating convergence difficulties and enhancing the solution accuracy especially in appropriately resolving the wall boundary layers where the rapid solution change exists along the perpendicular direction of the fluid flow.

The next aspect concerning the cell shape deals with grid *distortion* or *skewness*, which relates to the angle θ between the grid lines, as indicated in Fig. 4.19. It is normally desirable that the grid lines should be optimized in such a way that the angle θ is approximately 90° (orthogonal). If the angle $\theta < 45°$ or $\theta > 135°$, the mesh contains these highly skewed cells and often exhibits a deterioration of the computational results or leads to numerical instabilities. A typical example of highly skewed cells can be seen in Fig. 4.11 for the structured non-orthogonal body-fitted grid filling the interior circular cylinder. For some complicated geometries, there is a high probability that the generated mesh may contain cells that are just bordering the *skewness* angle limits. The convergence behavior of such a mesh may be hampered due to the significant influence of additional terms in the discretized form of the transformed equations. Such a case may be remedied through the use of *under-relaxation factors* to increase the diagonal dominance in the matrix solver, thereby gradually improving the solution between iterations. It is also necessary to avoid non-orthogonal cells near the geometry walls. The angle between the grid lines and the boundary of the computational domain (especially the wall, inlet, or outlet boundaries) should maintain as close as possible to 90°. The reader should pay special attention to this requirement and it is stronger than the requirement given for the grid lines in the flow field far away from the domain boundaries.

If an unstructured mesh is adopted, special care needs to be taken to ensure that the *warp angles* measuring between the surfaces normal to the triangular parts of the faces are not greater than 75°, as indicated by the angle β in Fig. 4.20. Cells with large deviations from the co-planar faces can lead to serious convergence problems and deterioration in the computational results. In many grid generation packages the problem can be overcome by a grid smoothing algorithm to improve the element *warp angles*. Whenever possible, the use of tetrahedral elements should be avoided in wall boundary layers. Prismatic

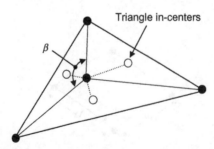

Fig. 4.20 A triangular cell having an angle of β between the surfaces normal to the triangular parts of the faces connected to two adjacent triangles.

or hexahedral cells are preferred because of their regular shape and their ability to adjust in accordance with the near-wall turbulence model requirements, which will be further discussed in a later section.

Thirdly, special grid design features such as the H-grid, O-grid, or C-grid and careful consideration to locate block interfaces in a sensible manner can assist *en masse* to improve the overall quality of a block-structured mesh. The presence of arbitrary mesh coupling, non-matching cell faces, or extended changes of element types at block interfaces should always be avoided in critical regions of high flow gradients or high shear. Wherever possible, finer and more regular mesh in these critical regions that may also include significant changes in the geometry or, where suggested, by error estimates should be employed. In all cases it is recommended that the CFD user checks the assumptions made when setting up the grid with regards to the critical regions of high flow gradients and large changes agreeing with the result of the computation and proceed to rearrange the grid nodal points if found to be necessary.

Finally, it is essential that a grid-independent study is performed to analyze the suitability of the mesh and to yield an estimate of the numerical errors in the simulation for each class of problem. Ideally, at least three significant different grid resolutions should be employed. Strictly speaking, *grid independency* could be examined by doubling the grid twice in each direction and then making sure later by applying the Richardson extrapolation (Roache, 1997). If this is not feasible, selective local refinement of the grid in critical flow regions of the domain can be applied. Otherwise, the reader may attempt to compare different orders of spatial discretizations on the same mesh, which will be further expounded in Section 6.5.

4.9 SUMMARY

Most engineering flows involve complex geometries. In principle, the mesh being considered is subject to constraints imposed by the *discretization* method. For a regular geometry, designing a mesh system is relatively simple since it is evident that the grid lines usually follow the coordinate directions. The use of structured and to a certain extent the curvilinear meshes complies with this requirement. For an irregular geometry, the choice of a mesh system conforming to the complicated shape is, however, not entirely trivial. The application of unstructured meshes, which allows the natural filling of cells of general shapes such as quadrilaterals or triangles in two dimensions or hexahedra or tetrahedral in three dimensions, provides the ease and flexibility of meshing such geometry. The meshing of complicated geometry using polyhedral cells has also attracted much attention and is becoming rather popular in recent times.

In spite of the different types of cells that could be realized either via structured, curvilinear, or unstructured mesh to fill the specific geometry of the CFD problem, the resolution of specific flow features requires particular attention. Some practical guidelines are summarized in the following. The use of local refinement via stretched or embedded meshes could assist in better resolving the flow gradients that exist for wall-bounded flows. Solution adaption via adaptive mesh redistribution or refinement allows the flexibility of adding or concentrating cells to better capture the presence of large flow gradients. The use of overlapping meshes permits the handling of moving bodies in stagnant fluids to be treated by resolving the fluid flow around the objects. Nevertheless, these practical guidelines that have been provided are by no means exhaustive. In essence, this chapter has been particularly written with the primary aim of exposing the reader to some basic understating of different mesh structures as well as recommending some specific guidelines for flow problems that are frequently encountered in the context of CFD.

The means of obtaining a solution to the governing equations are discussed in the next chapter. We will present some of the basic computational techniques that can be employed to solve such partial differential equations. The many discretization procedures predicting the fluid motion and heat transfer processes will also be further discussed and elaborated.

REVIEW QUESTIONS

4.1 What are some of the benefits of a well-designed grid?

4.2 What are some of the advantages of a structured mesh?

4.3 Why is it difficult to write CFD programs that involve a structured mesh for complex geometries?

4.4 Discuss some of the advantages of an unstructured grid.

4.5 What are some of the difficulties that arise regarding the programming of CFD problems for an unstructured mesh?

4.6 What conditions and constraints apply if you had to use a structured mesh for the geometry below? What about an unstructured mesh? Discuss the advantages and disadvantages of this case.

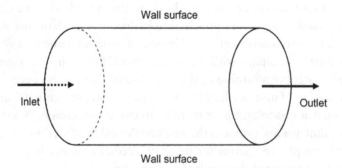

4.7 For the geometry below, discuss how using a block-structured mesh has advantages over a single structured or unstructured mesh.

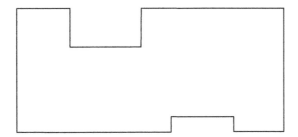

4.8 Why is it more favorable to start off with a coarse mesh when solving a CFD problem?

4.9 What is the aspect ratio of a mesh element? Why should a large aspect ratio be avoided, especially in important regions within the flow field? (See diagram below.)

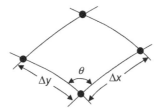

4.10 What is the skewness of a mesh element? Why is it best to avoid highly skewed elements?

4.11 Discuss why tetrahedral elements are a poor choice for meshing near walls and boundaries.

4.12 What techniques can be used in the solver when highly skewed and high aspect ratio elements exist within the mesh?

4.13 A computational domain with different boundary types for the flow around a hydrofoil is shown below. Show where a fine mesh should be appropriately located.

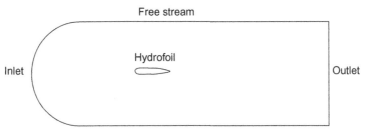

4.14 The stretched grid technique is an example of a local refinement technique. What is one problem associated with the application of this technique?

4.15 Describe how solution adaptation uses an adaptive or moving mesh for grid refinement.

CHAPTER 5

CFD Techniques – The Basics

5.1 INTRODUCTION

Some of the basic computational techniques that are required to solve the governing equations of fluid dynamics are examined in this chapter. Here, the authors will endeavor to demonstrate how these techniques are employed to obtain an approximate solution for the governing equations of flow problems with appropriate boundary conditions applied for the specific problem considered.

The process of obtaining the computational solution consists of two stages. The *first stage* involves the conversion of the partial differential equations and auxiliary (boundary and initial) conditions into a system of discrete algebraic equations. This *first stage* is commonly known as the *discretization* stage.

In Chapter 3 some analytical solutions have been derived for the partial differential equations of the Navier-Stokes to some simple one-dimensional flow problems. There, the closed-form expressions of u, v, w, p, etc., as functions of one of the spatial locations x, y, and z, were used to provide the desired values of the flow-field variables. However, real fluid flows are generally three-dimensional in nature; analytical relationships are not easily attainable. Even for a simplified three-dimensional flow to a two-dimensional problem, this can be difficult. We have seen in Chapter 2 how CFD can be employed to solve a simple two-dimensional channel flow problem. Instead of closed-form expressions, the respective values of u, v, w, p, etc., were obtained at the discrete locations within the flow domain by the CFD solver and the original Navier-Stokes equations were approximated by algebraic derivatives. The partial differential equations, totally replaced by a system of algebraic equations, solved the discrete values of the flow-field variables. For such a solution, the original partial differential equations were considered to be discretized in order to yield the values at the discrete locations.

All numerical methods in the main stream of CFD to solve the fluid flow equations require some form of *discretization*.

Firstly, the *finite difference* method is regarded as the oldest method within the CFD community that uses Taylor series expansions to generate appropriate finite difference expressions to approximate the partial derivatives of the governing equations. These derivatives are replaced by finite difference approximations to yield an algebraic equation for the flow solution at each grid point. Because of the need to have a high degree of regularity, such a method is more suited for structured meshes.

Computational Fluid Dynamics
https://doi.org/10.1016/B978-0-323-93938-6.00005-1

Secondly, the *finite element* method requires the application of simple piecewise polynomial functions that are employed on local elements to describe the variations of the
unknown flow-field variables. The concept of weighted residuals is introduced to measure the errors associated with the approximate functions, which are minimized with successive solution iterations, up to the limits of accuracy achievable by the level of
geometric *discretization*. A set of non-linear algebraic equations for the unknown terms
of the approximating functions is solved and hence yields a flow solution. The attraction
of the *finite element* method over the *finite difference* method is the ability to handle complicated geometries (and boundaries) with relative ease. While the *finite difference* method
is restricted to handling predominantly rectangular shapes and simple variations, the handling of geometries in the *finite element* method is theoretically straightforward. Nevertheless, such a method tends to require greater computational resources and computer
processing power. One well-known method to solve the Navier-Stokes equations is
the discontinuous Galerkin method proposed by Bassi and Rebay (1997). This method
combines features of both *finite element* and *finite volume* methods and the solution is represented within each element as a polynomial approximation (*finite element* method) while
the inter-element advection terms are resolved with appropriate approximation of the
fluxes (*finite volume* method). The discontinuous Galerkin method is used to simulate
a wide range of flow regimes and has been adapted for use with compressible and incompressible, steady and unsteady, as well as laminar and turbulent flows. Further description
of the *finite element* method is given in Section 5.2.3.

Thirdly, the *spectral* method employs the same general approach as the *finite difference*
and *finite element* methods, where the unknowns of the governing equations are replaced
with a truncated series. The difference is that, where the previous two methods employ
local approximations, the *spectral method* uses global approximation that is either by means
of a truncated Fourier series or a series of Chebyshev polynomials for the entire flow
domain. The discrepancy between the exact solution and the approximation is dealt with
by using a weighted residuals concept similar to the *finite element* method. As a variant to
the *finite element* method, the *spectral element* method is a formulation of the *finite element*
method that utilizes high-degree piecewise polynomials. The *spectral element* method as
described in Patera (1984) employs high-degree piecewise polynomials based on orthogonal Chebyshev polynomials or very high-order Legendre polynomials over non-
uniformly spaced nodes. However, a significant disadvantage of the *spectral element*
method is its inherent difficulty in modeling complex geometries when compared to
the flexibility of the *finite element* method. Further description of the *spectral* method is
provided in Section 5.2.4.

Fourthly, the *finite volume* method like the *finite element* method has the capacity of
handling arbitrary geometries with ease. It can be applied to structured, body-fitted,
and unstructured meshes. More importantly, this method bears many similarities with
the *finite difference* method; it is thus simple to apply. Because of the many advantages,

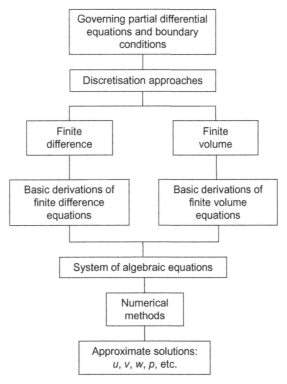

Fig. 5.1 Overview process of the computational solution procedure.

one of which is consistent with the concept of the control volume approach, the most common form of *discretization* in use today for finite volume applications will be thus emphasized.

The overview process of the computational solution procedure in illustrated in Fig. 5.1. Two main headings, namely, the *finite difference* and *finite volume* methods, constitute the most popular *discretization* approaches in CFD, which will be further elaborated on in Section 5.2. The authors will concentrate in detail on the basic derivations of the *finite difference* and *finite volume* methods within this book. The *finite difference* method is illustrated because of its simplicity in formulating the algebraic equations and it also forms the foundation for comprehending the essential basic features of *discretization*. The *finite volume* method is employed in the majority of all commercial CFD codes today. We believe the reader should familiarize themselves with this approach because of its ability to be applied to not only *structured* meshes but also *unstructured* meshes, which are gaining in popularity and usage in handling arbitrary geometrical shapes. A *structured* mesh is usually designated as a mesh containing cells having either a regular-shape element with four-nodal corner points in two dimensions or a hexahedral-shape element with eight-nodal corner points in three dimensions. An *unstructured* mesh commonly refers, however, to a mesh overlaying with cells that are in the form of either a triangle-shape element in two dimensions or a tetrahedron-shape element in three dimensions.

The *second stage* of the solution process involves the implementation of numerical methods to provide a solution to the system of algebraic equations. Appropriate methods for obtaining the numerical solution for the system of algebraic equations are discussed in Section 5.3.

Throughout this chapter, we will concentrate more on considering only systems of algebraic equations typically arising from the solution of steady-state flow problems. These governing equations contain only spatial derivatives that can be discretized by employing either the *finite difference* or the *finite volume* method. Nevertheless, some basic approximations to the time derivatives for unsteady flow problems, which in practice are exclusively discretized using the *finite difference* method, are briefly described in Section 5.2.1.

5.2 DISCRETIZATION OF GOVERNING EQUATIONS

5.2.1 Finite Difference Method

The finite difference method is the oldest of the methods for the numerical solution of partial differential equations. It is believed to have been developed by Euler in 1768 and was used to obtain numerical solutions to differential equations by hand calculation. At each nodal point of the grid used to describe the fluid flow domain, the Taylor series expansions are used to generate finite difference approximations to the partial derivatives of the governing equations. These derivatives, replaced by finite difference approximations, yield an algebraic equation for the flow solution at each grid point. In principle, finite difference can be applied to any type of grid system. However, the method is more commonly applied to structured grids since it requires a mesh having a high degree of regularity. The grid spacing between the nodal points need not be uniform, but there are limits on the amount of grid stretching or distortion that can be imposed, to maintain accuracy. Topologically, these finite difference structured grids must conform to the constraints of general coordinate systems such as Cartesian grids comprising six-sided computational domains. However, the use of an intermediate coordinate mapping such as the body-fitted coordinate system allows this major geometrical constraint to be relaxed, such that complex shapes can be modeled.

The finite difference allows the incorporation of higher-order differencing approximations on regular grids which provide a higher degree of accuracy in the solution. However, the main disadvantage is that the conservation property is not usually enforced unless special care is taken.

The first step toward obtaining a numerical solution involves the discretization of the geometric domain, that is, a numerical grid must be defined. In the finite difference method the grid is usually taken to be locally structured, which means that each grid node

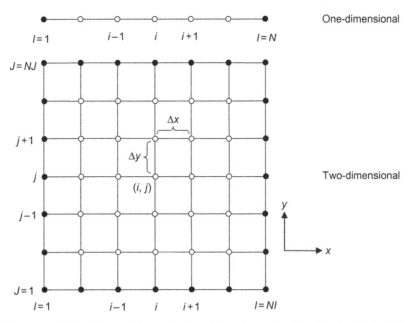

Fig. 5.2 A representation of a one-dimensional and two-dimensional uniformly distributed Cartesian grid for the finite difference method (full symbols denote boundary nodes and open symbols denote computational nodes).

may be considered the origin of a local coordinate system whose axes coincide with the grid lines. These two grid lines also imply that they do not intersect elsewhere and that any pair of grid lines belonging to the different families intersect only once at the grid point. In three dimensions three grid lines intersect at each node; none of these lines intersect each other at any other grid nodal point. Fig. 5.2 illustrates examples of one-dimensional and two-dimensional uniformly distributed Cartesian grids commonly used in the finite difference method. Within these two grid systems, each node is uniquely identified by a set of indices, which are indices of the grid lines that intersect at (i, j) in two-dimensional and (i, j, k) in three-dimensional. The neighboring nodes are defined by increasing or reducing one of the indices by unity.

Analytical solutions of partial differential equations involve closed-form expressions that provide the variation of the flow-field variables continuously throughout the domain. This is in contrast to numerical solutions, where they provide answers only at discrete points in the geometric domain, for example, at the grid points (open symbols) shown in Fig. 5.2. To illustrate the finite difference method, let us conveniently assume that the spacing of the grid points in the x direction is uniform and given by Δx and that the spacing of the points in the y direction is also uniform and given by Δy. The spacing of Δx or Δy need not necessarily be uniform. We could have easily dealt with totally unequal spacing in both directions. This uniform spacing does not have to correspond in the physical x-y space. As is frequently handled in CFD, the numerical calculations can be performed in a transformed computational space that has uniform spacing in the transformed independent variables but still corresponds to a non-uniform spacing in the physical space. In any event we assume uniform spacing in each of the coordinate directions in describing the basic techniques of the finite difference method.

The starting point to the representation of the partial derivatives in the governing equations is the Taylor series expansion. We have encountered the use of the Taylor series expansion in formulating the conservation equations in Chapter 3. Here, we are interested in replacing the partial derivative with a suitable algebraic difference quotient – a finite difference. For example, referring to Fig. 5.2, if at the indices (i, j), there exists a generic flow field variable ϕ, then the variable at point $(i + 1, j)$ can be expressed in terms of a Taylor series expanded about the point (i, j) as:

$$\phi_{i+1,j} = \phi_{i,j} + \left(\frac{\partial \phi}{\partial x}\right)_{i,j} \Delta x + \left(\frac{\partial^2 \phi}{\partial x^2}\right)_{i,j} \frac{\Delta x^2}{2} + \left(\frac{\partial^3 \phi}{\partial x^3}\right)_{i,j} \frac{\Delta x^3}{6} + \cdots \qquad (5.1)$$

Similarly, the variable at point $(i - 1, j)$ can also be expressed in terms of Taylor series about points (i, j) as:

$$\phi_{i-1,j} = \phi_{i,j} - \left(\frac{\partial \phi}{\partial x}\right)_{i,j} \Delta x + \left(\frac{\partial^2 \phi}{\partial x^2}\right)_{i,j} \frac{\Delta x^2}{2} - \left(\frac{\partial^3 \phi}{\partial x^3}\right)_{i,j} \frac{\Delta x^3}{6} + \cdots \qquad (5.2)$$

Eqs (5.1) and (5.2) are mathematically exact expressions for the respective variables $\phi_{i+1,j}$ and $\phi_{i-1,j}$ if the number of terms is infinite and the series converges and/or $\Delta x \to 0$. By subtracting these two equations, we obtain the approximation for the first derivative of ϕ:

$$\left(\frac{\partial \phi}{\partial x}\right) = \frac{\phi_{i+1,j} - \phi_{i-1,j} - (\Delta x^3/3)(\partial^3 \phi/\partial x^3)}{2\Delta x} + \cdots$$

or

$$\left(\frac{\partial \phi}{\partial x}\right) = \frac{\phi_{i+1,j} - \phi_{i-1,j}}{2\Delta x} + \underbrace{O(\Delta x^2)}_{\text{Truncation error}} \quad \text{Central difference} \qquad (5.3)$$

The term $O(\Delta x^n)$ signifies the truncation error of the finite difference approximation, which measures the accuracy of the approximation and determines the rate at which the error decreases as the spacing between points is reduced. Eq. (5.3) is taken to be second order accurate because the truncation error is of order 2. This is a major simplification and its validity depends on the size of Δx. The smaller Δx is, the better the agreement. This equation is called *central difference* since it depends equally on values to both sides of the node at x. It is also possible to form other expressions for the first derivative by invoking Eqs (5.1) and (5.2). These are

$$\left(\frac{\partial \phi}{\partial x}\right) = \frac{\phi_{i+1,j} - \phi_{i,j}}{\Delta x} + \underbrace{O(\Delta x)}_{\text{Truncation error}} \quad \textit{Forward difference} \qquad (5.4)$$

and

$$\left(\frac{\partial \phi}{\partial x}\right) = \frac{\phi_{i,j} - \phi_{i-1,j}}{\Delta x} + \underbrace{O(\Delta x)}_{\text{Truncation error}} \quad \textit{Backward difference} \qquad (5.5)$$

The above two equations are termed the *forward* and *backward differences*. They reflect their respective biases and both of these finite difference approximations are only first order accurate. It is expected that they will be less accurate in comparison to the central difference for a given value of Δx.

We further explore the idea behind finite difference approximations that is borrowed directly from the definition of derivatives. A geometric interpretation of Eqs (5.3), (5.4), and (5.5) is provided in Fig. 5.3. The first derivative $\partial \phi / \partial x$ at the point i in the direction of x is the slope of tangent to the curve $\phi(x)$ at that point and is the line marked *exact* in the figure. Its slope can be approximated by the slope of a line passing through the neighboring points, namely, $i + 1$ and $i - 1$, on the curve. The forward difference is evaluated by the slope BC between the points i and $i + 1$, while the backward difference is achieved by the slope AB between the points $i - 1$ and i. The line labeled central represents the approximation by a central difference that evaluates the slope AC. From the figure, it can be seen that some approximations are better than others. The line for the central difference, indicated by the slope AC, appears to be closer to the slope of the exact line; if the

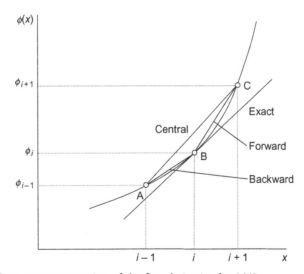

Fig. 5.3 Finite difference representation of the first derivative for $\partial \phi / \partial x$.

function $\phi(x)$ were a second-order polynomial and the points equally spaced in the x-direction, the slopes would match exactly. It also appears that the quality of approximation improves when additional points are made closer to the point; that is, as the grid is refined, the approximation improves.

Differences for the y derivatives are obtained in exactly the same fashion. The results are directly analogous to the previous equations for the x derivatives. They are given by

$$\left(\frac{\partial \phi}{\partial y}\right) = \frac{\phi_{i,j+1} - \phi_{i,j}}{\Delta y} + \underbrace{O(\Delta y)}_{\text{Truncation error}} \quad \text{Forward difference} \quad (5.6)$$

$$\left(\frac{\partial \phi}{\partial y}\right) = \frac{\phi_{i,j} - \phi_{i,j-1}}{\Delta y} + \underbrace{O(\Delta y)}_{\text{Truncation error}} \quad \text{Backward difference} \quad (5.7)$$

$$\left(\frac{\partial \phi}{\partial y}\right) = \frac{\phi_{i,j+1} - \phi_{i,j-1}}{2\Delta y} + \underbrace{O(\Delta y^2)}_{\text{Truncation error}} \quad \text{Central difference} \quad (5.8)$$

The second derivative can also be obtained through the Taylor series expansion as has been applied to approximate the first derivative. By summing Eqs (5.1) and (5.2), we have

$$\left(\frac{\partial^2 \phi}{\partial x^2}\right) = \frac{\phi_{i+1,j} - 2\phi_{i,j} + \phi_{i-1,j}}{\Delta x^2} + \underbrace{O(\Delta x^2)}_{\text{Truncation error}} \quad (5.9)$$

This equation represents the central finite difference for the second derivative with respect to x evaluated at the point (i, j). The approximation is second order accurate. An analogous expression can easily be obtained for the second derivative with respect to y, which results in

$$\left(\frac{\partial^2 \phi}{\partial y^2}\right) = \frac{\phi_{i,j+1} - 2\phi_{i,j} + \phi_{i,j-1}}{\Delta y^2} + \underbrace{O(\Delta y^2)}_{\text{Truncation error}} \quad (5.10)$$

The Taylor series expansion for time derivatives can also be obtained similar to those of the Taylor series expansion for space, Eq. (5.1). Since the numerical solution is more likely to be marched continuously in the discrete time interval of Δt, the finite approximation derived for the first-order spatial derivatives applies equally for the first-order time derivative. For a forward difference approximation in time,

$$\left(\frac{\partial \phi}{\partial t}\right) = \frac{\phi_{i,j}^{n+1} - \phi_{i,j}^{n}}{\Delta t} + \underbrace{O(\Delta t)}_{\text{Truncation error}} \quad \text{Forward difference} \quad (5.11)$$

The above equation introduces a truncation error of $O(\Delta t)$. More accurate approximations to the time derivative can be obtained through the consideration of additional discrete values of $\phi_{i,j}$ in time.

5.2.2 Finite Volume Method

The finite volume method discretizes the integral form of the conservation equations directly in the physical space. It was initially introduced by researchers such as McDonald (1971) and MacCormack and Paullay (1972) for the solution of two-dimensional time-dependent Euler equations and was later extended to three-dimensional flows by Rizzi and Inouye (1973). The computational domain is sub-divided into a finite number of contiguous control volumes, where the resulting statements express the exact conservation of relevant properties for each of the control volumes. At the centroid of each of the control volumes, the variable values are calculated. Interpolation is used to express variable values at the control volume surface in terms of the center values and suitable quadrature formulae are applied to approximate the surface and volume integrals. An algebraic equation for each of the control volumes can be obtained, in which a number of the neighboring nodal values appear.

As the finite volume method works with control volumes and not the grid intersection points, it has the capacity to accommodate any type of grid. Here, instead of structured grids, unstructured grids can be employed that allow a large number of options for the definition of the shape and location of the control volumes. Since the grid defines only the control volume boundaries, the method is conservative so long as the surface integrals that are applied at these boundaries are the same as the control volumes sharing the boundary. One disadvantage of this method compared to the finite difference schemes is that higher-order differencing approximations greater than the second order are more difficult to develop in three dimensions. This is because of the requirement for two levels of approximation, which are interpolation and integration. However, the finite volume method has more advantages than disadvantages. One important feature of the method is that a "finite element" type mesh can be used, in which the mesh can be formed by the combination of triangles or quadrilaterals in the case of two dimensions or tetrahedra and hexahedra in three dimensions. This type of unstructured mesh offers greater flexibility for handling complex geometries. Another attractive feature is the method requires no transformation of the equations in terms of body-fitted coordinate system as is required in the finite difference method.

As with the finite difference method, a numerical grid must be initially defined to discretize the physical flow domain of interest. For the finite volume method, we now have the flexibility of representing the grid by either structured or unstructured mesh. For illustration purposes of the finite volume method, we consider a typical representation of structured (quadrilateral) and unstructured (triangle) finite volume

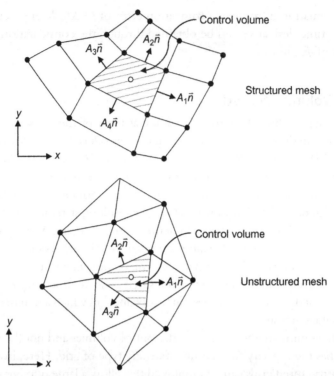

Fig. 5.4 A representation of structured and unstructured mesh for the finite volume method (full symbols denote element vertices and open symbols at the center of the control volumes denote computational nodes).

elements in two-dimensional shape as shown in Fig. 5.4 for the discretization of the partial differential equations. The cornerstone of the finite volume method is the *control volume integration*. In a control volume the bounding surface areas of the element are directly linked to the discretization of the first and second derivatives for ϕ (the generic flow field variable). Here, the surface areas in the normal direction $(\vec{n})$ to the volume surfaces as indicated in Fig. 5.4 are resolved with respect to the Cartesian coordinate directions to yield the projected areas A_i^x and A_i^y in the x and y directions, respectively. The projected areas are positive if their outward normal vectors from the volume surfaces are directed in the same directions as the Cartesian coordinate system; otherwise, they are negative.

By applying Gauss' divergence theorem to the volume integral, the first derivative of ϕ in two dimensions, for example, the term along the x direction represented in Eq. (3.13), can be approximated by

$$\left(\frac{\partial\phi}{\partial x}\right) = \frac{1}{\Delta V}\int_V \frac{\partial\phi}{\partial x}\,dV = \frac{1}{\Delta V}\int_A \phi\,dA^x \approx \frac{1}{\Delta V}\sum_{i=1}^{N}\phi_i A_i^x \qquad (5.12)$$

where ϕ_i are the variable values at the elemental surfaces and N denotes the number of bounding surfaces on the elemental volume. Eq. (5.12) applies to any type of finite volume element that can be represented within the numerical grid. For a quadrilateral element in two dimensions for the structured mesh, as seen in Fig. 5.4, N has the value of 4 since there are four bounding surfaces of the element. In three dimensions, for a hexagonal element, N becomes 6. Similarly, the first derivative for ϕ in the y direction is obtained in exactly the same fashion, which can be written as

$$\left(\frac{\partial \phi}{\partial y}\right) = \frac{1}{\Delta V} \int_{\Delta V} \frac{\partial \phi}{\partial y} dV = \frac{1}{\Delta V} \int_A \phi dA^y \approx \frac{1}{\Delta V} \sum_{i=1}^N \phi_i A_i^y \qquad (5.13)$$

EXAMPLE

Example 5.1

Consider the conservation of mass as described in Chapter 3. Determine the discretized form of the two-dimensional continuity equation $\dfrac{\partial u}{\partial x} + \dfrac{\partial v}{\partial y} = 0$ by the finite volume method in a structured uniform grid arrangement.

Solution: An elemental control volume of the two-dimensional structured grid is shown in Fig. 5.1.1. The centroid of the control volume is indicated by the point P, which is surrounded by the adjacent control volumes having their respective centroids indicated by the points: east, E; west, W; north, N; and south, S. The control volume face between points P and E is denoted by the area A_e^x. Subsequently, the rest of the control volume faces are respectively: A_w^x, A_n^y, and A_s^y.

We begin the analysis by introducing the control volume integration, which forms the key step of the finite volume method. Applying Eqs (5.12) and (5.13) yields the following expressions, which is applicable to both structured and unstructured grids:

$$\frac{1}{\Delta V}\int_V \frac{\partial u}{\partial x} dV = \frac{1}{\Delta V}\int_A u dA^x \approx \frac{1}{\Delta V}\sum_{i=1}^4 u_i A_i^x = \frac{1}{\Delta V}\left(u_e A_e^x - u_w A_w^x + u_n \overset{=0}{A_n^x} - u_s \overset{=0}{A_s^x}\right)$$

$$\frac{1}{\Delta V}\int_V \frac{\partial v}{\partial y} dV = \frac{1}{\Delta V}\int_A v\, dA^y \approx \frac{1}{\Delta V}\sum_{i=1}^4 v_i A_i^y = \frac{1}{\Delta V}\left(v_e \overset{=0}{A_e^y} - v_w \overset{=0}{A_w^y} + v_n A_n^y - v_s A_s^y\right)$$

For the structured uniform grid arrangement, the projection areas A_n^x and A_s^x in the x direction and the projection areas A_e^y and A_w^y in the y direction are zero. One important aspect demonstrated here by the finite volume method is that it allows direct discretization in the physical domain (or in a body-fitted conformal grid) without the need of transforming the continuity equation from the physical domain to a computational domain.

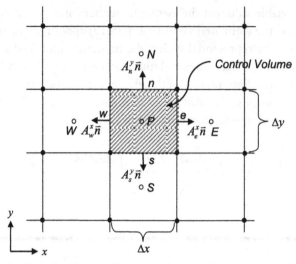

Fig. 5.1.1 Control volume for the two-dimensional continuity equation problem.

Since the grid has been taken to be uniform, the face velocities u_e, u_w, v_n, and v_s are located midway between each of the control volume centroids, which allows us to determine the face velocities from the values located at the centroids of the control volumes. Thus

$$u_e = \frac{u_P + u_E}{2}; \quad u_w = \frac{u_P + u_W}{2}; \quad v_n = \frac{v_P + v_N}{2}; \quad v_s = \frac{v_P + v_S}{2}$$

By substituting the above expressions to the discretized form of the velocity first derivatives, the final form of the discretized continuity equation becomes

$$\left(\frac{u_P + u_E}{2}\right) A_e^x - \left(\frac{u_P + u_W}{2}\right) A_w^x + \left(\frac{v_P + v_N}{2}\right) A_n^y - \left(\frac{v_P + v_S}{2}\right) A_s^y = 0$$

From Fig. 5.1.1, $A_e^x = A_w^x = \Delta y$ and $A_n^y = A_s^y = \Delta x$, the above equation can then be expressed by

$$\left(\frac{u_P + u_E}{2}\right) \Delta y - \left(\frac{u_P + u_W}{2}\right) \Delta y + \left(\frac{v_P + v_N}{2}\right) \Delta x - \left(\frac{v_P + v_S}{2}\right) \Delta x = 0$$

and reduced to

$$\left(\frac{u_E - u_W}{2}\right) \Delta y + \left(\frac{v_N - v_S}{2}\right) \Delta x = 0$$

or in another form

$$\frac{u_E - u_W}{2\Delta x} + \frac{v_N - v_S}{2\Delta y} = 0$$

For a uniform grid arrangement, the distances between P and E, and W and P are equivalent to Δx. Similarly, the distances between P and N, and S and P are given by Δy. If the finite difference method is applied to discretize the continuity equation through the central difference scheme, we obtain the same discretized form of the equation at point P as compared to the form derived through the finite volume method. The accuracy obtained is of second order, as inferred from the finite difference central difference scheme.

Discussion: The purpose of this example is to demonstrate the use of the finite volume method to discretize the two-dimensional continuity equation and compare its form to the finite difference approximation. We observed that the exact representation of the discretized form for the continuity equation can be obtained by applying either the finite volume or the finite difference method for a uniform grid arrangement. Nevertheless, there are two major advantages that the finite volume method holds over the finite difference method. Firstly, it has good conservation properties from the physical viewpoint and secondly, it allows the accommodation of complicated physical domains to be discretized in a simpler way rather than the need to transform the equation to generalized coordinates in the computational domain.

From the above example, the first derivatives that appear in the continuity equation have been discretized through the finite volume method. The discretization of the second derivatives is no different from that performed on the first derivatives. The second derivative along the x direction, for example, the diffusion terms represented in Eqs (3.24) and (3.25), can be evaluated by

$$\left(\frac{\partial^2 \phi}{\partial x^2}\right) = \frac{1}{\Delta V} \int_{\Delta V} \frac{\partial^2 \phi}{\partial x^2} dV = \frac{1}{\Delta V} \int_A \frac{\partial \phi}{\partial x} dA^x \approx \frac{1}{\Delta V} \sum_{i=1}^N \left(\frac{\partial \phi}{\partial x}\right)_i A_i^x \tag{5.14}$$

An analogous expression can be also easily obtained for the second derivative with respect to y, which is given by

$$\left(\frac{\partial^2 \phi}{\partial y^2}\right) = \frac{1}{\Delta V} \int_{\Delta V} \frac{\partial^2 \phi}{\partial y^2} dV = \frac{1}{\Delta V} \int_A \frac{\partial \phi}{\partial y} dA^y \approx \frac{1}{\Delta V} \sum_{i=1}^N \left(\frac{\partial \phi}{\partial y}\right)_i A_i^y \tag{5.15}$$

From Eqs (5.14) and (5.15), it can be seen that in order to approximate the second derivatives, the respective first derivatives of $(\partial \phi / \partial x)_i$ and $(\partial \phi / \partial y)_i$ appearing in the equations are required to be evaluated at the elemental surfaces of the control volume.

The approximation of the first derivatives at the control volume faces is usually determined from the discrete ϕ values of the surrounding elements. For example, in a structured mesh arrangement, as shown in Fig. 5.5, where the central control volume (shaded) is surrounded by only one adjacent control volume at each face, the first derivatives could be approximated by a piecewise linear gradient profile between the central and adjacent nodes. If needed, higher-order quadratic profiles could also be employed to attain higher accuracy for the numerical solution, which require more surrounding elemental volumes within the mesh system.

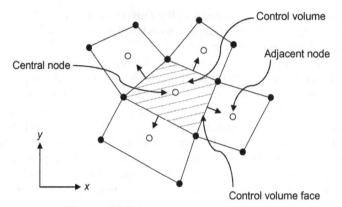

Fig. 5.5 A representation of a structured mesh arrangement (open symbols at the center of the control volumes denote computational nodes) for the evaluation of the face first derivatives.

5.2.3 Finite Element Method

In many aspects the finite element method, which was developed initially as an *ad hoc* engineering procedure for determining the stress and strain displacement solutions in structural analysis, is similar to the finite volume method. Both the finite element and finite volume methods are suitable for irregular computational domains, meaning that they can accommodate complex geometries.

Nevertheless, one distinguishing feature of the finite element method is that the governing equations are first approximated by multiplication with the so-called shape functions before they are integrated over the entire computational domain. For the domain that is divided into a set of finite elements, the generic variable ϕ can be approximated by

$$\phi = \sum_{j=1}^{n} \phi_j \psi_j(x, y, z) \tag{5.16}$$

where n represents the number of discrete nodal unknowns ϕ_j and $\psi_j(x, y, z)$ are the shape functions. For the consideration of linear shape functions, they can be constructed simply from the values at the corners of the elements. As a general guide, the use of linear shape functions generates solutions of about the same accuracy as those of the second-order finite difference method. This approximation is then substituted into the integral of the weighted residual over the computation domain that is taken to be equal to zero,

$$\iiint W_m(x, y, z) \, R \, dx \, dy \, dz = 0 \tag{5.17}$$

in order to generate a system of algebraic equations for ϕ_j, which can be normally solved via numerical methods. In Eq. (5.17) R is referred to as the equation residual, while W_m represents the weight functions. Different choices of W_m give rise to different methods. For example, if the Galerkin method is adopted, the weight functions are chosen to be the same as the shape functions. More details on the application of the finite element method to fluid mechanics can be found in Thomasset (1981), Baker (1983), and Fletcher (1984).

5.2.4 Spectral Method

The finite difference, finite volume, and finite element methods are generally considered *local* methods. Consequently, these methods can be identified in terms of the discrete nodal unknowns. In contrast, the spectral method is a *global* method, which makes it more difficult to implement in practice. Nevertheless, the method allows approximate solutions with a high degree of accuracy; the number of grid points required to achieve the desired precision can therefore be very low.

The spectral method is actually built on the same principal ideas as the finite element method discussed in the previous section. One main difference between these two methods is that the spectral method approximates the solution with a linear combination of continuous functions that are generally non-zero throughout the whole domain (Fourier series, Legendre polynomials and Chebyshev polynomials), while the finite element method approximates the solution of piecewise functions that are non-zero on localized sub-domains. Thus the spectral method works best when the solution is smooth.

Like the finite element method, the starting point for the spectral method is the introduction of the approximate solution of ϕ which can be assumed to be

$$\phi = \sum_{j=1}^{n} a_j(t)\, \varphi_j(x, y, z) \tag{5.18}$$

where $a_j(t)$ are the unknown coefficients and $\varphi_j(x, y, z)$ are the trial functions. The choice of trail functions is dependent on the particular consideration of the flow problem being solved. Fourier series are particularly suited for periodic boundary conditions, while Chebyshev polynomials are more appropriate for non-periodic boundary conditions. Substitution of Eq. (5.18) into the transport equation yields the residual of the equation. The unknown coefficients are now determined via the integration of the vanishing weighted residual over the computational domain, that is, applying Eq. (5.17). Here again, if the Galerkin method is adopted, the weight functions are automatically chosen from the same family as the trial functions. More details on the application of the spectral method to fluid mechanics can be found in Fletcher (1984) and Canuto et al. (1987).

5.3 CONVERTING GOVERNING EQUATIONS TO ALGEBRAIC EQUATION SYSTEM

The basic derivations of the finite difference and finite volume methods and some basic descriptions of the finite element and spectral methods have been provided. Primarily concentrating on the finite difference and finite volume methods, the next task is to consider the appropriate application of these various discretization techniques to numerically solve the governing partial differential equations. A numerical method is developed by considering the three transport processes associated with: (1) pure diffusion in steady state, (2) steady convection-diffusion, and (3) unsteady convection-diffusion. We further simplify the problem by systematically working through a one-dimensional equation in order to demonstrate the application of the discretization techniques to finally attain the algebraic form of the governing equation. It can be seen later that this discretized form of the equation is easily extended and accommodated for two- and three-dimensional diffusion problems.

Pure diffusion process: Let us consider the steady-state diffusion of the generic variable ϕ in a one-dimensional domain. The equation that governs such a process is given by

$$\frac{\partial}{\partial x}\left(\Gamma\frac{\partial \phi}{\partial x}\right) + S_\phi = 0 \tag{5.19}$$

where Γ is the diffusion coefficient and S_ϕ is the source term. This equation is typical of a one-dimensional heat conduction process, which we will further investigate in detail through worked examples. For this problem, the applications of the finite difference and finite volume methods to discretize the equation are illustrated.

5.3.1 Finite Difference Method

The use of the finite difference method is explored here. The first step in developing a numerical solution for this method involves dividing the geometric domain into discrete nodal points. Let us consider a general nodal point P and its surrounding neighboring nodal points to the west and east, W and E, respectively, for the one-dimensional geometry as demonstrated in Fig. 5.6. A uniform grid spanning the three nodal points W, P, and E is produced.

The next step in the discretization process is to discretize Eq. (5.19) around the nodal point P. To derive a suitable expression for the finite difference method, Eq. (5.19) is required to be expanded into its non-conservative form, which is given by

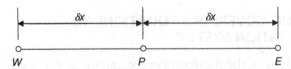

Fig. 5.6 A schematic representation of the uniform grid spacing along the *x* direction for the one-dimensional geometry.

$$\frac{\partial \Gamma}{\partial x} \frac{\partial \phi}{\partial x} + \Gamma \frac{\partial^2 \phi}{\partial x^2} + S_\phi = 0 \qquad (5.20)$$

By applying the central differencing of the first derivative Eq. (5.8) and second derivative Eq. (5.9), the discretized form of Eq. (5.20) is obtained as

$$\frac{(\Gamma_E - \Gamma_W)}{2\delta x} \frac{(\phi_E - \phi_W)}{2\delta x} + \Gamma_P \frac{(\phi_E - 2\phi_P + \phi_W)}{\delta x^2} + S_\phi = 0 \qquad (5.21)$$

After some re-arrangement, Eq. (5.21) can be alternatively expressed as

$$\frac{2\Gamma_P}{\delta x^2} \phi_P = \left[\frac{(\Gamma_E - \Gamma_W)}{4\delta x^2} + \frac{\Gamma_P}{\delta x^2} \right] \phi_E + \left[-\frac{(\Gamma_E - \Gamma_W)}{4\delta x^2} + \frac{\Gamma_P}{\delta x^2} \right] \phi_W + S_\phi \qquad (5.22)$$

Identifying the coefficients of ϕ_E and ϕ_W in Eq. (5.19) as a_E and a_W and the coefficient of ϕ_P as a_P, the equation can be written in a simple form:

$$a_P \phi_P = a_E \phi_E + a_W \phi_W + b \qquad (5.23)$$

where

$$a_E = \frac{(\Gamma_E - \Gamma_W)}{4\delta x^2} + \frac{\Gamma_P}{\delta x^2}; \quad a_W = -\frac{(\Gamma_E - \Gamma_W)}{4\delta x^2} + \frac{\Gamma_P}{\delta x^2}; \quad a_P = \frac{2\Gamma_P}{\delta x^2}; \quad b = S_\phi$$

5.3.2 Finite Volume Method

When the finite volume method is applied, we have to consider the physical domain as being divided into finite control volumes surrounding the nodal points W, P, and E. Fig. 5.7 shows a control volume surrounding the nodal point P. The distances between the nodes W and P, and between nodes P and E, are identified by the respective notations δx_W and δx_E. For this one-dimensional case, the control volume width surrounding the nodal point P is Δx since Δy and Δz have dimensions of unit length.

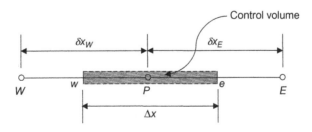

Fig. 5.7 A schematic representation of a control volume around a node P in a one-dimensional domain using the finite volume method.

To apply the finite volume discretization, the gradient term in Eq. (5.19) can be approximated by the use of Eq. (5.12). This gives

$$\frac{\partial}{\partial x}\left(\Gamma\frac{\partial \varphi}{\partial x}\right) = \frac{1}{\Delta V}\int_{\Delta V}\frac{\partial}{\partial x}\left(\Gamma\frac{\partial \phi}{\partial x}\right)dV = \frac{1}{\Delta V}\int_{A}\left(\Gamma\frac{\partial \phi}{\partial x}\right)dA^x \approx \frac{1}{\Delta V}\sum_{i=1}^{2}\left(\Gamma\frac{\partial \phi}{\partial x}\right)_i A_i^x$$

(5.24)

Here, the projected areas A_i^x for the one-dimensional case are given by $A_1^x = -A_W$ and $A_2^x = A_E$. Eq. (5.24) can thus be written as

$$\left(\Gamma\frac{\partial \phi}{\partial x}\right)_e A_E - \left(\Gamma\frac{\partial \phi}{\partial x}\right)_w A_w$$

(5.25)

For the remaining term in the equation, the source term is approximated as

$$\frac{1}{\Delta V}\int_{\Delta V}S_\phi dV = S_\phi$$

(5.26)

where S_ϕ is assumed to be constant within ΔV, which is the finite control volume. The final form of the discretized equation becomes

$$\frac{1}{\Delta V}\left(\Gamma\frac{\partial \phi}{\partial x}\right)_e A_E - \frac{1}{\Delta V}\left(\Gamma\frac{\partial \phi}{\partial x}\right)_w A_w + S_\phi = 0$$

(5.27)

To express an algebraic form for Eq. (5.27) with the nodal points W, E, and P, approximations to the gradients $\partial \phi/\partial x$ at the west ("w") and east ("e") faces of the control volume are required. We will assume the piecewise linear gradient profiles spanning the nodal points between W and P and between P and E to sufficiently approximate the first derivatives at "w" and "e"; the diffusive fluxes are evaluated as

$$\left(\Gamma\frac{\partial \phi}{\partial x}\right)_e A_E = \Gamma_e A_E\left(\frac{\phi_E - \phi_P}{\delta x_E}\right)$$

(5.28)

$$\left(\Gamma\frac{\partial \phi}{\partial x}\right)_w A_W = \Gamma_w A_W\left(\frac{\phi_P - \phi_W}{\delta x_W}\right)$$

(5.29)

Substitution of Eqs. (5.28) and (5.29) into Eq. (5.27) gives

$$\frac{\Gamma_e A_E}{\Delta V}\left(\frac{\phi_E - \phi_P}{\delta x_E}\right) - \frac{\Gamma_w A_W}{\Delta V}\left(\frac{\phi_P - \phi_W}{\delta x_W}\right) + S_\phi = 0$$

(5.30)

Eq. (5.30) presents a very attractive feature of the finite volume method. This discretized equation possesses a clear physical interpretation. It states that the difference between the diffusive fluxes of ϕ at the east and west faces of the control volume is equal to the generation of ϕ and constitutes a balance equation for ϕ over the control volume. Eq. (5.30) can be re-arranged as

$$\frac{1}{\Delta V}\left(\frac{\Gamma_e A_E}{\delta x_E} + \frac{\Gamma_w A_W}{\delta x_W}\right)\phi_P = \frac{1}{\Delta V}\left(\frac{\Gamma_e A_E}{\delta x_E}\right)\phi_E + \frac{1}{\Delta V}\left(\frac{\Gamma_w A_W}{\delta x_W}\right)\phi_W + S\phi \qquad (5.31)$$

As above, by identifying the coefficients of ϕ_E and ϕ_W in Eq. (5.31) as a_E and a_W and the coefficient of ϕ_P as a_P, the algebraic form can be written as

$$a_P\phi_P = a_E\phi_E + a_W\phi_W + b \qquad (5.32)$$

where

$$a_E = \frac{\Gamma_e A_E}{\Delta V \delta x_E}; \quad a_W = \frac{\Gamma_w A_W}{\Delta V \delta x_W}; \quad a_P = a_E + a_W; \quad b = S_\phi$$

Eq. (5.32) represents the discretized form through the finite volume method for Eq. (5.19). For the one-dimensional problem considered here, the face areas A_E and A_W are unity since Δy and Δz have dimensions of unit length; the finite control volume ΔV is therefore the width Δx.

5.3.3 Comparison of the Finite Difference and Finite Volume Discretizations

Although the same algebraic form of equation for the steady-state one-dimensional diffusion process is obtained, different expressions for the coefficients of a_E, a_W, and a_P are derived as seen in Eq. (5.23) for the finite difference method and Eq. (5.32) for the finite volume method. Nevertheless, let us consider for a special case where the diffusion coefficient is spatially invariant and the mesh is uniformly distributed. The coefficients in the algebraic Eq. (5.23) for the finite difference method reduce to

$$a_E = \frac{\Gamma}{\delta x^2}; \quad a_W = \frac{\Gamma}{\delta x^2}; \quad a_P = a_E + a_W; \quad b = S_\phi$$

while for the finite volume method, the coefficients in Eq. (5.32) become

$$a_E = \frac{\Gamma}{\delta x^2}; \quad a_W = \frac{\Gamma}{\delta x^2}; \quad a_P = a_E + a_W; \quad b = S_\phi$$

where the control volume is $\Delta x = \delta x$ (uniform grid and Δy, Δz is unity). From the example above for the discretization of the continuity equation, the resultant algebraic equations are again exactly the same whether adopting either the finite difference or finite volume discretization.

It should be noted that the finite difference method generally requires a uniformly distributed mesh in order to apply the first and second derivative approximations to the governing equation. For a non-uniform grid distribution, some mathematical manipulation (e.g. transformation functions) is required to transform Eq. (5.20) into a computational domain in generalized coordinates before applying the finite difference approximations. This requirement is, however, not a prerequisite for the finite volume method. Because of the availability of different control volume sizes, any non-uniform grid could therefore be easily accommodated.

In comparison to the finite difference method that is mathematically derived from the Taylor series, the finite volume method ensures that the property is conserved and it retains this physical significance throughout the discretization process. Almost all commercial CFD codes adopt the finite volume discretization of the Navier-Stokes equation to obtain numerical solutions for complex fluid flow problems, as the mesh is not restricted to structured-type elements but can include a variety of unstructured-type elements of different shapes and sizes.

EXAMPLE

Example 5.2

Consider the problem of a steady heat conduction problem in a large brick plate with a uniform heat generation. The faces A and B as shown in Fig. 5.2.1 below are maintained at constant temperatures. The governing equation is of the generic form presented in Eq. (5.19). The diffusion coefficient Γ governing the heat conduction problem becomes the thermal conductivity k of the material. For a given thickness $L = 2$ cm with constant thermal conductivity $k = 5$ W/m $\cdot$ K, determine the steady-state distribution in the plate. Temperatures at T_A and T_B are respectively 100°C and 400°C and heat generation q is 500 kW/m^3.

Solution: Assuming that the dimensions in the y direction and z direction are so large that the temperature gradients are only significant in the x direction, we can reduce the problem to a one-dimensional analysis. We apply the finite volume method to obtain the solution of this simple heat conduction problem.

Let us divide the domain into four control volumes (see Fig. 5.2.2), giving $\delta x = 0.005$ m. There are four nodal points, each representing the central location for the four control volumes. For illustration purposes, a unit area is considered in the y-z plane.

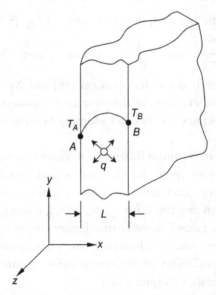

Fig. 5.2.1 Schematic representation of the large brick plate with heat generation.

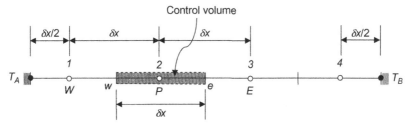

Fig. 5.2.2 Finite volume discretization of the large brick plate domain.

Since the thermal conductivity is constant, we can define $k = k_e = k_w$ and the volumetric source term S_ϕ equal to q, the general discretized form of the equation at point P (node 2), as from Eq. 5.32, is given as

$$a_P T_P = a_E T_E + a_W T_W + b$$

where the coefficients are

$$a_E = \frac{k}{\delta x^2}; \quad a_W = \frac{k}{\delta x^2}; \quad a_P = a_E + a_W; \quad b = q$$

This algebraic equation is also valid for the control volume of node 3.

For the control volumes of nodes 1 and 4, we apply a linear approximation for the temperatures between the boundary points and its adjacent nodal point of the control volume (Fig. 5.2.3). At the west face of the control volume at the fixed end, the temperature is known by T_A. We begin by re-visiting Eq. (5.27).

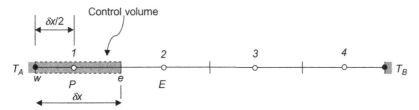

Fig. 5.2.3 Finite volume discretization of the large brick plate domain, showing the first control volume.

For a constant thermal conductivity and uniform heat generation, the equation becomes

$$k\left(\frac{\partial T}{\partial x}\right)_e - k\left(\frac{\partial T}{\partial x}\right)_w + q\delta x = 0$$

Introducing linear approximations to the gradients at the east and west faces of the control volume 1 gives

$$k\left(\frac{T_E - T_P}{\delta_x}\right) - k\left(\frac{T_P - T_A}{\delta x/2}\right) + q\delta x = 0$$

The above equation can be re-arranged to yield the discretized equation for node 1:

$$a_P T_P = a_E T_E + a_W T_W + b$$

where the coefficients are

$$a_E = \frac{k}{\delta x}; \quad a_W = 0; \quad a_P = a_E + a_W + \frac{2k}{\delta x}; \quad b = q\delta x + \frac{2k}{\delta x} T_A$$

The area $a_W = 0$, since it is a fixed end. At node 4, the temperature of the east face of the control volume is known. The node is similarly treated and we obtain

$$k\left(\frac{\partial T}{\partial x}\right)_e - k\left(\frac{\partial T}{\partial x}\right)_w + q\delta x = 0$$

and

$$k\left(\frac{T_B - T_P}{\delta x/2}\right) - k\left(\frac{T_P - T_W}{\delta x}\right) + q\delta x = 0$$

with the discretized equation for node 4 given as:

$$a_P T_P = a_E T_E + a_W T_W + b$$

where

$$a_E = 0; \quad a_W = \frac{k}{\delta x}; \quad a_P = a_E + a_W + \frac{2k}{\delta x}; \quad b = q\delta x + \frac{2k}{\delta x} T_B$$

Substitution of numerical values for the thermal conductivity $k = 5$ W/m^2.K, heat generation $q = 500$ kW/m^3, $\delta x = 0.005$ m, and unit area $A = 1$ m^2 throughout provides the coefficients of the discretized equation summarized in Table 5.2.1.

Table 5.2.1 The Coefficients at Each Node of the Control Volumes.

Node	a_E	a_W	a_P	b
1	1000	0	3000	2500 + 2000 T_A
2	1000	1000	2000	2500
3	1000	1000	2000	2500
4	0	1000	3000	2500 + 2000 T_B

The resulting set of algebraic equations for this example is

$$3000 T_1 = 1000 T_2 + 2000 T_A + 2500$$
$$2000 T_2 = 1000 T_1 + 1000 T_3 + 2500$$
$$2000 T_3 = 1000 T_2 + 1000 T_4 + 2500$$
$$3000 T_4 = 1000 T_3 + 2000 T_B + 2500$$

This set of equations can be re-arranged in matrix form as

$$
\begin{bmatrix}
3000 & -1000 & 0 & 0 \\
-1000 & 2000 & -1000 & 0 \\
0 & -1000 & 2000 & -1000 \\
0 & 0 & -1000 & 3000
\end{bmatrix}
\begin{bmatrix}
T_1 \\ T_2 \\ T_3 \\ T_4
\end{bmatrix}
=
\begin{bmatrix}
2000\,T_A + 2500 \\
2500 \\
2500 \\
2000\,T_B + 2500
\end{bmatrix}
$$

The above set of equations yields the steady state temperature distribution for the given situation.

Discussion: For a one-dimensional steady heat conduction process, we obtained the algebraic equations in matrix form. Because the problem only involves a small number of nodes, the matrix can be solved directly with a software package such as MATLAB (*The Student Edition of MATLAB*, The Math Works Inc., 1992). Analytically the matrix can be solved by methods such as Gaussian elimination. This matrix algorithm will be discussed in the next section.

In most CFD problems, however, the complexity and size of a set of equations depends on the dimensionality of the problem and the number of grid nodes, where the system can consist of up to 100,000 or 1 million equations. For such problems, the available computer resources set a powerful constraint and the Gaussian elimination may not be the most economical way to obtain the solution of the discretized equations. We will further explore other algorithms that are currently available to efficiently obtain the solution to the algebraic equations in Section 5.3.

Steady convection-diffusion process: In the absence of sources the equation governing the steady convection and diffusion process of a property ϕ in a given one-dimensional flow field u from Eq. (3.53) simplifies to

$$\frac{\partial}{\partial x}(\rho u \phi) = \frac{\partial}{\partial x}\left(\Gamma \frac{\partial \phi}{\partial x}\right) \tag{5.33}$$

The mass conservation for the one-dimensional convection-diffusion process is enforced through

$$\frac{\partial(\rho u)}{\partial x} = 0 \tag{5.34}$$

By applying the finite volume method based on the elemental volume described in Fig. 5.7, the algebraic form of Eq. (5.33) can be expressed as

$$(\rho u \phi)_e A_E - (\rho u \phi)_w A_W = \left(\Gamma \frac{\partial \phi}{\partial x}\right)_e A_E - \left(\Gamma \frac{\partial \phi}{\partial x}\right)_w A_W \tag{5.35}$$

Assuming piecewise linear gradient profiles spanning the nodal points between W and P and between P and E, the first derivatives at "w" and "e" of the diffusive fluxes can be similarly approximated using the expressions formulated in Eqs (5.28) and (5.29) to yield

$$(\rho u\phi)_e A_E - (\rho u\phi)_w A_W = \Gamma_e A_E \left(\frac{\phi_E - \phi_P}{\delta x_E}\right) - \Gamma_w A_W \left(\frac{\phi_P - \phi_W}{\delta x_W}\right) \qquad (5.36)$$

It seems rather sensible for the convective term to be approximated in a similar fashion. Using the Taylor series expansion, the interface values ϕ_w and ϕ_e at the respective cell faces w and e can be obtained via linear interpolation between the neighboring points as

$$\phi_w = \frac{1}{2}(\phi_W + \phi_P) \qquad (5.37)$$

$$\phi_e = \frac{1}{2}(\phi_P + \phi_E) \qquad (5.38)$$

The above approximation is second order accurate and the scheme does not exhibit any bias on the flow direction. Using these relationships, Eq. (5.36) thus becomes

$$(\rho u)_e A_E \frac{1}{2}(\phi_P + \phi_E) - (\rho u)_w A_W \frac{1}{2}(\phi_W + \phi_P)$$

$$= \Gamma_e A_E \left(\frac{\phi_E - \phi_P}{\delta x_E}\right) - \Gamma_w A_W \left(\frac{\phi_P - \phi_W}{\delta x_W}\right) \qquad (5.39)$$

Eq. (5.39) can be re-arranged as

$$\left(\frac{\Gamma_e A_E}{\delta x_E} + \frac{(\rho u)_e A_E}{2} + \frac{\Gamma_w A_W}{\delta x_W} - \frac{(\rho u)_w A_W}{2}\right)\phi_P$$

$$= \left(\frac{\Gamma_e A_E}{\delta x_E} - \frac{(\rho u)_e A_E}{2}\right)\phi_E + \left(\frac{\Gamma_w A_W}{\delta x_W} + \frac{(\rho u)_w A_W}{2}\right)\phi_W \qquad (5.40)$$

By identifying the coefficients of ϕ_E and ϕ_W in Eq. (5.40) as a_E and a_W and the coefficient of ϕ_P as a_P, the algebraic form can be written as

$$a_P \phi_P = a_E \phi_E + a_W \phi_W \qquad (5.41)$$

where

$$a_E = \frac{\Gamma_e A_E}{\delta x_E} - \frac{(\rho u)_e A_E}{2}; \quad a_W = \frac{\Gamma_w A_W}{\delta x_W} + \frac{(\rho u)_w A_W}{2};$$

$$a_P = a_E + a_W + (\rho u)_e A_E - (\rho u)_w A_W$$

One renowned inadequacy of the central differencing schemes in a strongly convective flow is its inability to identify the flow direction. It is well recognized that the above treatment usually results in large "undershoots" and "overshoots," eventually causing the numerical procedure to diverge. Increasing the mesh resolution for the flow domain with very small grid spacing could probably overcome the problem but such an approach usually precludes practical flow calculations to be carried out robustly and effectively in practice.

Consider for a moment the flow moving from the upstream point W (left) to the downstream point E (right). Through the central differencing approximation, the interface values of ϕ are always assumed to be equally weighted by the influence of the available variables at the neighboring grid nodal points. This implies that the downstream values of ϕ_P and ϕ_E are prevailing during the evaluation of ϕ_w and ϕ_e. In the majority of flow cases these values are not known *a priori*; the remedy is therefore to design a numerical solution to recognize the direction of the flow by exerting an unequal weighting influence based on the available variables located at the surrounding grid nodal points to appropriately determine the interface values. This is essentially the hallmark of the *upwind* or *donor-cell* concept. The first-order upwind scheme is described herein. With reference to Fig. 5.7, if the interface velocities $u_w > 0$ and $u_e > 0$, respectively, the interface values ϕ_w and ϕ_e according to the *donor-cell* concept are now approximated according to their upstream neighboring counterparts as

$$\phi_w = \phi_W \tag{5.42}$$

$$\phi_e = \phi_P \tag{5.43}$$

Using the above relationships, Eq. (5.36) can be expressed as

$$(\rho u)_e A_E \phi_P - (\rho u)_w A_W \phi_W = \Gamma_e A_E \left(\frac{\phi_E - \phi_P}{\delta x_E} \right) - \Gamma_w A_W \left(\frac{\phi_P - \phi_W}{\delta x_W} \right) \tag{5.44}$$

and the coefficients a_E, a_W, and a_P, as described by Eq. (5.41), are given by

$$a_E = \frac{\Gamma_e A_E}{\delta x_E}; \quad a_W = \frac{\Gamma_w A_W}{\delta x_W} + (\rho u)_w A_W; \quad a_P = a_E + a_W + (\rho u)_e A_E - (\rho u)_w A_w$$

Similarly, if the interface velocities $u_w < 0$ and $u_e < 0$, respectively, the interface values ϕ_w and ϕ_e are conversely evaluated by

$$\phi_w = \phi_P \tag{5.45}$$

$$\phi_e = \phi_E \tag{5.46}$$

Using the above relationships, Eq. (5.38) can be written as

$$(\rho u)_e A_E \phi_E - (\rho u)_w A_W \phi_P = \Gamma_e A_E \left(\frac{\phi_E - \phi_P}{\delta x_E} \right) - \Gamma_w A_W \left(\frac{\phi_P - \phi_W}{\delta x_W} \right) \tag{5.47}$$

and the coefficients a_E, a_W, and a_P are now given by

$$a_E = \frac{\Gamma_e A_E}{\delta x_E} - (\rho u)_e A_E; \quad a_W = \frac{\Gamma_w A_W}{\delta x_W}; \quad a_P = a_E + a_W + (\rho u)_e A_E - (\rho u)_w A_w$$

A more in-depth look at the different discretization schemes based on upwind and central differencing schemes can be described by three fundamental properties: *transportiveness*, *boundedness*, and *conservativeness*.

Transportiveness concerns the direction of the flow. Central differencing scheme introduces the influence at node P from all directions from its neighboring nodes to calculate the convective and diffusive fluxes. Hence this particular scheme does not recognize the direction or the strength of the convection relative to diffusion and does not possess the transportiveness property. Conversely, the upwind differencing scheme accounts for the flow direction and satisfies the transportiveness property. Fig. 5.8 illustrates the preferential direction of the flow moving across three control volumes from the upstream node W (left) to the downstream node E (right) for interface velocities $u_w > 0$ and $u_e > 0$ as well as in the opposite direction from upstream node E (right) to downstream node W (left) for interface velocities $u_w < 0$ and $u_e < 0$, respectively.

Boundedness concerns the condition whereby the resulting matrix of the coefficients is diagonally dominant to ensure numerical convergence. It can be demonstrated that the matrix coefficients for central differencing can become negative, which greatly violates the requirement of boundedness and may lead to unphysical solutions. For coefficients a_W and a_E, it is highly possible that they can become negative if advection dominates. For a_W and a_E to be positive, they must satisfy the following conditions:

$$\frac{(\rho u)_w}{\Gamma_w/\delta x_W} = Pe_w < 2 \quad \text{and} \quad \frac{(\rho u)_e}{\Gamma_e/\delta x_E} = Pe_e < 2 \tag{5.48}$$

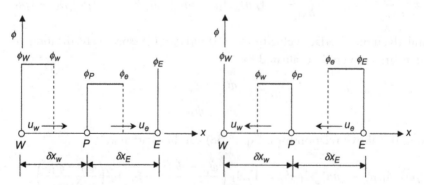

Fig. 5.8 A schematic representation of the upwind differencing scheme.

If the local Peclet numbers Pe_w and Pe_e are greater than 2, a_W and a_E become negative and these coefficients violate the requirement of boundedness and may result in a physically impossible solution. However, all coefficients in the upwind differencing scheme are positive and the coefficient matrix is diagonally dominant, which satisfy the requirement of boundedness.

In addition, Spalding (1972) developed a scheme that combines the central and upwind differencing schemes by employing piecewise formulae based on the Peclet number. This so-called *hybrid differencing* retains a second-order accuracy for small Peclet numbers due to central differencing but reverts to the upwind differencing for large Peclet numbers. For the case where the interface velocities are positive, that is, $u_w > 0$ and $u_e > 0$, the interface values ϕ_w and ϕ_e according to the hybrid differencing formulae are given by

$$\phi_w = \frac{1}{2}(\phi_W + \phi_P); \quad \phi_e = \frac{1}{2}(\phi_P + \phi_E); \quad \text{for} \quad Pe_w, Pe_e < 2 \qquad (5.49)$$

$$\phi_w = \phi_W; \quad \phi_e = \phi_P \quad \text{for} \quad Pe_w, Pe_e \geq 2 \qquad (5.50)$$

where

$$Pe_w = \frac{(\rho u)_w}{\Gamma_w / \delta x_W} \quad \text{and} \quad Pe_e = \frac{(\rho u)_e}{\Gamma_e / \delta x_E}$$

Similar considerations can be obtained for the respective interface values of ϕ_w and ϕ_e when the interface velocities u_w and u_e are in the opposite direction or negative. Like upwind differencing, this scheme is highly stable, satisfies transportiveness, and produces physically realistic solutions.

Another popular scheme that is considered to yield better results than the hybrid scheme is the *power-law differencing scheme* of Patankar (1980). Here, the upwind differencing becomes effective only when $Pe > 10$. For example, the interface values ϕ_w and ϕ_e according to the power-law differencing formulae for the respective interface velocities $u_w > 0$ and $u_e > 0$ can be determined as

$$\phi_w = (1 - \chi_w)\phi_W + \chi_w\phi_P; \quad \phi_e = (1 - \chi_e)\phi_P + \chi_e\phi_E \quad \text{for} \quad 0 < Pe_w, Pe_e < 10 \qquad (5.51)$$

$$\text{where} \quad \chi_w = (1 - 0.1Pe_w)^5 / Pe_w \quad \text{and} \quad \chi_e = (1 - 0.1Pe_e)^5 / Pe_e$$

$$\phi_w = \phi_W; \quad \phi_e = \phi_P \quad \text{for} \quad Pe_w, Pe_e \geq 10 \qquad (5.52)$$

The power-law differencing scheme possesses similar properties to the hybrid scheme. It has also enjoyed much extensive usage in practical flow calculations and can be used as an alternative to the hybrid scheme.

Conservativeness concerns the fluxes being represented in a reliable manner. The upwind and central differencing schemes use consistent expressions to evaluate the convective and diffusive fluxes at the control volume faces. This meant that fluxes cancel out when summed and overall conservation is satisfied.

It should be noted that despite the favorable properties of the upwind differencing scheme in satisfying the above fundamental properties and promoting numerical stability, it is widely known that it generally causes unwanted numerical diffusion in space. In order to reduce these numerical errors, high-order approximations such as the second-order upwind and third-order QUICK scheme that are widely applied in many CFD problems have been proposed. The formulation of these schemes is further described in Appendix B.

Unsteady convection-diffusion process: The equation governing the unsteady convection and diffusion process of a property ϕ in a given one-dimensional flow field u is given by

$$\frac{\partial(\rho\phi)}{\partial t} + \frac{\partial}{\partial x}(\rho u \phi) = \frac{\partial}{\partial x}\left(\Gamma \frac{\partial \phi}{\partial x}\right) + S_\phi \tag{5.53}$$

For the purpose of illustration, let us assume that the fluid is incompressible (i.e. density is constant) and Eq. (5.53) can be rearranged as

$$\frac{\partial \phi}{\partial t} + \frac{\partial}{\partial x}(u\phi) = \frac{1}{\rho}\frac{\partial}{\partial x}\left(\Gamma \frac{\partial \phi}{\partial x}\right) + \frac{S_\phi}{\rho} \tag{5.54}$$

The mass conservation becomes

$$\frac{\partial u}{\partial x} = 0 \tag{5.55}$$

Assuming central differencing for the convective and diffusive terms, the partial algebraic form of Eq. (5.53) via the finite volume method is given by

$$\frac{\partial \phi}{\partial t} + u_e A_E \frac{1}{2}(\phi_W + \phi_P) - u_w A_W \frac{1}{2}(\phi_P + \phi_E)$$

$$= \Gamma_e A_E \frac{1}{\rho}\left(\frac{\phi_E - \phi_P}{\delta x_E}\right) - \Gamma_w A_W \frac{1}{\rho}\left(\frac{\phi_P - \phi_W}{\delta x_W}\right) + \frac{S_\phi}{\rho}\Delta V \tag{5.56}$$

In the majority of cases the time derivative can be approximated by applying the first-order forward difference scheme as above:

$$\frac{\partial \phi}{\partial t} = \frac{\phi^{n+1} - \phi^n}{\Delta t} = \frac{\phi_P^{n+1} - \phi_P^n}{\Delta t} \tag{5.57}$$

where Δt is the incremental time step and the superscripts n and $n + 1$ denote the previous and current time levels, respectively. It should be clear from above that a suitable time-marching procedure needs to appropriately update the property ϕ at the central point P

and the neighboring points through time. Let us first examine the implication due to an *explicit* approach. For this approach, Eq. (5.56) can be written as:

$$\frac{\phi_P^{n+1} - \phi_P^n}{\Delta t} + u_e^n A_E \frac{1}{2}\left(\phi_W^n + \phi_P^{n+1}\right) - u_w^n A_W \frac{1}{2}\left(\phi_P^{n+1} + \phi_E^n\right)$$

$$= \Gamma_e^n A_E \frac{1}{\rho}\left(\frac{\phi_E^n - \phi_P^{n+1}}{\delta x_E}\right) - \Gamma_w^n A_W \frac{1}{\rho}\left(\frac{\phi_P^{n+1} - \phi_W^n}{\delta x_W}\right) + \frac{S_\phi^n}{\rho}\Delta V \qquad (5.58)$$

Here again, by identifying the coefficients of ϕ_E and ϕ_W in Eq. (5.58) as a_E and a_W and the coefficient of ϕ_P as a_P, the algebraic form can be written as

$$a_P \phi_P^{n+1} = a_E \phi_E^n + a_W \phi_W^n + b \qquad (5.59)$$

where

$$a_E = \frac{\Gamma_e^n A_E}{\rho \delta x_E} - \frac{u_e^n A_E}{2}; \quad a_W = \frac{\Gamma_w^n A_W}{\rho \delta x_W} + \frac{u_w^n A_W}{2};$$

$$a_P = a_E + a_W + u_e^n A_E - u_w^n A_W + \frac{1}{\Delta t}; \quad b = \frac{S_\phi^n}{\rho}\Delta V + \frac{1}{\Delta t}\phi_P^n$$

Casting our attention to the Eq. (5.58) and the sketch in Fig. 5.9, we see a straightforward mechanism of evaluating the unknown ϕ_P^{n+1} (indicated by the square at time level $n+1$) that is calculated directly from all the values obtained from the indicated circles of the property ϕ at the previous time level n. By definition, in an *explicit* approach, each difference equation contains only one unknown and therefore can be solved *explicitly* for this unknown in a simple manner.

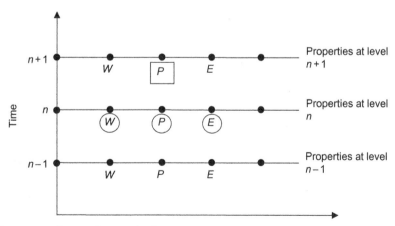

Fig 5.9 Illustration of an explicit method.

On the other hand, let us consider a time-matching procedure that requires the solution for all the variables at the time level $n + 1$. Eq. (5.58) can be rewritten as

$$\frac{\phi_P^{n+1} - \phi_P^n}{\Delta t} + u_e^{n+1} A_E \frac{1}{2}\left(\phi_W^{n+1} + \phi_P^{n+1}\right) - u_w^{n+1} A_W \frac{1}{2}\left(\phi_P^{n+1} + \phi_E^{n+1}\right)$$

$$= \Gamma_e^{n+1} A_E \frac{1}{\rho}\left(\frac{\phi_E^{n+1} - \phi_P^{n+1}}{\delta x_E}\right) - \Gamma_w^{n+1} A_W \frac{1}{\rho}\left(\frac{\phi_P^{n+1} - \phi_W^{n+1}}{\delta x_W}\right) + \frac{S_\phi^{n+1}}{\rho}\Delta V \tag{5.60}$$

In its algebraic form Eq. (5.60) can be recast as

$$a_P \phi_P^{n+1} = a_E \phi_E^{n+1} + a_W \phi_W^{n+1} + b \tag{5.61}$$

where

$$a_E = \frac{\Gamma_e^{n+1} A_E}{\rho \delta x_E} - \frac{u_e^{n+1} A_E}{2}; \quad a_W = \frac{\Gamma_w^{n+1} A_W}{\rho \delta x_W} + \frac{u_w^{n+1} A_W}{2};$$

$$a_P = a_E + a_W + u_e^{n+1} A_E - u_w^{n+1} A_W + \frac{1}{\Delta t}; \quad b = \frac{S_\phi}{\rho}\Delta V + \frac{1}{\Delta t}\phi_P^n$$

As sketched in Fig. 5.10, the property ϕ at the central point P and the properties at the neighboring points are required to be solved simultaneously, possibly coupling with other flow variables such as pressure and temperature appearing in the source term S_ϕ at the current time level within the same difference equation. This is an example of the *fully implicit* approach. By definition, an *implicit* approach is one where the unknowns must be obtained by means of a simultaneous solution of the difference equations applied at *all* grid nodal points at a given time level. Implicit methods usually involve the manipulation of large matrices because of the need to solve large systems of algebraic equations.

The above explicit and implicit schemes are generally considered methods having first order in time. Similar to the first order in space, these methods may also cause unwanted

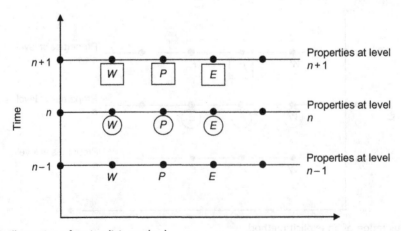

Fig. 5.10 Illustration of an implicit method.

numerical diffusion in time. In order to reduce these numerical errors, second-order approximations such as the explicit Adams-Bashford and semi-implicit Crank-Nicolson schemes have been proposed. The formulation of these schemes is further described in Appendix C.

5.4 NUMERICAL SOLUTIONS TO ALGEBRAIC EQUATIONS

The various discretization methods for the partial differential equations have been described. Through this process, we obtain a system of linear or non-linear algebraic equations that needs to be solved by some numerical methods. The complexity and size of this set of equations depends on the dimensionality and geometry of the physical problem. Whether the equations are linear or non-linear, efficient and robust numerical methods are required to solve the system of algebraic equations. There are essentially two families of numerical methods: *direct methods* and *iterative methods*.

Previously, we have highlighted a direct method such as the Gaussian elimination, which can be used to solve the resultant matrix form of a simple one-dimensional steady heat conduction process. It is noted that there are other direct methods that may be employed, such as Cramer's rule for matrix inversion and the Thomas algorithm (tri-diagonal matrix algorithm). For finite difference or finite volume discretization on structured mesh, the resultant matrix of the algebraic equations is typically sparse; most of the elements are zero and the non-zero terms are close to the diagonal. For one-dimensional situations, the Thomas algorithm or tri-diagonal matrix algorithm is actually a direct method that takes advantage of this particular matrix structure. It is computationally inexpensive and requires a minimum amount of storage in the core memory and as a result is extensively used in CFD programs. The Thomas algorithm is further described in the next section.

Iterative methods are, however, based on the repeated applications of an algorithm leading to its eventual convergence after a number of repetitions. They are generally much more economical and only non-zero terms of the algebraic equations are required to be stored in the core memory. For non-linear problems, they are used out of necessity but are just as valuable for sparse linear systems. Well-known point-by-point methods such as Jacobi and Gauss-Seidel are described in Section 5.4.2 in order to provide the reader with some basic understanding of iterative methods. Other variants from these two iterative methods will also be described, particularly those algorithms that are used in solving CFD problems.

5.4.1 Direct Methods

One of the most basic methods for solving linear systems of algebraic equations is *Gaussian elimination*. The algorithm derives from the basis of systematic reduction of large systems of equations to smaller ones. Let us suppose that the systems of equations can be written in the form:

$$A\phi = B \qquad (5.62)$$

where ϕ is the unknown nodal variables. Matrix A contains non-zero coefficients of the algebraic equations as illustrated below:

$$A = \begin{bmatrix} A_{11} & A_{12} & A_{13} & \cdots & A_{1n} \\ A_{21} & A_{22} & A_{23} & \cdots & A_{2n} \\ A_{31} & A_{32} & A_{33} & \cdots & A_{3n} \\ \vdots & \vdots & \vdots & \ddots & \vdots \\ A_{n1} & A_{n2} & A_{n3} & & A_{nn} \end{bmatrix} \qquad (5.63)$$

while B comprises known values of ϕ, for example, that are given by the boundary conditions or source/sink terms.

It can be observed that the diagonal coefficients of matrix A are represented by the entries of $A_{11}, A_{22}, \ldots, A_{nn}$. The heart of the algorithm is to eliminate the entries below the diagonal to yield a lower triangle of zeros. This means eliminating the elements of $A_{21}, A_{31}, A_{32}, \ldots, A_{nn-1}$ by replacing them with zeros. We begin the elimination process by considering the first column elements of $A_{21}, A_{31}, \ldots, A_{n1}$ in matrix A. By multiplying the first row of the matrix by A_{21}/A_{11} and subtracting these values from the second row, all the elements in the second row are subsequently modified, which includes the terms in B on the right-hand side of the equations. The other elements $A_{31}, A_{41}, \ldots, A_{n1}$ in the first column of matrix A are treated similarly and by repeating this process down the first column, all the elements below A_{11} are reduced to zero. The same procedure is then applied for the second column (for all elements below A_{22}) and so forth until the process reaches the $n-1$ column.

After this process is complete, the original matrix A becomes an *upper triangular matrix* that is given by:

$$U = \begin{bmatrix} A_{11} & A_{12} & A_{13} & \cdots & A_{1n} \\ 0 & A_{22} & A_{23} & \cdots & A_{2n} \\ 0 & 0 & A_{33} & \cdots & A_{3n} \\ \vdots & \vdots & \vdots & \ddots & \vdots \\ 0 & 0 & 0 & & A_{nn} \end{bmatrix} \qquad (5.64)$$

All the elements in matrix U except the first row differ from those in the original matrix A. It is therefore more efficient to store the modified elements in place of the original ones. This process is called the *forward elimination* process. The upper triangular system of equations can now be solved by the *back substitution* process. It is observed that the entry of the matrix U contains only one variable, ϕ_n, and is solved:

$$\phi_n = \frac{B_n}{U_{nn}} \tag{5.65}$$

The entry in matrix U just above Eq. (5.65) contains only ϕ_{n-1} and ϕ_n and, once ϕ_n is known, it can be solved for ϕ_{n-1}. By proceeding in an upward fashion, each of the variables ϕ_i is solved in turn. The general form of the equation for ϕ_i can be expressed as:

$$\phi_i = \frac{B_i - \sum_{j=i+1}^{n} A_{ij}\phi_j}{A_{ii}} \tag{5.66}$$

It is not difficult to see that the bulk of the computational effort is in the *forward elimination* process; the back substitution process requires fewer arithmetic operations and is thus much less costly. Gaussian elimination can be expensive especially for a full matrix containing a large number of unknown variables to be solved but it is as good as any other methods that are currently available.

As observed in Example 5.2, we obtained a matrix of the form typically found from the application of the finite difference or finite volume method. This special form of matrix can be solved by the Thomas algorithm. We observed that the non-zero elements (neighboring entries) lie close to the main diagonal entries; it is useful to consider variants of Gaussian elimination that take advantage of this particular banded structure of the matrix (tri-diagonal) to maximize computational resources and reduce the arithmetic operations. Let us consider the tri-diagonal form of a system of algebraic equations as:

$$
\begin{bmatrix}
A_{11} & A_{12} & & & & & \\
A_{21} & A_{22} & A_{23} & & & & \\
\cdots & \cdots & \cdots & & & & \\
& A_{ii-1} & A_{ii} & A_{ii+1} & & & \\
& & \cdots & \cdots & \cdots & & \\
& & & A_{nn-2} & A_{n-1n-1} & A_{n-1n} & \\
& & & & A_{nn-1} & A_{nn}
\end{bmatrix}
\begin{bmatrix}
\phi_1 \\
\phi_2 \\
\cdots \\
\phi_i \\
\cdots \\
\phi_{n-1} \\
\phi_n
\end{bmatrix}
=
\begin{bmatrix}
B_1 \\
B_2 \\
\cdots \\
B_i \\
\cdots \\
B_{n-1} \\
B_n
\end{bmatrix}
\tag{5.67}
$$

The *Thomas algorithm* like the Gaussian elimination solves the system of equations above in two parts: *forward elimination* and *back substitution*. For the *forward elimination* process, the neighboring banded entries are eliminated below the diagonal to yield zero entries. This means replacing the elements of A_{21}, A_{32}, A_{43},....., A_{nn-1} with zeros. For the first row, the diagonal entry A_{11} is normalized to unity and the neighboring entry A_{12} and matrix B term B_1 are modified according to

$$A'_{12} = \frac{A_{12}}{A_{11}}, \qquad B'_1 = \frac{B_1}{A_{11}} \tag{5.68}$$

Like the Gaussian elimination, by multiplying the first row of the matrix by A_{21} and subtracting it from the second row, all the elements in the second row are subsequently modified, which also include the terms in B on the right-hand side of the equations. Applying the same procedure to the rest of the rows of the matrix, the neighboring element entries and the matrix B terms in general form are:

$$A'_{ii+1} = \frac{A_{ii+1}}{A_{ii} - A_{ii-1}A'_{i-1i}}, \qquad B'_i = \frac{B_i - A_{ii-1}B'_{i-1}}{A_{ii} - A_{ii-1}A'_{i-1i}} \tag{5.69}$$

The matrix containing the non-zero coefficients is therefore manipulated into:

$$\begin{bmatrix} 1 & A'_{12} & & & & & \\ & 1 & A'_{23} & & & & \\ & & \cdots & \cdots & & & \\ & & & 1 & A'_{ii+1} & & \\ & & & & \cdots & \cdots & \\ & & & & & 1 & A'_{n-1n} \\ & & & & & & 1 \end{bmatrix} \begin{bmatrix} \phi_1 \\ \phi_2 \\ \cdots \\ \phi_i \\ \cdots \\ \phi_{n-1} \\ \phi_n \end{bmatrix} = \begin{bmatrix} B'_1 \\ B'_2 \\ \cdots \\ B'_i \\ \cdots \\ B'_{n-1} \\ B'_n \end{bmatrix} \tag{5.70}$$

The second stage simply involves the *back substitution* process, which involves evaluating:

$$\phi_n = B'_n \quad \text{and} \quad \phi_i = B'_i - \phi_{i+1}A'_{ii+1} \tag{5.71}$$

It can be seen that the Thomas algorithm is more economical than the Gaussian elimination because of the absence of arithmetic operations (multiplication and divisions) in obtaining ϕ_i during back substitution. Nevertheless, in order to prevent ill-conditioning (and hence round-off error) for the two direct methods, it is necessary that

$$|A_{ii}| > |A_{ii-1}| + |A_{ii+1}| \tag{5.72}$$

This means that the diagonal coefficients are required to be much larger than the sum of the neighboring coefficients.

EXAMPLE

Example 5.3
A steady heat conduction problem in a large brick plate with a uniform heat generation is presented in Example 5.2. With the boundary temperatures of T_A and T_B given respectively as 100°C and 400°C, determine the discrete nodal temperatures across the brick plate using the Thomas algorithm.

Solution: In this worked example we will illustrate the arithmetic operations of the Thomas algorithm to solve the resultant system of equations for the one-dimensional steady heat conduction process. The system of equations in matrix form that was previously derived is given as

$$
\begin{bmatrix}
3000 & -1000 & 0 & 0 \\
-1000 & 2000 & -1000 & 0 \\
0 & -1000 & 2000 & -1000 \\
0 & 0 & -1000 & 3000
\end{bmatrix}
\begin{bmatrix} T_1 \\ T_2 \\ T_3 \\ T_4 \end{bmatrix}
=
\begin{bmatrix} 2000\,T_A + 2500 \\ 2500 \\ 2500 \\ 2000\,T_B + 2500 \end{bmatrix}
$$

Substituting the temperatures of $T_A = 100°C$ and $T_B = 400°C$ into the right-hand side of the forcing terms, we have

$$
\begin{bmatrix}
3000 & -1000 & 0 & 0 \\
-1000 & 2000 & -1000 & 0 \\
0 & -1000 & 2000 & -1000 \\
0 & 0 & -1000 & 3000
\end{bmatrix}
\begin{bmatrix} T_1 \\ T_2 \\ T_3 \\ T_4 \end{bmatrix}
=
\begin{bmatrix} 202,500 \\ 2500 \\ 2500 \\ 802,500 \end{bmatrix}
$$

In the Thomas algorithm the first step of the *forward elimination* process involves eliminating the lower triangular coefficients below the diagonal coefficients to yield zero entries. By applying Eqs (5.68) and (5.69), the matrix is reduced to

$$
\begin{bmatrix}
1 & -0.333 & 0 & 0 \\
0 & 1 & -0.6 & 0 \\
0 & 0 & 1 & -0.714 \\
0 & 0 & 0 & 1
\end{bmatrix}
\begin{bmatrix} T_1 \\ T_2 \\ T_3 \\ T_4 \end{bmatrix}
=
\begin{bmatrix} 67.5 \\ 42.0 \\ 31.8 \\ 365.0 \end{bmatrix}
$$

The second stage of the Thomas algorithm simply involves the *back substitution* process. Using Eq. (5.71), the solution to the above system is

$$
\begin{bmatrix} T_1 \\ T_2 \\ T_3 \\ T_4 \end{bmatrix}
=
\begin{bmatrix} 140.0 \\ 217.4 \\ 292.4 \\ 365.0 \end{bmatrix}
$$

Discussion: For this simple problem, we could have obtained the solution using the Gaussian elimination instead of the Thomas algorithm. For such a small matrix, the additional arithmetic operations required for the Gaussian elimination to perform on the zero entries may not be as significant compared to the Thomas algorithm. Nevertheless, this is not true when a number of grid points are used to better predict the temperature distribution across the plate. This is because of the additional and more cumbersome numerical computations (multiplication and divisions) that have to be performed on the matrix entries. The algorithm degenerates and becomes inefficient once the order of the matrix becomes higher (>10).

5.4.2 Iterative Methods

Direct methods such as Gaussian elimination can be employed to solve any system of equations. Unfortunately, in most CFD problems that usually result in a large system of non-linear equations, the cost of using this method is generally quite high. It has been demonstrated through the worked example in the previous section that the Thomas algorithm is particularly economical in obtaining the solution for a one-dimensional steady-state heat conduction problem because of the inherent banded matrix structure (tri-diagonal). For multi-dimensional situations, the nature of the solver, however, cannot be readily extended to solve such problems.

This therefore leaves the option of employing iterative methods. In an iterative method one guesses the solution and uses the equation to systematically improve the solution until it reaches some level of convergence. If the number of iterations is small in achieving convergence, an iterative solver may cost less to use than a direct method. This is usually the case for CFD problems.

Jacobi and Gauss-Seidel methods: The simplest method from the various classes of iterative methods is the Jacobi method. Let us revisit the system of equations, $A\phi = B$, as described in the previous section; the general form of the algebraic equation for each unknown nodal variable of ϕ can be written as:

$$\sum_{j=1}^{i-1} A_{ij}\phi_j + A_{ii}\phi_i + \sum_{j=i+1}^{n} A_{ij}\phi_j = B_i \tag{5.73}$$

In Eq. (5.73) the Jacobi method requires that the nodal variables ϕ_j (non-diagonal matrix elements) are assumed to be known at iteration step k and the nodal variables ϕ_i are treated as unknown at iteration step $k + 1$. Solving for ϕ_i, we have

$$\phi_i^{(k+1)} = \frac{B_i}{A_{ii}} - \sum_{j=1}^{i-1} \frac{A_{ij}}{A_{ii}} \phi_j^{(k)} - \sum_{j=i+1}^{n} \frac{A_{ij}}{A_{ii}} \phi_j^{(k)} \tag{5.74}$$

The iteration process begins with an initial guess of the nodal variables ϕ_j ($k = 0$). After repeated application of Eq. (5.74) to all the n unknowns, the first iteration, $k = 1$, is completed. We proceed to the next iteration step, $k = 2$, by substituting the iterated values at $k = 1$ into Eq. (5.74) to estimate the new values at the next iteration step. This process is continuously repeated for as many iterations as required to converge to the desired solution.

An immediate improvement to the Jacobi method is provided by the Gauss-Seidel method, in which the updated nodal variables $\phi_j^{(k+1)}$ are immediately used on the right-hand side of Eq. (5.74) as soon as they are available. In such a case the previous values of $\phi_j^{(k)}$ that appear in the second term of the right-hand side of Eq. (5.74) are replaced by the current values of $\phi_j^{(k)}$, due to which the equivalent of Eq. (5.74) becomes

$$\phi_i^{(k+1)} = \frac{B_i}{A_{ii}} - \sum_{j=1}^{i-1} \frac{A_{ij}}{A_{ii}} \phi_j^{(k+1)} - \sum_{j=i+1}^{n} \frac{A_{ij}}{A_{ii}} \phi_j^{(k)} \tag{5.75}$$

Comparing the above two iterative procedures, the Gauss-Seidel iteration is typically twice as fast as the Jacobi iteration. After repeated applications of Eqs (5.74) and (5.75), convergence can be gauged in a number of ways. One convenient condition to terminate the iteration process is to ensure that the maximum difference $\phi_j^{(k+1)} - \phi_j^{(k)}$ falls below some predetermined value of acceptable error. The smaller the acceptable error, the more accurate the solution will be, but it is noted that this is achieved at the expense of more number of iterations.

EXAMPLE

Example 5.4
Based on the same worked example of the steady heat conduction problem in a large brick plate with a uniform heat generation, as previously presented in Example 5.2, determine the discrete nodal temperatures across the brick plate using
(a) The Jacobi method.
(b) The Gauss-Seidel method.
Solution:
(a) To illustrate the *Jacobi method*, the resulting set of algebraic equations as previously derived in Example 5.2 is rewritten

$$3000T_1 + 1000T_2 + 0 \times T_3 + 0 \times T_4 = 2500 + 2000\,T_A$$
$$1000T_1 + 2000T_2 + 1000T_3 + 0 \times T_4 = 2500$$
$$0 \times T_1 + 1000T_2 + 2000T_3 + 1000T_4 = 2500$$
$$0 \times T_1 + 0 \times T_2 + 1000T_3 + 3000T_4 = 2500 + 2000T_B$$

Substituting the boundary temperatures of $T_A = 100°C$ and $T_B = 400°C$, we have

$$3000T_1 + 1000T_2 + 0 \times T_3 + 0 \times T_4 = 202,500$$
$$1000T_1 + 2000T_2 + 1000T_3 + 0 \times T_4 = 2500$$
$$0 \times T_1 + 1000T_2 + 2000T_3 + 1000T_4 = 2500$$
$$0 \times T_1 + 0 \times T_2 + 1000T_3 + 3000T_4 = 802,500$$

The above set of equations can be reorganized so that the required variable is on the left-hand side of the equation.

$$T_1 = 0.333\,T_2 + 0 \times T_3 + 0 \times T_4 + 67.5$$
$$T_2 = 0.5\,T_1 + 0.5\,T_3 + 0 \times T_4 + 1.25$$
$$T_3 = 0 \times T_1 + 0.5\,T_2 + 0.5\,T_4 + +1.25$$
$$T_4 = 0 \times T_1 + 0 \times T_2 + 0.333\,T_3 + 267.5$$

By employing initial guesses, that is $T_1^{(0)} = T_2^{(0)} = T_3^{(0)} = T_4^{(0)} = 100$, the nodal temperatures for the first iteration are determined as:

$$T_1^{(1)} = 0.333(100) + 67.5 = 100.8$$
$$T_2^{(1)} = 0.5(100) + 0.5(100) + 1.25 = 101.25$$
$$T_3^{(1)} = 0.5(100) + 0.5(100) + 1.25 = 101.25$$
$$T_4^{(1)} = 0.333(100) + 267.5 = 300.8$$

The above first iteration values of $T_1^{(1)} = 100.8$, $T_2^{(1)} = 101.25$, $T_3^{(1)} = 101.25$, and $T_4^{(1)} = 300.8$ are substituted back into the system of equations; the second iteration yields

$$T_1^{(2)} = 0.333(101.25) + 67.5 = 101.2$$
$$T_2^{(2)} = 0.5(100.8) + 0.5(101.25) + 1.25 = 102.3$$
$$T_3^{(2)} = 0.5(101.25) + 0.5(300.8) + 1.25 = 202.3$$
$$T_4^{(2)} = 0.333(101.25) + 267.5 = 301.2$$

After repeated applications of the iterative process up to 10 and 20 iterations, the nodal temperatures have advanced to

$$\begin{bmatrix} T_1^{(10)} \\ T_2^{(10)} \\ T_3^{(10)} \\ T_4^{(10)} \end{bmatrix} = \begin{bmatrix} 135.2 \\ 207.9 \\ 282.2 \\ 360.5 \end{bmatrix} \text{ and } \begin{bmatrix} T_1^{(20)} \\ T_2^{(20)} \\ T_3^{(20)} \\ T_4^{(20)} \end{bmatrix} = \begin{bmatrix} 139.7 \\ 217.1 \\ 292.0 \\ 364.7 \end{bmatrix}$$

From the previous Example 5.3, we obtained the exact direct solution by the Thomas algorithm, which are

$$\begin{bmatrix} T_1^{exact} \\ T_2^{exact} \\ T_3^{exact} \\ T_4^{exact} \end{bmatrix} = \begin{bmatrix} 140.0 \\ 217.4 \\ 292.4 \\ 365.0 \end{bmatrix}$$

It is observed that the nodal temperatures after 20 iterations are edging closer toward the exact nodal temperature values.

(b) Let us now employ the iterative *Gauss-Seidel method* to the system of algebraic equations. We begin as in the Jacobi method with the set of equations

$$T_1 = 0.333\,T_2 + 67.5$$
$$T_2 = 0.5\,T_1 + 0.5\,T_3 + 1.25$$
$$T_3 = 0.5\,T_2 + 0.5\,T_4 + 1.25$$
$$T_4 = 0.333\,T_3 + 267.5$$

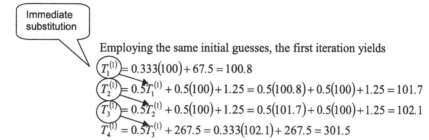

Immediate substitution

Employing the same initial guesses, the first iteration yields

$$T_1^{(1)} = 0.333(100) + 67.5 = 100.8$$
$$T_2^{(1)} = 0.5\,T_1^{(1)} + 0.5(100) + 1.25 = 0.5(100.8) + 0.5(100) + 1.25 = 101.7$$
$$T_3^{(1)} = 0.5\,T_2^{(1)} + 0.5(100) + 1.25 = 0.5(101.7) + 0.5(100) + 1.25 = 102.1$$
$$T_4^{(1)} = 0.5\,T_3^{(1)} + 267.5 = 0.333(102.1) + 267.5 = 301.5$$

After performing 10 iterations, the nodal temperatures have advanced to

$$\begin{bmatrix} T_1^{(10)} \\ T_2^{(10)} \\ T_3^{(10)} \\ T_4^{(10)} \end{bmatrix} = \begin{bmatrix} 139.4 \\ 216.7 \\ 291.9 \\ 364.7 \end{bmatrix}$$

The temperature values obtained through the Gauss-Seidel method at this present stage are comparable to the values obtained by the Jacobi method at 20 iterations.

Discussion: We can infer from this example that the Gauss-Seidel iteration is twice as fast as the Jacobi iteration. Convergence is achieved quicker by the Gauss-Seidel method because of the *immediate substitution* of the temperatures to the right-hand side of the equations whenever they are made available. Thus far, we have not discussed the issue of terminating the iteration process for this particular problem. The degree to which you wish convergence to be achieved is entirely up to you. If the absolute maximum difference $|\phi_j^{(k+1)} - \phi_j^{(k)}|$ is chosen as the condition for the termination process, the accuracy of the solution depends on the targeted number of significant figures you wish to obtain for the temperatures. The smaller the acceptable error, the higher the number of iterations, but this will achieve greater accuracy.

Sometimes the convergence to a solution can be enhanced by utilizing the numerical technique called *successive overrelaxation*. The basic idea behind this technique is an extrapolation procedure where the intermediate nodal variables $\phi_j^{(k+1)}$ are further advanced by a weighted average of the current values of $\phi_j^{(k+1)}$ with the previous values of $\phi_j^{(k)}$. The extrapolated values for $\phi_j^{(k+1)}$ is obtained as follows:

$$\overline{\phi}_i^{(k+1)} = (1 - \lambda)\phi_i^{(k)} + \lambda\phi_i^{(k+1)} \tag{5.76}$$

These extrapolated values are continuously used in the system of equations as the iteration process progresses. In the above equation λ is a relaxation factor whose value is usually found by trial-and-error experimentation for a given problem. Generally, the value of λ is bounded between $0 < \lambda < 2$ in order to ensure convergence. If the *successive overrelaxation* is used in conjunction with the Gauss-Seidel method, for a value λ in the range $1 < \lambda < 2$, a significant improvement to the nodal temperatures obtained at each iteration step is realized and hence convergence is achieved at a faster rate.

Right up to this moment, we have only discussed the application of the iterative methods on one-dimensional problems. For multi-dimensional problems, with a larger number of grid points, thus a larger number of equations to be solved, the Jacobi and Gauss-Seidel methods, despite their simplicity, may prove rather expensive especially since they generally require a large number of iterations to reach convergence. Successive overrelaxation described above though provides a way of accelerating the iteration process; however, the difficulty in determining the optimum values of λ precludes its wide application in tackling CFD problems. In fact, the method of under-relaxation, which limits the amount by which a variable changes from the previous iteration to the next one, is routinely used in CFD. Under-relaxation slows down the convergence rate but increases the stability. Nevertheless, the primary aim of this section is to introduce the reader to some basic understanding of iterative methods and demonstrate with some worked examples of their numerical computations.

Other methods: A practice often applied to multi-dimensional problems is the use of iterative matrices that correspond to lower-dimensional problems. One commonly used method is the ADI (alternating direction implicit), introduced by Peaceman and Rachford (1955), that is used to reduce multi-dimensional problems, whether they are two-dimensional or three-dimensional, to a sequence of a one-dimensional problem. The resulting matrix is of a tri-diagonal form in which the Thomas algorithm is applied in each of the coordinate directions; this procedure solves the nodal variables for the lines in one direction and repeats for the lines in other directions. Another iterative method for solving multi-dimensional discretization equations, particularly for structured mesh, is the strongly implicit procedure (SIP) proposed by Stone (1968). The basic idea of this method involves approximating the matrix A, Eq. (5.31), by an incomplete LU (lower-upper) factorization to yield an iteration matrix M. Unlike other methods, SIP is a good iterative technique in its own right. It has been used in some commercial CFD codes as the standard solver for non-linear equations. It also provides a good basis for acceleration techniques such as the conjugate gradient methods and multigrid methods.

5.5 PRESSURE-VELOCITY COUPLING – SIMPLE SCHEME

The incompressible form of the conservation equations governing the fluid flow is derived in Chapter 3 and summarized in Section 3.6. Because of the incompressible assumption, solution to the governing equations is complicated by the lack of an independent equation for pressure. In each of the momentum equations the fluid flow is driven by the contribution of the pressure gradients. With the additional equation provided by the continuity equation, this system of equations is self-contained; there are four equations for four dependents u, v, w, and p but no independent transport equation for pressure. The implication here is that the continuity and momentum equations are all required to solve for the velocity and pressure fields in an incompressible flow. For such a flow, the continuity equation is a kinematic constraint on the velocity field rather than a dynamic equation. In order to link the pressure with the velocity for an incompressible flow, one possible way is to construct the pressure field so as to guarantee conservation of the continuity equation.

Within this section, we describe the basic philosophy behind one of the most popular schemes of pressure-velocity coupling for an incompressible flow. It belongs to the class of iterative methods, which is embodied in a scheme called SIMPLE, where the acronym stands for Semi-Implicit Method for Pressure-Linkage Equations. This scheme was developed for practical engineering solutions by Patankar and Spalding (1972). Ever since their pioneering work, it has found widespread application in the majority of commercial CFD codes. In this scheme a guessed pressure field is used to solve the momentum equations. A pressure correction equation, deduced from the continuity equation, is then solved to obtain a pressure correction field, which in turn is used to update the velocity and pressure fields. These guessed fields are progressively improved through the iteration process until convergence is achieved for the velocity and pressure fields. The salient features of the SIMPLE scheme and the assembly of the complete iterative procedure will be discussed later.

Variable arrangement on the grid: Before describing the SIMPLE scheme, the choice of arrangement on the grid requires some consideration. Among the many arrangements, the two most popular ones that have gained wide acceptance are the *staggered* and *collocated* grid arrangements.

The aim of having a *staggered* grid arrangement for CFD computations is to evaluate the velocity components at the control volume faces while the rest of the variables governing the flow field, such as the pressure, temperature, and turbulent quantities, are stored at the central node of the control volumes. A typical arrangement is depicted in Fig. 5.11 on a structured finite volume grid and it can be demonstrated that the discrete values of the velocity component, u, from the x-momentum equation are evaluated and stored at the east, e, and west, w, faces of the control volume. By evaluating the other velocity components using the y-momentum and z-momentum equations on the rest of the control volume faces, these velocities allow a straightforward evaluation of the mass fluxes that are used in the pressure correction equation. This arrangement therefore provides a strong coupling between the velocities and pressure, which helps to avoid some types of convergence problems and oscillations in the pressure and velocity fields. Historically, staggered grid arrangement enjoyed its dominance within the CFD framework between the 1960s and 1980s. However, as the use of non-orthogonal grids became

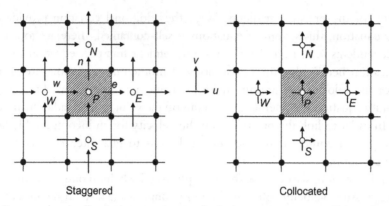

Staggered   Collocated

Fig. 5.11 Staggered and collocated arrangements of velocity components on a finite volume grid (full symbols denote element vertices and open symbols at the center of the control volumes denote computational nodes for the storage of other governing variables).

commonplace, because of the need to handle complex geometries, alternative grid arrangements had to be explored because of some inherent difficulties in the staggered approach. In particular, if the staggered approach is used in generalized coordinates, curvature terms are required to be introduced into the equations that are usually difficult to treat numerically and may create non-conservative errors when the grid is not smooth.

Nowadays, the alternative grid arrangement that is frequently adopted in many commercial CFD codes is the *collocated* grid arrangement. Here, all the flow–field variables including the velocities are stored at the same set of nodal points. For the finite volume grid, as shown in Fig. 5.12, they are stored at the central node of the control volumes (open symbols). The collocated arrangement offers significant advantages in complicated

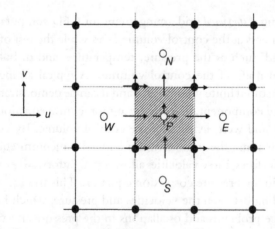

Fig. 5.12 Arrangement of velocity components on a control volume element of a structured grid at the central node, element faces, and element vertices.

domains, especially the capability of accommodating slope discontinuities or boundary conditions that may be discontinuous. Furthermore, if multigrid methods are used, the collocated arrangement allows the ease of transfer of information between various grid levels for all the variables. This grid arrangement was out of favor for incompressible flow computation for a substantial period because of the difficulties faced in coupling the pressure with the velocity and the occurrence of oscillations in the pressure. Nevertheless, the widespread use of the collocated grid arrangement became prominent once again through significant developments of pressure-velocity coupling algorithms such as the well-known Rhie and Chow (1983) interpolation scheme. This scheme, which has provided physically sensible solutions on structured collocated meshes, generated much interest in unstructured meshing applications. For the vast majority of general flows, this treatment ties together the pressure fields to yield smooth solutions while only minimally affecting the mass fluxes. More details of this interpolation scheme are left to the interested reader.

Pressure correction equation and its solution: The SIMPLE scheme is essentially a guess-and-correct procedure for the calculation of pressure through the solution of a pressure correction equation. The method is illustrated by considering a two-dimensional steady laminar flow problem in a structured grid, as shown in Fig. 5.11.

The SIMPLE scheme provides a robust method of calculating the pressure and velocities for an incompressible flow. When coupled with other governing variables such as temperature and turbulent quantities, the calculation needs to be performed sequentially since it is an iterative process. The sequence of operations in a typical CFD iterative process that embodies the SIMPLE scheme is given in Fig. 5.13, with more details of each iterative step elaborated below.

Step 1. The iterative SIMPLE calculation process begins by guessing the pressure field, p^*. During the iterative process, the discretized momentum equations are solved using the guessed pressure field. Applying the finite volume method, the equations for the x-momentum and y-momentum that yield the velocity components, u^* and v^*, can be expressed in the same algebraic form as previously derived in Eq. (5.30), which can be recast into

$$a_P^u u_P^* = \sum a_{nb}^u u_{nb}^* - \frac{\partial p^*}{\partial x} \Delta V + b' \qquad (5.76)$$

$$a_P^v v_P^* = \sum a_{nb}^v v_{nb}^* - \frac{\partial p^*}{\partial y} \Delta V + b' \qquad (5.77)$$

where ΔV is the finite control volume. Here, we simplify the above expressions by introducing a_{nb} to represent the presence of the neighboring coefficients and u_{nb}^* and v_{nb}^* to denote the neighboring nodal velocity components. The pressure gradient terms appearing in the above two equations are taken out from the original source term b of the momentum equations, while the other terms governing the fluid flow are left in the source term b'.

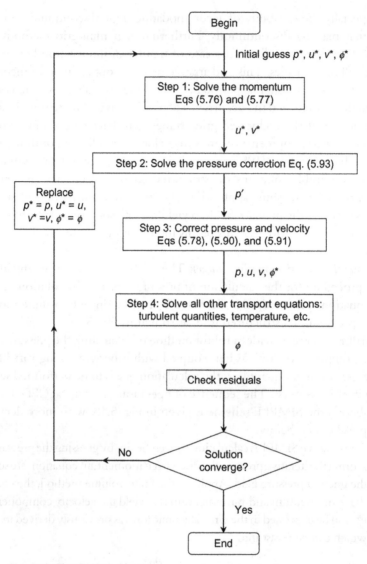

Fig. 5.13 The SIMPLE scheme.

Step 2. If we define the correction p' as the difference between the correct pressure field and the guessed pressure field p^*, we get

$$p = p^* + p' \qquad (5.78)$$

Similarly, we can also define the corrections u' and v' to relate the correct velocities u and v to the guessed velocities u^* and v^*:

$$u = u* + u' \tag{5.79}$$

$$v = v* + v' \tag{5.80}$$

The algebraic form of the correct velocities u and v can also be similarly expressed as presented in Eqs (5.76) and (5.77) so that

$$a_p^u u_p = \sum a_{nb}^u u_{nb} - \frac{\partial p}{\partial x} \Delta V + b' \tag{5.81}$$

$$a_p^v v_p = \sum a_{nb}^v v_{nb} - \frac{\partial p}{\partial y} \Delta V + b' \tag{5.82}$$

Subtracting Eqs (5.81) and (5.82) from Eqs (5.76) and (5.77), we obtain

$$a_P^u \left(u_p - u_P^*\right) = \sum a_{nb}^u (u_{nb} - u_{nb}^*) - \frac{\partial (p - p^*)}{\partial x} \Delta V \tag{5.83}$$

$$a_P^u \left(v_p - v_P^*\right) = \sum a_{nb}^u (v_{nb} - v_{nb}^*) - \frac{\partial (p - p^*)}{\partial y} \Delta V \tag{5.84}$$

Using the correction formulae (5.78)–(5.80), the above equations may be rewritten as follows:

$$a_p^u u'_p = \sum a_{nb}^u u'_{nb} - \frac{\partial p'}{\partial x} \Delta V \tag{5.85}$$

$$a_p^v v'_p = \sum a_{nb}^v v'_{nb} - \frac{\partial p'}{\partial y} \Delta V \tag{5.86}$$

The SIMPLE scheme approximates the above two Eqs (5.85) and (5.86) by the omission of the terms $\sum a_{nb}^u u'_{nb}$ and $\sum a_{nb}^v v'_{nb}$. It is reminded that this scheme is an iterative approach, and there is no reason why the formula designed to predict p' needs to be physically correct. Hence we are allowed to construct a formula for p' that is simply a numerical artifice with the aim to expedite the convergence of the velocity field to a solution that satisfies the continuity equation. This is the essence of the algorithm. Once the pressure correction field is known, the correct velocities u and v can be updated through the guessed velocities $u*$ and $v*$ from the simplified equations of (5.85) and (5.86):

$$u_p = u_p^* - D^u \frac{\partial p'}{\partial x} \tag{5.87}$$

$$v_p = v_p^* - D^v \frac{\partial p'}{\partial y} \tag{5.88}$$

$$\text{where } D^u = \frac{\Delta V}{a_p^u} \text{ and } D^v = \frac{\Delta V}{a_p^v} \tag{5.89}$$

Although the above Eqs (5.87)–(5.89) have been developed to correct the velocities from the guessed velocities at the central node of the control volume, these correction formulae can also be generally applied to any location where the velocity components reside within the computational grid (as shown in Fig. 5.9, the velocities may be located at central node P or at the control volume faces or at the vertices of the control volume). The general form of the velocity correction formulae, by removing the subscript P, can be expressed as:

$$u = u^* - D^u \frac{\partial p'}{\partial x} \tag{5.90}$$

$$v = v^* - D^v \frac{\partial p'}{\partial y} \tag{5.91}$$

The derivation of a pressure correction equation utilizes the above two equations. Differentiating Eq. (5.90) by the Cartesian direction x and Eq. (5.91) by the Cartesian direction y and summing them together yields

$$-\frac{\partial}{\partial x}\left(D^u \frac{\partial p'}{\partial x} \right) - \frac{\partial}{\partial y}\left(D^v \frac{\partial p'}{\partial y} \right) + \underbrace{\frac{\partial u^*}{\partial x} + \frac{\partial v^*}{\partial y}}_{\text{guessed velocity gradients}} = \underbrace{\frac{\partial u}{\partial x} + \frac{\partial v}{\partial y}}_{\text{correct velocity gradients}} {}^{= 0} \tag{5.92}$$

By invoking the continuity equation, it is shown that the term represented by the source term of the right-hand side for the Eq. (5.84) is zero, and Eq. (5.84) can be re-arranged as

$$\frac{\partial}{\partial x}\left(D^u \frac{\partial p'}{\partial x} \right) + \frac{\partial}{\partial y}\left(D^v \frac{\partial p'}{\partial y} \right) = \underbrace{\left(\frac{\partial u^*}{\partial x} + \frac{\partial v^*}{\partial y} \right)}_{\text{mass residual}} \tag{5.93}$$

Interestingly, Eq. (5.93) behaves like a steady-state diffusion process in a two-dimensional domain. It is a Poisson equation, one of the well-known equations from classical physics and mathematics. The solution to this Poisson equation can be achieved through some efficient numerical solvers (conjugate gradient and multigrid methods) as previously discussed above to accelerate its convergence.

Step 3. Once the pressure correction p' field is obtained, the pressure and velocity components are subsequently updated through the correction formulae of Eqs (5.78), and (5.90) and (5.91), respectively. If the solution only concerns a laminar CFD flow problem, the iteration process proceeds directly to check the convergence of the solution. If the solution is not converged, the process is repeated by returning to Step 1. The source term appearing in the pressure correction Eq. (5.93), commonly known as the *mass residual*, is normally used in CFD computations as a criterion to terminate the iteration procedure. As the mass residual continues to diminish, the pressure correction p' will tend toward zero, thereby yielding a converged solution of $p^* = p$, $u^* = u$, and $v^* = v$.

Step 4. This step is executed if the CFD flow problem is turbulent or it may involve the transfer of heat or mass exchanges between different flow phases. Additional transport equations governing such a flow system need to be solved before convergence is checked. If the solution is not converged, the iterative process returns to Step 1 and repetitive calculations are carried out until convergence is reached.

The application of this SIMPLE scheme can be best illustrated by solving the Chapter 2 CFD problem of a steady two-dimensional incompressible laminar flow in a channel, which is described by the worked example below.

EXAMPLE

Example 5.5

Consider the case for a steady two-dimensional incompressible, laminar flow between two stationary parallel plates (Fig. 5.5.1) as in Chapter 2. By obtaining the solution from a CFD code using the finite volume method, track the progress of the intermediate values of u, v, p, p' and the mass residual during the iterative process at a computational nodal point at the center of the channel.

Solution: The problem is described as follows:

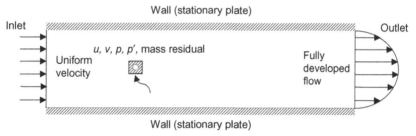

Fig. 5.5.1 Two-dimensional laminar flow in a channel with a monitoring point located at the center of the channel.

To demonstrate the robustness of the SIMPLE scheme, the iterative process begins by employing the initial guesses: $p^* = 0$, $u^* = 0$, and $v^* = 0$. The discretized equations governing the momentum and pressure correction are solved using the default iterative solvers provided in the commercial CFD code. The inlet, outlet, and wall conditions remain the same as applied in Chapter 2.

Based on Eqs (5.63), (5.64), (5.65), (5.77), (5.78), and (5.80), the calculated values of the pressure p, pressure correction p', velocities u and v, and mass residual for the first iteration at the *monitoring point* are:

$$\begin{bmatrix} p_{monitor}^{(1)} \\ p'^{(1)}_{monitor} \\ u_{monitor}^{(1)} \\ v_{monitor}^{(1)} \end{bmatrix} = \begin{bmatrix} 0.02043 \\ 0.06812 \\ 0.01033 \\ -0.1246\times10^{-4} \end{bmatrix} \qquad \boxed{mass\ residual \Rightarrow \quad 1.2\times10^{-4}\ \text{kg}/\text{s}}$$

The solution of the first iteration from above is subsequently used as intermediate values for the next iteration step; the second iteration yields

$$
\begin{bmatrix} p^{(2)}_{monitor} \\ p'^{(2)}_{monitor} \\ u^{(2)}_{monitor} \\ v^{(2)}_{monitor} \end{bmatrix} = \begin{bmatrix} 4.774 \times 10^{-3} \\ -0.0522 \\ 0.01181 \\ 1.104 \times 10^{-5} \end{bmatrix} \quad \boxed{\begin{array}{l} \textit{mass residual} \Rightarrow \\ 2.0494 \times 10^{-4} \text{ kg/s} \end{array}}
$$

After repeated applications of the iterative process, the respective values for the pressure p, pressure correction p', velocities u and v, and mass residual (Figs. 5.5.2–5.5.5) after 10 and 20 iterations are:

$$
\begin{bmatrix} p^{(10)}_{monitor} \\ p'^{(10)}_{monitor} \\ u^{(10)}_{monitor} \\ v^{(10)}_{monitor} \end{bmatrix} = \begin{bmatrix} 6.83 \times 10^{-4} \\ -4.949 \times 10^{-4} \\ 0.01419 \\ -2.963 \times 10^{-6} \end{bmatrix} \quad \boxed{\begin{array}{l} \textit{mass residual} \Rightarrow \\ 1.8639 \times 10^{-5} \text{ kg/s} \end{array}}
$$

and

$$
\begin{bmatrix} p^{(20)}_{monitor} \\ p'^{(20)}_{monitor} \\ u^{(20)}_{monitor} \\ v^{(20)}_{monitor} \end{bmatrix} = \begin{bmatrix} 6.626 \times 10^{-4} \\ 3.793 \times 10^{-6} \\ 0.01482 \\ 2.444 \times 10^{-7} \end{bmatrix} \quad \boxed{\begin{array}{l} \textit{mass residual} \Rightarrow \\ 5.061 \times 10^{-7} \text{ kg/s} \end{array}}
$$

\From a theoretical viewpoint, the vertical velocity v is zero at the monitoring point and the iterative process confirms the trend of the prediction toward the zero value. It is seen during the iterative process that the intermediate values of this velocity are much smaller than the rest of the other governing variables; the convergence history plot of this velocity component is therefore omitted since no quantitative comparison can be realized against the other variable convergence histories. The convergence histories for the rest of the governing variables that include the pressure p, pressure correction p', horizontal velocity u, and mass residual are illustrated below.

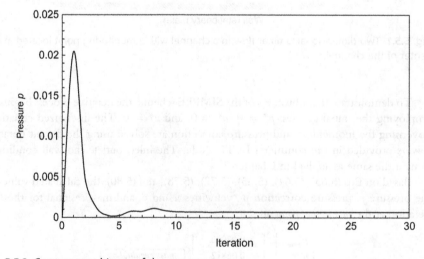

Fig. 5.5.2 Convergence history of the pressure p.

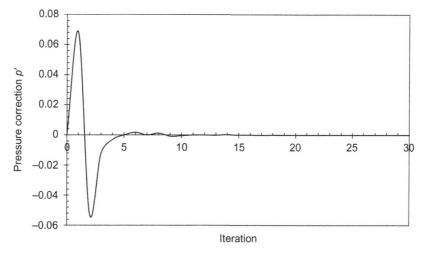

Fig. 5.5.3 Convergence history of the pressure correction p'.

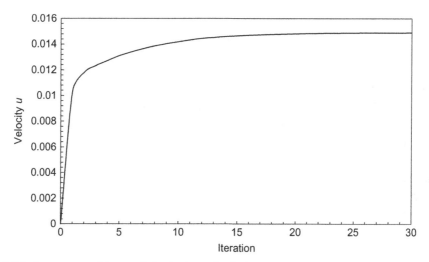

Fig. 5.5.4 Convergence history of the horizontal velocity u.

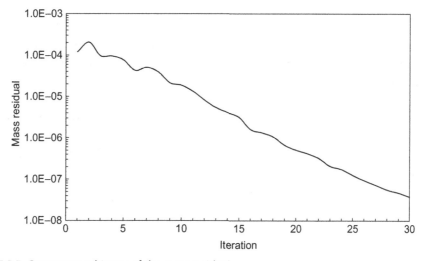

Fig 5.5.5 Convergence history of the mass residual.

Discussion: From this worked example of a channel flow, the SIMPLE scheme provides an efficient iterative procedure for obtaining the velocity and pressure fields for an incompressible flow. The SIMPLE scheme is a robust method that produces rapid stabilization of the velocity and pressure as seen by their respective convergence histories after 5 iterations. The mass residual that appears as a source term in the pressure correction Eq. (5.80) continues to diminish during the iteration process, thus reaffirming conservation in the continuity equation. Subsequently, the pressure correction p' is seen to be approaching zero. Hence the corrections that are required to update the velocity field are also approaching zero. The trend of the convergence histories favors the likelihood of a converged steady-state solution.

The reader should be aware of other types of pressure-velocity coupling algorithms that employ a similar philosophy to the SIMPLE algorithm that are employed by CFD users or adopted in commercial CFD codes. These variant SIMPLE algorithms have been formulated with the aim of better improving the robustness and convergence rate of the iterative process. We do not intend to provide the reader with all the details of the available algorithms but to briefly indicate and describe the modifications made to the original SIMPLE algorithm.

The SIMPLEC (SIMPLE-Consistent) algorithm by Van Doormal and Raithby (1984) follows the same iterative steps as in the SIMPLE algorithm. The main difference between SIMPLEC and SIMPLE is that the discretized momentum equations are manipulated so that the SIMPLEC velocity correction formulae omit terms that are less significant than those omitted in SIMPLE. Another pressure correction procedure that is also commonly employed is the PISO (Pressure Implicit with Splitting of Operators) algorithm proposed by Issa (1986). This pressure-velocity calculation procedure was originally developed for the non-iterative computation of unsteady compressible flows. Nevertheless, it has been adapted successfully for the iterative solution of steady-state problems. PISO is simply recognized as an extension of SIMPLE with an additional corrector step that involves an additional pressure correction equation to enhance the convergence. The SIMPLER (SIMPLE-Revised) algorithm developed by Patankar (1980) also falls within the framework of two corrector steps like in PISO. Here, a discretized equation for the pressure provides the intermediate pressure field before the discretized momentum equations are solved. A pressure correction is later solved where the velocities are corrected through the correction formulae as similarly derived in the SIMPLE algorithm.

There are other SIMPLE-like algorithms such as SIMPLEST (SIMPLE-ShorTened) by Spalding (1980) or SIMPLEX by Van Doormal and Raithby (1985) or SIMPLEM (SIMPLE-Modified) by Acharya and Moukalled (1989) that share the same essence in their derivations. More details of all the above pressure-velocity coupling algorithms are left to keen and interested readers.

5.6 MULTIGRID METHOD

For the multigrid method, it can be categorized into two types: geometric and algebraic. The former, also known as the full approximation scheme (FAS) multigrid, involves a hierarchy of meshes (cycling between fine and coarse grids) and the discretized equations are evaluated on every level, while in the latter, the coarse level equations are generated without any geometry or re-discretization on the coarse levels, a feature that makes algebraic multigrid particularly attractive for the use on unstructured meshes.

The multigrid method is ideal for solving the Poisson-like pressure or pressure-correction equation such as the SIMPLE method, which will be further discussed in the next section. Conceptually, the multigrid method can be described in the following. Focusing on the system of equations, $A\phi = B$, intermediate solutions of ψ are obtained if this system is solved with an iterative method after some predetermined number of iterations. The residuals R can be defined as:

$$A\ \psi = B - R \tag{5.94}$$

By also defining the errors as the difference between the true and intermediate solutions:

$$E\ = \phi - \psi \tag{5.95}$$

and subtracting Eq. (5.95) from Eq. (5.94), the following relationship between the errors and residuals is:

$$A\ E = R \tag{5.96}$$

During the multigrid cycle, the matrix of coefficients of A and the residuals as described by Eq. (5.94) are transferred from a finer grid to a coarser grid through volume-weighted *restrictions*. After obtaining the converged solutions of the errors, *prolongations* of computed corrections on the coarse grids are transferred to the next fine grids through tri-linear interpolations. The simplest choice of a multigrid cycle can be described by the typical V-cycle with five different grid levels described as shown in Fig. 5.14 (see further details in Appendix F and references given within). Another strategy, for example, a W-cycle such as that depicted in Fig. 5.15, may also be used for cycling between coarse and fine grids. Efficiency may be improved by the decision to switch from one grid to another on the rate of convergence through the combination of V-W cycles or other possible combinations. The optimum choice of parameters is problem dependent but their effect on performance is not as dramatic as for the single-grid method.

In theory, the advantage of geometric multigrid over algebraic multigrid is that the former should perform better for non-linear problems since non-linear properties in the system are propagated down to the coarse levels through the re-discretization while for the latter, once linearization is performed on the system of equations, the non-linear

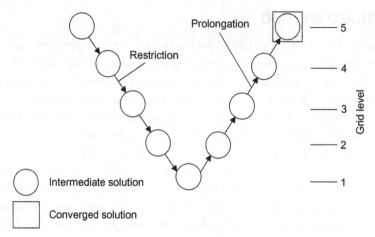

Fig. 5.14 Schematic representation of a multigrid method using a V-cycle.

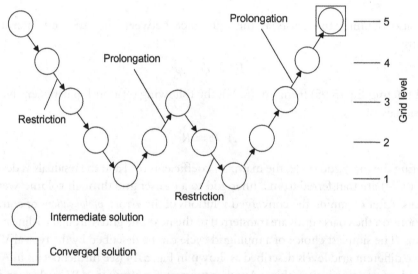

Fig. 5.15 Schematic representation of a multigrid method using a W-cycle.

properties are not "felt" by the solver until the fine level operator is updated. The multi-grid approach is more of a strategy than a particular method. More details of these acceleration techniques are left to the interested reader.

5.7 SUMMARY

Let us review some of the basic computational techniques that have been examined in this chapter to solve the governing equations of fluid dynamics.

The first stage of obtaining the computational solution involves the conversion of the governing equations into a system of algebraic equations. This is usually known as the *discretization stage*. We have discussed some of the *discretization tools* such as the finite difference and finite volume methods, which form the foundation of understanding the basic features of *discretization*. Both of these methods are found in many CFD applications.

The second stage involves numerically solving the system of algebraic equations, which can be achieved by either the *direct methods* or *iterative methods*. Basic direct methods such as the Gaussian elimination and the Thomas algorithm have been described, of which the latter is exceedingly economical for a tri-diagonal matrix system and is a standard algorithm for the solution of fluid flow equations in a structured mesh. Simple iterative methods such as the point-by-point Jacobi and Gauss-Seidel methods are also described. Nevertheless, CFD problems are generally multi-dimensional and comprise a large system of equations to be solved. Efficient iterative methods such as the ADI or Stone's SIP are often applied to solve such a system of equations. To further enhance the convergence of the computational solution, precondition conjugate gradient methods or multigrid methods are employed to accelerate the iteration process.

The reader may return to Fig. 5.1 to view how these two stages fit within the overview process of the computational solution procedure. Within the block that comprises numerical methods, an iterative algorithm for the calculation of pressure and velocity fields based on the SIMPLE scheme is presented for an incompressible flow. The basic philosophy behind this popular scheme is to initially guess a pressure field in the discretized momentum equations to yield the intermediate velocities. The continuity equation in the form of a pressure correction is subsequently solved, which is then used to correct the velocity and pressure fields. These guessed fields are continuously improved until convergence is reached. The reader may refer back to Fig. 5.13 for a more exhaustive description of the iterative steps that are involved within the SIMPLE scheme.

Finally, we have not discussed in depth the assessment of *convergence*. In practice, the algebraic equations that result from the discretization process yield the flow solution at each nodal point on a finite grid layout. It is expected that, from the truncation errors given in Section 5.2.1, more accurate solutions can be obtained by refining the grid. For an unsteady problem, this can be achieved by employing smaller time intervals. However, for a given required *solution accuracy*, it may be more economical to solve higher-order approximations of the first and second derivative equations governing the fluid flow on a coarse grid rather than using a low-order approximation on a finer grid. This leads to the concept of *computational efficiency*. Other issues such as the *solution consistency* and *stability* of the numerical procedure are also important considerations for the *convergence* of the computational solution. All of these will be investigated in the next chapter.

REVIEW QUESTIONS

5.1 What are the differences between solving a fluid flow problem analytically and solving numerically? What are the advantages and disadvantages of each method?

5.2 What are the main advantages and disadvantages of discretization of the governing equations through the finite difference method?

5.3 Is finite difference more suited for structured or unstructured mesh geometries? Why?

5.4 Consider the following finite difference formulation for a simplified flow: $\dfrac{\phi_{i-1} - 2\phi_i + \phi_{i+1}}{\Delta x^2} = 0$. Is the flow steady or transient? Is it one-, two-, or three-dimensional? Is the nodal spacing constant or variable?

5.5 Using finite difference, show that the steady one-dimensional heat conduction equation, $k\dfrac{\partial^2 T}{\partial x^2} = 0$, can be expressed as $\dfrac{T_{i-1} - 2T_i + T_{i+1}}{\Delta x^2} = 0$.

5.6 What is the second term in the central difference approximation for a first derivative (given below) called, and what does it measure? $\left(\dfrac{\partial \phi}{\partial x}\right) = \dfrac{\phi_{i+1,j} - \phi_{i-1,j}}{2\Delta x} + O(\Delta x^2)$

5.7 Which of the following is most accurate, and why: *forward difference, backward difference,* and *central difference*?

5.8 What are the main advantages and disadvantages of discretization of the governing equations through the finite volume method?

5.9 Is the finite volume method more suited for structured or unstructured mesh geometries? Why?

5.10 What is the significance of the integration of the governing equations over a control volume during the finite volume discretization?

5.11 For the control volume below, show how the one-dimensional steady-state diffusion term $\dfrac{\partial}{\partial x}\left(\Gamma \dfrac{\partial \phi}{\partial x}\right)$ is discretized to obtain its discretized equation $\left(\Gamma \dfrac{\partial \phi}{\partial x}\right)_e A_E - \left(\Gamma \dfrac{\partial \phi}{\partial x}\right)_w A_w$ for central grid nodal point P.

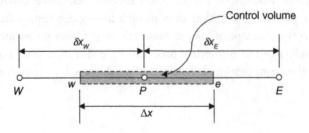

5.12 In a finite difference scheme data is resolved at nodal points; how is this different to the finite volume scheme?

5.13 How is a steady convective-diffusion process different from a pure diffusion process?

5.14 Why are upwind schemes important for strongly convective flow?

5.15 Why are higher-order upwind schemes more favorable than the first-order upwind scheme?

5.16 For the unsteady convection-diffusion process, what is the difference between *explicit* and *implicit* time-marching approaches?

5.17 What is the difference between using a direct method and an iterative method to solve the discretized equations?

5.18 Is the direct method or iterative method more suitable in solving for a large system of non-linear equations?

5.19 Why does the Gauss-Seidel iterative method converge to a solution quicker than the Jacobi method?

5.20 What is the technique associated with *successive overrelaxation* and why is it used?

5.21 Where are the flow field variables located in collocated grids? How is this different to the locations in a staggered grid?

5.22 Write down the formulation of the central difference scheme for u velocity in the x direction. What is its truncation error in terms of $\triangle x$? And state the order of this discretization scheme?

5.23 What is the purpose of the SIMPLE scheme? Does it give us a direct solution or depend on the iterative concept?

5.24 What is the Gaussian elimination method based on? Can this method be used to solve a system of non-linear algebraic equations?

5.25 Solve the following set of equations by Gaussian elimination:

$$\begin{bmatrix} 100 & 100 & 0 \\ 200 & 100 & - \\ 300 & - & - \\ - & 200 & 300 \end{bmatrix} \begin{bmatrix} T_1 \\ T_2 \\ T_3 \end{bmatrix} = \begin{bmatrix} 400 \\ 100 \\ -300 \\ 400 \end{bmatrix}$$

5.26 Solve the following set of equations by the Thomas algorithm:

$$\begin{bmatrix} 100 & - & 200 & - \\ 200 & - & 300 & - \\ 100 & 100 & 100 & 0 \\ 100 & - & 400 & 300 \end{bmatrix} \begin{bmatrix} T_1 \\ T_2 \\ T_3 \\ T_4 \end{bmatrix} = \begin{bmatrix} 800 \\ -2000 \\ -200 \\ 400 \end{bmatrix}$$

5.27 Solve the following set of equations using the Gauss-Seidel method:

$$\begin{bmatrix} -1000 & -100 & 200 \\ -100 & -1100 & -100 \\ 200 & -100 & 1000 \\ 0 & 300 & -100 \end{bmatrix} \begin{bmatrix} T_1 \\ T_2 \\ T_3 \end{bmatrix} = \begin{bmatrix} 600 \\ -2500 \\ 1100 \\ 1500 \end{bmatrix}$$

5.28 For the same matrix given in question 5.27, use the Jacobi method to solve the set of equations. Compare the number of iterations for convergence between the Jacobi method and the Gauss-Seidel method.

5.29 Solve the following set of equations using the Gauss-Seidel method.

(a)
$$\begin{aligned} 3x_1 &- x_2 + 3x_3 &= 0 \\ -x_1 &+ 2x_2 + x_3 &= 3 \\ 2x_1 &- x_2 - x_3 &= 2 \end{aligned}$$

(b)
$$\begin{aligned} 10x_1 &- x_2 + 2x_3 & &= 6 \\ -x_1 &+ 11x_2 - x_3 &+ 3x_4 &= 25 \\ 2x_1 &- x_2 - 10x_3 &- x_4 &= -11 \\ & 3x_2 - x_3 &+ 8x_4 &= 15 \end{aligned}$$

5.30 Following the grid arrangement below, derive the following expression:

$$\left(\frac{\partial^2 u}{\partial x \partial y}\right)_{i,j} = \frac{u_{i+1,j+1} - u_{i+1,j-1} - u_{i-1,j+1} + u_{i-1,j-1}}{4\Delta x \Delta y}$$

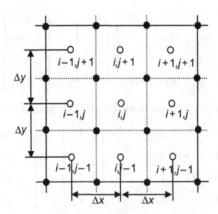

CHAPTER 6

CFD Solution Analysis – Essentials

6.1 INTRODUCTION

Analyzing a *computational solution* represents an integral part in the use of CFD. In Chapter 5 some basic discretization techniques are introduced to allow the reader to familiarize with common methodologies of converting the governing partial differential equations into a system of algebraic equations. This system of algebraic equations is subsequently solved through numerical methods to provide *approximate solutions* to the governing equations. It is these *approximate solutions* that we can interchangeably refer to as *computational solutions*.

In the context of CFD some of the primary concerns regarding the *computational solutions* are whether the solution can be guaranteed to approach the exact solution of the partial differential equations and if so under what circumstances. This can be (superficially) achieved by forcing the computational solution to converge to an exact solution as the finite quantities shrink to zero. We recall from Chapter 5 that the finite quantities, in time Δt and in space Δx, Δy, and Δz, are prevalent in the system of algebraic equations as a result of the discretization of the partial differential equations. Nevertheless, *convergence* can neither be straightforward nor directly established. Indirect considerations of convergence need, however, to be inferred from aspects such as *consistency* and *stability*. First, it is required that the formulation of system of algebraic equations through the discretization process should be consistent with the original partial differential equations. The implication of *consistency* here is the recovery of the governing equations by reversing the discretization process through a Taylor series expansion. Secondly, for any chosen numerical algorithm adopted to solve the algebraic equations, *stability* shares a part of the platform and, with the *consistency* criteria, ensures *convergence*.

The *accuracy* of the *computational solutions* can be affected by errors and uncertainties in the numerical calculations. These errors and uncertainties can be generated in either the conceptual modeling or during the computational design phase, which needs to be measured or bounded. The credibility of the *computational solutions* is strongly dependent on whether the errors and uncertainties are identified and qualified, irrespective of their sources. Systematic reduction of errors and uncertainties leads to better representation of real physical flow problems and thus increases the confidence in the use of the CFD simulation code. We will provide a pragmatic approach for estimating these errors and uncertainties in CFD through the *verification* and *validation* procedures.

Computational Fluid Dynamics
https://doi.org/10.1016/B978-0-323-93938-6.00015-4
209

To close this chapter, indicative case studies are selected to demonstrate the relevance and credibility of the computational solutions by addressing the essentials of *consistency*, *stability*, *convergence*, and *accuracy* that concern a CFD solution analysis. Through discussing these essentials, the reader may take it upon himself/herself to realize and appreciate various numerical aspects that are involved in solving the particular flow problems. Each of these embraces its own physical significance; it is assumed that the reader already possesses some basic knowledge of the fluid flow and heat transfer processes in order to better understand the physical considerations of the numerical simulations.

6.2 CONSISTENCY

The property of *consistency* appears rather rhetorical in hindsight. Nevertheless, it is an important property and concerns the discretization of the partial differential equations where the approximation performed should diminish or become exact if the finite quantities, such as the time step Δt and mesh spacing Δx, Δy, and Δz, tend to zero. In Section 5.2.1 of the finite difference method, the concept of *truncation error* measures the discrete approximation obtained through a Taylor series expansion about a single nodal point. Essentially, the *truncation error* represents the difference between the discretized equation and the exact one. As a result, the original partial differential equation is recovered by the addition of a remainder, the *truncation error*. This error basically measures the accuracy of the approximation and determines the rate at which the error decreases as the time step and/or mesh spacing are reduced.

For any numerical method to be *consistent*, the *truncation error* must become zero when the time step $\Delta t \to 0$ and/or mesh spacing Δx, Δy, and $\Delta z \to 0$. This error is usually proportional to a power of nth for the finite quantities. If the most important term is proportional to $(\Delta t)^n$ or $(\Delta x_i)^n$, the numerical method results in an nth-order approximation for $n > 0$. Ideally, all terms in the governing equations should be discretized with the approximation of the same order of accuracy. However, in practice, some terms (e.g., advection terms for high Reynolds number flows) may be particularly dominant and a high-order approximation may be required to treat them with more accuracy than others.

Some basic ideas of *consistency* are further elucidated through the illustrative worked examples below.

EXAMPLES

Example 6.1
Consider the discretized form of the incompressible steady-state two-dimensional continuity equation, $\frac{\partial u}{\partial x} + \frac{\partial v}{\partial y} = 0$, in a structured uniform grid arrangement as in Example 4.1. Discuss the remainder or truncation error associated with the original form of the partial differential equation.

Solution: An elemental control volume of the two-dimensional structured grid is shown in Fig. 6.1.1. The centroid of the control volume is indicated by the point P, which is surrounded by adjacent control volumes having their respective centroids indicated by the points: east, E; west, W; north, N; and south, S. The control volume having its centroid at P has respective faces indicated by: east, e; west, w; north, n; and south, s.

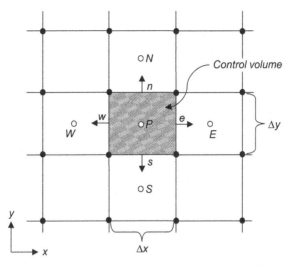

Fig. 6.1.1 Control volume for the two-dimensional continuity equation problem.

The discretized form obtained through the finite volume method is expressed by

$$\frac{u_e - u_w}{\Delta x} + \frac{v_n - v_s}{\Delta y} = 0$$

The face velocities u_e, u_w, v_n, and v_s are located midway between each of the control volume centroids, which allows us to determine the face velocities through interpolation of the centroid values. Thus

$$u_e = \frac{u_P + u_E}{2}; \quad u_w = \frac{u_P + u_W}{2}; \quad v_n = \frac{v_P + v_N}{2}; \quad v_s = \frac{v_P + v_S}{2}$$

By substituting the above expressions to the discretized form of the velocity first-order derivatives, we get

$$\frac{u_E - u_W}{2\Delta x} + \frac{v_N - v_S}{2\Delta y} = 0$$

To recover the original form of the partial differential equation, the above equation can be rewritten in terms of the truncation errors obtained through Taylor series expansion described in Section 5.2.1 of Eqs (5.3) and (5.8) as

$$\frac{u_E - u_W}{2\Delta x} + \frac{v_N - v_S}{2\Delta y} + \left[\underbrace{O\left(\Delta x^2, \Delta y^2\right)}_{\text{Truncation error}} \right]_P = 0$$

The numerical method results in an *n*th-order approximation of 2. The approximation is therefore second order accurate at the grid nodal point *P*.

Discussion: The original partial differential equation is recovered (satisfying *consistency*) from the discretized equation that includes the truncation error, as the mesh spacing Δx and $\Delta y \rightarrow 0$. For the above second-order scheme, halving the mesh spacing results in a reduction in the truncation errors by a factor of four.

Example 6.2

Consider the discretized form of the one-dimensional transient diffusion equation. Assuming that the thermal diffusivity α ($= k/\rho C_p$) is constant, discuss the remainder or truncation error associated with the DuFort-Frankel (DuFort and Frankel, 1953) differencing of the transient heat conduction equation.

Solution: The one-dimensional transient diffusion equation can be expressed as

$$\frac{\partial \phi}{\partial t} - \alpha \frac{\partial^2 \phi}{\partial x^2} = 0 \tag{6.2A}$$

Applying the DuFort and Frankel differencing to the above equation yields

$$\frac{\phi_i^{n+1} - \phi_i^{n-1}}{2\Delta t} - \frac{\alpha}{\Delta x^2}\left(\phi_{i+1}^n - \phi_i^{n+1} - \phi_i^{n-1} + \phi_{i-1}^n\right) = 0$$

The left-hand side term of Eq. (6.2A) that represents the time derivative about the *i*th node is analogous to the truncation error of the second-order central differencing in space, *viz.*,

$$\frac{\partial \phi}{\partial t} = \frac{\phi_i^{n+1} - \phi_i^{n-1}}{2\Delta t} + \underbrace{O(\Delta t^2)}_{\text{Truncation error}}_i$$

The right-hand side term of Eq. (6.2A) can be represented according to the second-order derivative in space of Eq. (5.9), yielding the following truncation error:

$$\alpha \frac{\partial^2 \phi}{\partial x^2} = \frac{\alpha}{\Delta x^2}\left(\phi_{i+1}^n - \phi_i^{n+1} - \phi_i^{n-1} + \phi_{i-1}^n\right) + \underbrace{O(\Delta x^2)}_{\text{Truncation error}}_i$$

A Taylor series expansion of the exact solution substituted into the one-dimensional transient heat conduction equation, neglecting higher order terms, is thus given by

$$\left[\frac{\partial \phi}{\partial t} - \alpha \frac{\partial^2 \phi}{\partial x^2} + \alpha\left(\frac{\Delta t}{\Delta x}\right)^2 \frac{\partial^2 \phi}{\partial t^2}\right]_i + \left[\underbrace{O(\Delta t^2, \Delta x^2)}_{\text{Truncation error}}\right]_i = 0$$

Discussion: As demonstrated in Fletcher (1991), $\Delta t/\Delta x$ must $\rightarrow 0$ at the same rate as $\Delta t, \Delta x \rightarrow 0$ to achieve *consistency*. It is also required that $\Delta t << \Delta x$ for *consistency* or else the scheme becomes inaccurate (i.e. if $(\Delta t/\Delta x)^2$ is large). From a practical viewpoint, there is effectively a restriction on the size of Δt when using the DuFort and Frankel scheme.

Consider the Dirichlet boundary conditions set at the opposite ends as $\phi\,(t, x = 0) = 1$ and $\phi\,(t, x = 1) = 0$. With a constant thermal diffusivity of $\alpha = 0.5$ and $\Delta x = 1/10$, the steady-state solutions to Eq. (6.2A) subjected to time step $\Delta t = 1/10$ (blue) and time step $\Delta t = 50$ (red) are shown in Fig. 6.2.1. The exact solution is given in Fig. 6.2.2. It is clearly seen that the solution for large $(\Delta t/\Delta x)$ is inconsistent with the exact solution when compared with the result obtained for a smaller $(\Delta t/\Delta x)$.

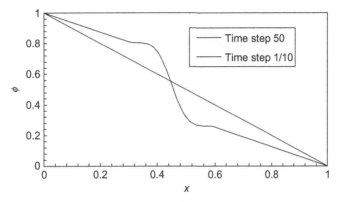

Fig. 6.2.1 Steady-state solution for $\Delta t = 1/10$ and $\Delta t = 50$ with a fixed grid step size $\Delta x = 1/10$.

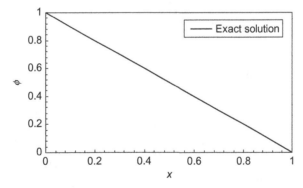

Fig. 6.2.2 The exact solution for Eq. (6.2A).

From the worked examples above, it is clear that the property of *consistency* is necessary if the approximate solution is to converge to the solution of the partial differential equation. Nevertheless, this property alone is not a sufficient condition. Even though the discretized equations might be equivalent to the partial differential equation as the finite quantities shrink to zero, it may not necessarily mean that the solution of the discretized equations follows the exact solution of the partial differential equation. The latter is evidenced by the inaccurate solution as a result of large $(\Delta t/\Delta x)$ in Example 6.2

6.3 STABILITY

In addition to *consistency*, another property that also strongly governs the numerical solution method is *stability*. This property concerns the growth or decay of errors introduced at any stage during the computation. It is noted that the errors being referred to here are not those produced by incorrect logic but those that occur because of rounding-off at every step of computation due to the finite number of significant figures the computer hardware can accommodate as well as a poor initial guess. A numerical solution method is therefore considered to be stable if it does not magnify the errors that appear in the course of the numerical solution process. *Stability* in temporal problems guarantees that the method yields a bounded solution whenever the exact solution is bounded. *Stability* in the context of iterative methods ensures that the solution does not diverge.

The property of *stability* can be rather difficult to investigate. The problem is further exacerbated when boundary conditions and non-linearities are present. For this reason, the *stability* aspect of a numerical method is commonly investigated with constant coefficients without boundary conditions. The results obtained in this way can often be applied to more complicated problems, albeit with some notable exceptions. However, when solving complex, non-linear, and coupled equations with complex boundary conditions, there are few stability studies we can employ. In this circumstance we have to rely on experience and intuition to ensure *stability* of the numerical procedure. A number of solution schemes require that the time step is set below a certain limit or promote under-relaxation in the system of algebraic equations. We shall discuss these issues and provide guidelines for the appropriate selection of time step size and suitable under-relaxation factors in Section 6.4.3.

For linear problems, the two most common methods of stability analysis are the matrix method and the von Neumann method. Both methods are based on predicting whether there will be a growth of error between the true solution of the numerical method and the actual computed solution, which also includes the round-off contamination. A worked example will be demonstrated on the application of the von Neumann stability method on a convection-type equation. Other related stability issues will also be demonstrated by additional numerical examples below.

EXAMPLES

Example 6.3

Consider the one-dimensional convection-type equation: $\frac{\partial \phi}{\partial t} + u \frac{\partial \phi}{\partial x} = 0$. Demonstrate the use of the von Neumann stability method to analyze the stability properties of the linear partial differential equation.

Solution: The exact form of the stability criterion depends on the particular differencing approximation applied to the equation. Using the finite difference method, let us approximate the time and spatial derivatives with forward and central differences, where the discretized form of the convection-type equation becomes

$$\frac{\phi_i^{n+1} - \phi_i^n}{\Delta t} + u\frac{\phi_{i+1}^n - \phi_{i-1}^n}{2\Delta x} = 0 \qquad (6.3A)$$

where u is the velocity. To analyze the stability of the above equation, consider the errors introduced at every grid point as

$$\xi_i^n = \phi_i^n - {}^*\phi_i^n \qquad (6.3B)$$

where ϕ_i^n is the true solution of the numerical method and ${}^*\phi_i^n$ is the actually computed solution. For the discretized equation, we are actually calculating

$$\frac{{}^*\phi_i^{n+1} - {}^*\phi_i^n}{\Delta t} + u\frac{{}^*\phi_{i+1}^n - {}^*\phi_{i-1}^n}{2\Delta x} = 0$$

Substituting Eq. (6.3B) into the above, followed by the application of Eq. (6.3A), yields

$$\frac{\xi_i^{n+1} - \xi_i^n}{\Delta t} + u\frac{\xi_{i+1}^n - \xi_{i-1}^n}{2\Delta x} = 0 \qquad (6.3C)$$

For a linear computational algorithm, the error ξ_i^n in the majority of textbooks (Fletcher, 1991; Anderson, 1995) deals with just one term of the finite complex Fourier series, which is given as

$$\xi_i^n = e^{at}e^{ik_m x} \qquad (6.3D)$$

where a is a constant and k_m is the wave number. Substituting the above into Eq. (6.3C), we obtain

$$\frac{e^{a(t+\Delta t)}e^{ik_m x} - e^{at}e^{ik_m x}}{\Delta t} + u\frac{e^{at}e^{ik_m(x+\Delta x)} - e^{at}e^{ik_m(x-\Delta x)}}{2\Delta x} = 0$$

After some arithmetic manipulation and applying trigonometric identities, the above equation reduces to

$$e^{a\Delta t} = 1 - iC\sin(k_m\Delta x)$$

where $C = u\Delta t/\Delta x$.

The modulus of the amplification factor, $\left|e^{a\Delta t}\right|$, is

$$\left|e^{a\Delta t}\right| = \sqrt{1 + C^2\sin^2(k_m\Delta x)}$$

For any non-zero value of C would produce $\left|e^{a\Delta t}\right| > 1$, and therefore the method is *unconditionally unstable*.

Alternatively, let us replace the time variable ϕ_i^n in Eq. (6.3A) as an average value between grid points $i + 1$ and $i - 1$, that is,

$$\phi_i^n = \frac{1}{2}\left(\phi_{i-1}^n + \phi_{i+1}^n\right)$$

Substituting the above into Eq. (6.3A), the discretized form becomes

$$\frac{\phi_i^{n+1} - \frac{1}{2}\left(\phi_{i-1}^n + \phi_{i+1}^n\right)}{\Delta t} + u\frac{\phi_{i+1}^n - \phi_{i-1}^n}{2\Delta x} = 0 \qquad (6.3E)$$

We get a similar error equation in the form of Eq. (6.3C) as

$$\xi_i^{n+1} = \frac{\xi_{i-1}^n + \xi_{i+1}^n}{2} - C\frac{\xi_{i+1}^n - \xi_{i-1}^n}{2}$$

By substituting Eq. (6.3D) into the above and after some arithmetic manipulation, the modulus of the amplification factor becomes

$$\left|e^{a\Delta t}\right| = \sqrt{1 + (C^2 - 1)\sin^2(k_m\Delta x)}$$

Here, the von Neumann stability requirement of $\left|e^{a\Delta t}\right| \leq 1$ is met as long as the parameter $C \leq 1$.

Discussion: The von Neumann stability analysis performed on a simple linear equation provided some fundamental insights into the application of various differencing schemes to achieve stability. The forward differencing in time employed in Eq. (6.3A) fails to satisfy the stability requirement of $\left|e^{a\Delta t}\right| \leq 1$. However, by cleverly replacing the time derivative with a first-order difference where the variable $\phi(t)$ is represented by an average value between neighboring grid points as illustrated in Eq. (6.3E), the stability requirement of $\left|e^{a\Delta t}\right| \leq 1$ can be met for $C \leq 1$. The differencing used to represent the time derivative is called the *Lax method*, named after the mathematician Peter Lax who first proposed it. The recurring parameter C in this example is commonly called the Courant number. It means that $\Delta t \leq \Delta x/u$ for the numerical solution to be stable. Moreover, it is also commonly called the *Courant-Friedrichs-Lewy* condition, generally written as the CFL condition. It is an important stability criterion for convection-type equations.

Example 6.4

Consider again the one-dimensional transient diffusion equation as described in Example 6.2. By applying the finite difference discretization to the equation, discuss the stability of the numerical solution using the explicit Euler method at two different time step sizes of $\Delta t = 1/100,000$ and $\Delta t = 1/1000$ with a fixed grid step size of $\Delta x = 1/100$. Approximate the time and spatial derivatives according to the first-order forward and second-order central differences. The initial condition is set according to $\phi\,(t = 0, x) = 1 - x + \sin(2\pi x)$, where $0 \leq x \leq 1$. The Dirichlet boundary conditions are respectively $\phi\,(t, x = 0) = 1$ and $\phi\,(t, x = 1) = 0$. The thermal diffusivity α is also assumed constant with a value of 0.5.

Solution: The finite difference discretized form of the diffusion Eq. (6.2A) using the explicit Euler method in this example can be expressed as:

$$\phi_i^{n+1} = \phi_i^n + \alpha\frac{\Delta t}{\Delta x^2}\left(\phi_{i+1}^n - 2\phi_i^n + \phi_{i-1}^n\right)$$

The transient results for ϕ are:

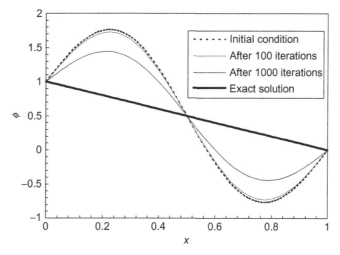

Fig. 6.4.1 Time-advancement with $\Delta t = 1/100{,}000$ and $\Delta x = 1/100$.

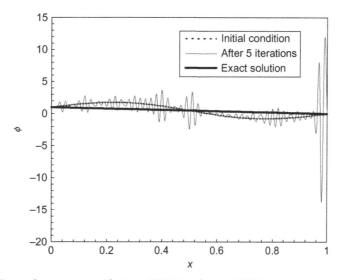

Fig. 6.4.2 Time-advancement with $\Delta t = 1/1000$ and $\Delta x = 1/100$.

Discussion: The sensitivity of the time step Δt to the time-advancement procedure is demonstrated for the explicit Euler method of a diffusion-type equation. For this particular example, the condition for stability is given by $\Delta t \leq \Delta x^2$. In Fig. 6.4.1, where the time step Δt, which is much smaller than the grid size Δx, the numerical procedure marches in a stable fashion and shows signs of convergence tendencies. After 1000 iterations, the intermediate result is gradually approaching the exact solution profile. Nevertheless, for larger time step such as that employed for the case in Fig. 6.4.2 (note the difference in scale along the vertical axis for ϕ), when the time step Δt is much greater than the grid size Δx, the numerical procedure exhibits strong signs of instability. This is evidenced even after 5 iterations.

Example 6.5

Consider again the convection-type equation as described in Example 6.3. Using the finite difference method to discretize the equation, discuss the stability of the numerical solution using the explicit Euler method with a fixed time step size of $\Delta t = 1$ at two different grid step sizes of $\Delta x = 1$ and $\Delta x = \frac{1}{2}$. Approximate the time and spatial derivatives according to the first-order forward and backward differences. The initial condition is set according to a Gaussian profile (similar to the initial value proposed in Hu et al., 1996):

$$\phi(t = 0, x) = \exp\left[-\ln(2)\left(\frac{x - 50}{3}\right)^2\right]$$

with the Dirichlet boundary condition of $\phi\ (t,\ x = 0) = 0$.

Solution: The finite difference discretization of the convection-type equation using the explicit Euler method is given by:

$$\phi_i^{n+1} = \phi_i^n - u\frac{\Delta t}{\Delta x}\left(\phi_i^n - \phi_{i-1}^n\right)$$

For stability, the CFL number must be less or equal to unity, that is, $C \leq 1$. Assuming a constant CFL number of unity, $\Delta t \leq \Delta x/u$. If the velocity u is taken to be 1 m/s, the approximation is stable when $\Delta t \leq \Delta x$. The transient results for ϕ are:

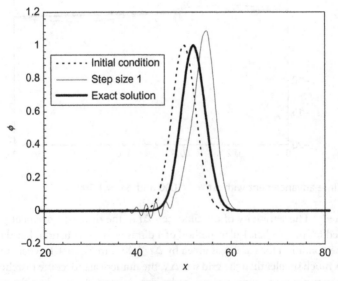

Fig. 6.5.1 Time-advancement with $\Delta t = 1$ and $\Delta x = 1$.

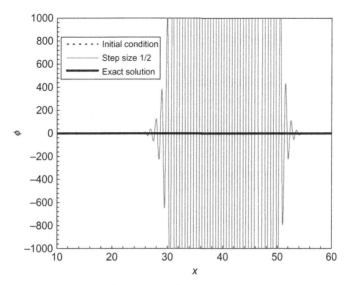

Fig. 6.5.2 Time-advancement with $\Delta t = 1$ and $\Delta x = 1/2$.

Discussion: The sensitivity of the grid size Δx to the time-advancement procedure, while maintaining a fixed time step Δt, is demonstrated for the explicit Euler method for a convection-type equation. This example illustrates the influence of the important CFL number on the stability of the numerically explicit marching procedure. For the case in Fig. 6.5.1, where the condition is $\Delta t = \Delta x$, the numerical procedure is relatively stable, though some unfavorable wiggles are attained during the numerical computations. The removal of these unwanted wiggles can be overcome by marching with smaller time step sizes. However, when $\Delta t > \Delta x$ as for the case depicted in Fig. 6.5.2 (note again the difference in scale along the vertical axis for ϕ), the time-advancing numerical procedure is evidently unstable. The case that is solved in Fig. 6.5.2 clearly accentuates the violation of the CFL number for explicit time-marching methods.

We have thus far discussed *stability* by predominantly focusing on explicit types of numerical procedures in the worked examples above. Explicit-type procedures can be considered *conditionally stable* since they are strongly influenced by the temporal resolution. For implicit-type procedures, they are usually *unconditionally stable*. This is because allowance is provided for the variable to be continuously updated within the time step instead of calculating from the previous time step values. The majority of commercial CFD codes employ implicit-type procedures due to the inherent stability they possess. Nevertheless, instability that arises in these codes does not depend on the temporal resolution but rather the adoption of the segregated approach where calculations of the transport variables are performed sequentially in the iterative process. In order to ensure *convergence*, the use of under-relaxation factors can assist in promoting the *stability* of the segregated iterative computations. These will be further discussed and explored in the next section.

6.4 CONVERGENCE

6.4.1 What Is Convergence?

If a numerical method can satisfy the two important properties of *consistency* and *stability*, we generally find that the numerical procedure is convergent. *Convergence* of a numerical process can therefore be stated as the solution of the system of algebraic equations approaching the true solution of the partial differential equations having the same initial and boundary conditions as the refined grid system (*grid convergence*). For initial value (marching) problems governed by the finite difference approximations of linear partial differential equations, Lax's equivalence theorem is given here without proof. It states that: "Given a properly posed linear initial valued problem and a finite difference approximation *consistency* and *stability* are necessary and sufficient conditions that need to be satisfied for *convergence*," that is, *consistency* + *stability* = *convergence*. We might add that most computational work for nonlinear partial differential equations, as used in CFD, proceeds as though this theorem applies, although it has not been proven directly for this general category of equations.

In the majority of commercial CFD codes the system of algebraic equations is usually solved iteratively. When dealing with these codes, there are three important aspects to abide by for *iterative convergence*. Firstly, all the discretized equations (momentum, energy, etc.) are deemed to be converged when they reach a specified tolerance at every nodal location. Secondly, the numerical solution no longer changes with additional iterations. Thirdly, overall mass, momentum, energy, and scalar balances are obtained. During the numerical procedure, the imbalances (errors) of the discretized equations are monitored and these defects are commonly referred to as the *residuals* of the system of algebraic equations; that is, they measure the extent of imbalances arising from these equations and terminate the numerical process when a specified tolerance is reached. For satisfactory convergence, the *residuals* should diminish as the numerical process progresses. In the likelihood that the imbalances grow, as reflected by increasing residual values, the numerical solution is thus classified as being unstable (divergent). It is noted that *iterative convergence* is not the same as *grid convergence*. *Grid convergence* seeks a grid-independent solution, which means approaching the exact solution. We will further discuss this later. Additionally, in Section 6.4.2, the concepts of *residuals* and *convergence tolerance* are discussed in the context of attaining a numerical solution.

EXAMPLE

Example 6.6

Based on the explicit Euler method as previously described in Example 6.4 for the one-dimensional transient diffusion equation with a time step size of $\Delta t = 1/100,000$ and a grid step size of $\Delta x = 1/100$, discuss the aspect of convergence for the numerical solution attained with identical initial and boundary conditions (Fig. 6.6.1). The thermal diffusivity α is also assumed constant with a value of 0.5.

Solution: The computational results showing the transient development for the variable ϕ are given by:

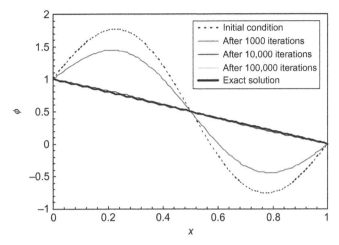

Fig. 6.6.1 Transient development of the variable ϕ with $\Delta t = 1/100,000$ and $\Delta x = 1/100$.

Discussion: The purpose of this simple example is to illustrate the condition of *consistency + stability = convergence*. The first aspect concerns *stability*; it is observed that no signs of instability are experienced during the course of the numerical calculations. The second aspect concerns *convergence*; after 100,000 cycles, the numerical result converges and collapses to the exact solution profile. Since the difference between the discretized equation and the exact one is negligible and thus the remainder or truncation error diminishes, *consistency* prevails.

6.4.2 Residuals and Convergence Tolerance

For any transport variable ϕ, the discretized form of the partial differential can be specifically written as:

$$a_P\phi_p = \sum a_{nb}\phi_{nb} + b_P \tag{6.1}$$

In Eq. (6.1) the central coefficient a_P and neighboring coefficients a_{nb} normally depend on the solution of other flow field variables including the time- and spatial-varying fluid flow properties. These coefficients are updated consecutively during the iterative procedure. At the start of each iteration step, the equality in Eq. (6.1) will not hold. We can therefore rewrite the above equation by introducing an imbalance variable called residual R_p, where Eq. (6.1) can be re-expressed as

$$R_P = \sum a_{nb}\phi_{nb} + b_P - a_P\phi_p \tag{6.2}$$

From the above equation, we introduce the concept of *residuals* as applied for each discretized equation of the system of transport equations. For a well-posed formulation, the *residuals* become *negligible* with increasing iterations. In CFD *residuals* are employed to monitor the behavior of the numerical process. Importantly, they implicate whether the solution shows a trend of convergence or divergence. It is noted that the concept of *mass residual* introduced in Section 5.5 and used in Example 5.5 is different from the concept of *residuals* defined herein. The former is a source term appearing in the pressure correction Eq. (4.80), while the latter pertains to the imbalances in Eq. (6.2) during the iterative procedure.

The residual that arises in Eq. (6.2) actually depicts the imbalance (error) at the nodal point *P* for one cell volume. For practical purposes, a *global* residual *R*, taken as the sum of each *local* residual R_p over all the grid nodal points, is monitored:

$$R = \sum_{grid\ points} |R_P| \tag{6.3}$$

Convergence is deemed to be achieved for the discretized Eq. (6.1) so long as the global residual *R* satisfies a specified tolerance, that is, $R \leq \varepsilon$ or $\sum_{grid\ points} |R_P| \leq \varepsilon$. The variable ε is usually referred to as the *convergence tolerance* for the system of algebraic equations. There is some practical guidance in selecting appropriate values for the *convergence tolerance*. It is noted that specifying appreciably small tolerance values will incur a large number of iteration steps in reaching convergence. On the other hand, large tolerance values constitute an early termination of the iteration process for which the numerical solution of the algebraic equations is considered to be rather coarse or not sufficiently converged. By default, the monitored residuals are usually scaled. Generally, a decrease of the residual by three orders of magnitude during the iteration process indicates at least *qualitative convergence*. Here, the major flow features are considered to be sufficiently established. Nevertheless, stricter convergence consideration is required for transport variables like energy and scalar species. It is recommended that the scaled energy residual decreases to a recommended convergence tolerance of 10^{-6} while the scaled scalar species may need to only decrease to a convergence tolerance of 10^{-5} to achieve energy and species balance, respectively. For *quantitative convergence*, changes are monitored for all considered flow field variables. During the monitoring of these residuals, the reader is also advised to ensure that property conservation is satisfied.

EXAMPLE

Example 6.7
Consider the two-dimensional CFD case of the incompressible laminar flow between two stationary parallel plates with the dimensions of height $H = 0.1$ m and length $L = 0.5$ m. Demonstrate the convergence behavior through monitoring the residuals of the transport

variables u and v in their algebraic form represented in Eq. (6.2) as well as the pressure correction p' from Eq. (4.80) where air (density $\rho = 1.2$ kg/m^3 and viscosity 4×10^{-5} kg/m · s) is the working fluid. The inlet velocity is fixed at $u_{in} = 0.01$ m/s. The outlet and wall conditions remain the same as applied in Chapter 2.

Solution: The schematic diagram of the channel flow is identical to the one that is used in Example 4.5. Here, the discretized equations governing the momentum and pressure correction are solved using default iterative solvers as provided by an in-house research CFD code.

Convergence histories of the velocities u and v and pressure correction p' are depicted in Fig. 6.7.1 below. The residuals for each of the transport variables are not scaled and the convergence tolerance ε has been set at 10^{-7} to terminate the numerical simulation.

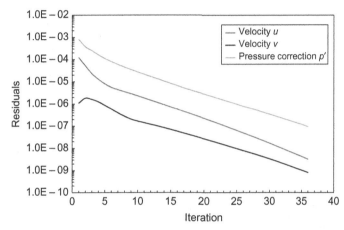

Fig. 6.7.1 Convergence histories of the horizontal velocity u, vertical velocity v, and pressure correction p'.

Discussion: In the previous Example 6.6 convergence is demonstrated and ascertained for an *explicit* methodology. Here, convergence is illustrated for an *implicit* methodology. The well-posed behavior of the numerical solution is evident from the diminishing values for the residuals of the velocities u and v and pressure correction p'. Convergence is achieved, thereby satisfying the condition of Eq. (6.3) when all the residuals fall below the convergence tolerance of $\varepsilon = 1 \times 10^{-7}$. Nevertheless, there are circumstances where divergence can occur during the simulation of a channel flow problem. Such behavior is typically exhibited by the ascending trend of the residuals leading to large catastrophic values during the iterative process, as shown in Fig. 6.7.2. This ill-posed (opposite to well-posed) numerical solution that is particularly designed for the purpose of illustration is due to the incorrect usage of the under-relaxation factors to control the numerical calculations. More discussions on the importance of under-relaxation factors generally required for implicit methods are elaborated in the next section.

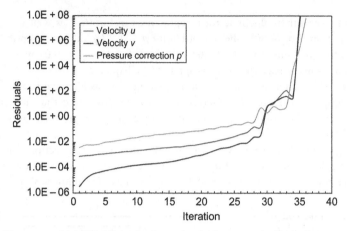

Fig. 6.7.2 Divergence histories of the horizontal velocity u, vertical velocity v, and pressure correction p'.

6.4.3 Convergence Difficulty and Using Under-Relaxation

In consideration of the channel flow example discussed in previous chapters and herein, the residuals for the continuity (pressure correction) and velocity components represent useful indicators that can be progressively evaluated to ascertain the convergence trend of the numerical calculations. Whether these residuals are being tracked locally at some grid nodal point within the flow system or globally through the sum of the local residuals, a convergence trend showing a diminishing residual value ensures the satisfaction of the conservation laws of mass and momentum (see above Example 6.7). Since these laws are physical statements of fluid dynamics, the removal of any imbalances is imperative. In some flow cases the local solution may be known and the converged solution can be gauged and assessed by directly comparing the computed results against the available solution values. There are also practical considerations during the numerical calculations where convergence can be assessed through the evaluation of some physical variables. For example, the calculation of the drag or lift coefficients for a flow of air over an aerofoil are useful indicators for determining convergence of the flow system and thus determining when to terminate the numerical procedure.

In any numerical calculation numerical instabilities can occur while solving the discretized equations. Poorly constructed meshes, improper solver settings, non-physical boundary conditions, and selection of inappropriate models are typically some of the factors that may cause the ill convergence of the numerical calculations. These are usually exhibited and amplified by the increasing (diverging) or "plateau" residual values throughout the iteration process (as seen above in Example 6.7). Diverging residuals clearly imply the increase of imbalances in the conservation equations and thus the physical laws of the fluid flow are vigorously violated. Computational results that are not converged are misleading.

There are, however, some practical corrective steps that can be undertaken to overcome the difficulties in achieving convergence. For a poor-quality mesh, the flow region can be re-meshed by increasing the number of grid nodal points. Such a strategy is particularly adopted to prevent the grid from having large aspect ratios or highly skewed cells. In cases where high-order approximations are required, it may be sensible to initially compute the solution using low-order approximations. The diffusive nature of the low-order approximations allows large imbalances to dissipate quickly and promote stability of the numerical procedure. Once the flow is established, greater accuracy of the numerical solution can be attained by switching to the high-order approximations.

Another strategy to promote convergence is through the use of *under-relaxation factors*. Poor initial guesses or unresolved steep gradients in the flow field may cause divergence in the iterative process. The incorporation of under-relaxation factors into the system of algebraic equations can significantly moderate the iteration process by limiting the change in each of the transport variables from one step to the next. By introducing an under-relaxation factor β, the change in the value of the transport variable ϕ at the central node of the cell volume between subsequent iteration steps can be expressed as:

$$\phi_P^{New} = \phi_P^{Old} + \beta\left(\phi_p^{New} - \phi_p^{Old}\right) \tag{6.4}$$

In CFD under-relaxation is often introduced to stabilize the numerical calculations of the governing equations that are generally non-linear and where the equation of one transport variable is dependent on the others, for example, temperature affecting the velocities in buoyancy flows. The under-relaxation factor β in Eq. (6.4) controls the advancement of the transport variable ϕ_P during the iteration process. For a specified under-relaxation factor of 0.5, a restriction of 50% change is implied for ϕ_P from the value determined at the previous iteration to the current iteration step. The advancement for ϕ_P through the iteration process will therefore be increasingly impeded as smaller under-relaxation factors are employed. For the numerical solution in Example 6.7, the advancement of pressure has been under-relaxed by a factor of 0.3, which is typically employed when applying the SIMPLE scheme (Ferziger and Perić, 1999), to ensure stability and convergence of the iterative process.

In the majority of commercial CFD codes the default settings of the under-relaxation factors are generally applicable to a wide range of problems. It is usually recommended to employ default factors in beginning the numerical calculations. Nevertheless, a more aggressive approach may be warranted by tightening the transport variable advancement through smaller under-relaxation factors to aid convergence. CFD users still constantly face many challenges in ascertaining optimal under-relaxation factors, usually not known *a priori*, to solve CFD problems. Settings of appropriate under-relaxation factors remain best learned from practical experience and application of CFD methodologies.

6.4.4 Accelerating Convergence

Some practical guidelines to attain quicker convergence of the computational solution are explored. In the majority of cases handling flow problems the supply of good initial or starting conditions is important, which leads to beneficial consequences in the iteration process. This can be achieved either by knowledge of similar physical conditions that can be imposed or by beginning the iteration from a previously converged solution. Inappropriate initial conditions generally lead to slower convergence but may result in some untenable situations that promote divergence tendencies. Good initial or starting conditions promote computational efficiency and reduce computational efforts and resources. In the previous section the use of under-relaxation factors was discussed to promote stability during the iteration process. Default settings of the under-relaxation factors or Courant number in commercial CFD codes are generally applicable to a wide range of flow problems. Nevertheless, there are special circumstances where depending on the flow problem, the under-relaxation factors or Courant number can be plausibly increased to accelerate the convergence. However, excessively high under-relaxation values can lead to unwanted instabilities. The reader may wish to adopt a strategy whereby to store intermediate solutions through incrementally increasing the under-relaxation factors. This can be carried out periodically before carrying out subsequent calculations. Such a procedure allows computations to be kept at a minimum but more importantly eliminates the need to re-compute the problem from the initial state. In Chapter 4 a number of accelerating techniques that are gaining prominence in solving CFD problems were described. In many commercial CFD codes multigrid solvers – a procedure to solve the algebraic equations by employing a combination of iterative solvers such as Jacobi, Gauss–Seidel or SIP, and direct solver cycling through different levels of grid densities (see Sections 5.5 and 9.2.4 for more detail description) – are offered as the default solvers in accelerating the convergence for the iteration process. In this case provisions are given for users to change, at their discretion, the settings of the multigrid solver. More often than not, the default settings provided are sufficiently robust and they do not necessarily need to be altered.

6.5 ACCURACY

The previous discussion of *convergence*, *consistency*, and *stability* has been primarily concerned with the solution behavior where the finite quantities, such as the time step Δt and mesh spacing Δx, Δy, and Δz, diminish. Since the discretized forms of the transport equations governing the flow and energy transfer are always solved numerically on a finite grid layout and the effects of turbulence are generally modeled through approximate theories, the solution obtained is always approximate. The corresponding issue of *accuracy* therefore becomes another important consideration.

In Section 6.2 the determination of consistency produces an explicit expression for the truncation error. As aforementioned, the truncation error represents the difference between the discretized equation and the exact one and it provides a means of evaluating the *accuracy* of the solution for the partial differential equations. The order of the truncation error coincides with the order of the solution error if the grid spacings are sufficiently small and if the initial and auxiliary boundary conditions are sufficiently smooth. It is commonly implied that an improvement in accuracy (from the truncation error) of high-order approximations can be achieved for a sufficiently fine grid. Refining the grid will often produce a superior accuracy for high-order approximations over low-order approximations. However, at an absolute accuracy level, justification for more expensive computations may not demonstrate the desired superior accuracy due to limited computing capacity.

One method where accuracy can be assessed for a particular algorithm on a finite grid is to apply it to a related but simplified problem that possesses an exact solution. However, accuracy is usually problem dependent; an algorithm that is accurate for one model problem may not necessarily be as accurate as for another more complicated problem. Another probable way for assessing accuracy is to obtain solutions on successively refined grids (*grid convergence*) and to check that, with successive refinements, the solution is not changing satisfying some predetermined accuracy. This technique assumes that the approximate solutions will converge to the exact solution as the finite quantities diminish and then the approximate solution on the finest grid can be used in place of the exact solution; grid independence solution is thus achieved. Assuming that the *accuracy* of this approximate solution can be assessed, it is important to consider the related question of how accuracy may be improved. At a specific level, the use of high-order approximation or grid refinement would be expected to produce more accurate solutions. Nevertheless, such choices are only meaningful if they are considered in conjunction with execution time and computational efficiency.

It is important for the reader to be aware that *a converged solution does not necessarily mean an accurate solution*. Some possible sources of solution errors resulting from the numerical calculations of the algebraic equations require analysis and this will be discussed in the next section. If these errors are to be minimized, some systematic steps to perform numerical analysis such as grid independence and verification and validation of numerical models are necessary.

6.5.1 Source of Solution Errors

Not only should the reader be aware of the existence of errors in *computational solutions*, but more importantly, the reader must attempt to distinguish one from another. This section serves to address the possible sources of errors that the reader is likely to encounter applying CFD methodologies. Errors are introduced because the numerical solutions of the fluid flow and heat transfer problems are only *approximate solutions*. Some prevalent sources of errors that occur in numerical solutions include the following classifications:

- *Discretization error*
- *Round-off error*

- *Iteration or convergence error*
- *Physical modeling error*
- *Human error*

Before we elaborate on the source of these errors in CFD, we would like to establish a clear and logical distinction between *error* and *uncertainty* that is based on the AIAA publication guide for the Verification and Validation of Computational Fluid Dynamics Simulations (AIAA, 1998). Error can be defined as *a recognizable deficiency that is not due to lack of knowledge* while uncertainty can be *defined as a potential deficiency that is due to lack of knowledge*. Although these definitions appear to be rather philosophical, they will become clearer as the origin of these errors in CFD is further explored below.

Discretization Error

These errors are due to the difference between the exact solution of the modeled equations and a numerical solution with a limited time and space resolution. They arise because an exact solution to the equation being solved is not obtained but numerically approximated. For a consistent discretization of the algebraic equations, the computed results are expected to become closer to the exact solution of the modeled equations as the number of grid cells is increased. However, the results are strongly affected by the density of the mesh and distribution of the grid nodal points.

We identify two types of discretization errors: local and global (or accumulated). To have an idea about the *local error* and *global error*, consider the finite difference formulation of the derivatives for the transport variable ϕ in space and time at a specified grid nodal point expressed through the Taylor series expansion,

$$\left(\frac{\partial \phi}{\partial x}\right) = \frac{\phi_{i+1,j} - \phi_{i,j}}{\Delta x} + \underbrace{O(\Delta x)}_{\text{Truncation error}} \quad \text{Spatial derivative} \qquad (6.5)$$

$$\left(\frac{\partial \phi}{\partial t}\right) = \frac{\phi_{i,j}^{n+1} - \phi_{i,j}^{n}}{\Delta t} + \underbrace{O(\Delta t)}_{\text{Truncateion error}} \quad \text{Time derivative} \qquad (6.6)$$

Termination of the Taylor series expansion in Eqs (6.5) and (6.6) results in the so-called truncation errors involved in the approximation.

The *local error* is the formulation associated with a single step and provides an idea about the accuracy of the method used. For this error, the accuracy of the numerical solution concerns mainly the approximation of the spatial derivative. The solution accuracy for a transient problem, however, focuses on the advancement of the transport variable ϕ through time usually characterized by the *global error*. The representation of the *local error* and *global error* is best illustrated in Fig. 6.1. We can observe that the smaller the mesh size or time step in transient problems, the smaller the error, and thus the more accurate the approximation.

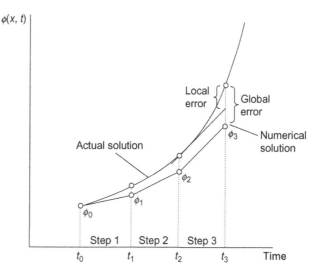

Fig. 6.1 The local and global discretization errors resulting from the finite difference method at a specified grid nodal point.

We shall illustrate the significance of the discretization error through a worked example below.

Example

Example 6.8

Consider the transient one-dimensional convection-type equation to further illustrate the aspect of discretization error. A fourth-order central difference is employed for the spatial derivative to attain higher accuracy. Using the Euler explicit method, demonstrate the discretization error that is associated with the numerical solution obtained through the first- and second-order approximations to the time derivative at a fixed time step size of $\Delta t = 1/128$ accompanied by the variation of two different grid step sizes of $\Delta x = 1$ and $\Delta x = \frac{1}{2}$.

Solution: Through the consideration of additional grid nodal points along the spatial direction x and applying the Taylor series expansion, the fourth-order finite difference approximation can be obtained as

$$\left(\frac{\partial \phi}{\partial x}\right) = \frac{-\phi_{i+2} + 8\phi_{i+1} - 8\phi_{i-1} + \phi_{i-2}}{12\Delta x} + \underbrace{O\left(\Delta x^4\right)}_{\text{Truncation error}}$$

Substituting the above approximation along with the first-order forward difference to the time derivative yields the following algebraic equation:

$$\frac{\phi_i^{n+1} - \phi_i^n}{\Delta t} - u\left(\frac{-\phi_{i+2} + 8\phi_{i+1} - 8\phi_{i-1} + \phi_{i-2}}{12\Delta x}\right) = 0$$

For the second-order approximation of the time derivative, the central difference is employed; in other words

$$\frac{\phi_i^{n+1} - \phi_i^{n-1}}{2\Delta t} - u\left(\frac{-\phi_{i+2} + 8\phi_{i+1} - 8\phi_{i-1} + \phi_{i-2}}{12\Delta x}\right) = 0$$

This newly developed formula is similar to the well-known *Leap Frog* method.

The computed results compared with the exact solution for the first-order and second-order time approximations against two different grid sizes are illustrated in Figs. 6.8.1 and 6.8.2.

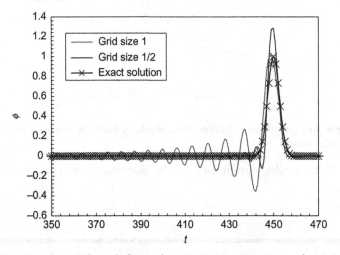

Fig. 6.8.1 Fourth-order grid and first-order time approximations of convection-type equation for the variable ϕ with a fixed time step of $\Delta t = 1/128$ and two grid sizes of $\Delta x = 1$ and $\Delta x = 1/2$.

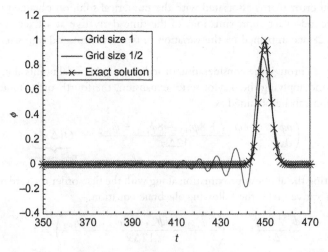

Fig. 6.8.2 Fourth-order grid and second-order time approximations of convection-type equation for the variable ϕ with a fixed time step of $\Delta t = 1/128$ and two grid sizes of $\Delta x = 1$ and $\Delta x = 1/2$.

Discussion: By increasing the approximation of the time derivative from first order to second order, the numerical simulation for the higher order time approximation at a grid size of $\Delta x = 1$ is shown to be less sensitive to the time step changes with the suppression of wiggles through time (compare the solutions between Figs. 6.8.1 and 6.8.2). By halving the grid size to $\Delta x = 1/2$, the numerical simulation is stabilized. The solution for the second-order time derivatives converges to the exact solution profile but the first-order solution retains some oscillatory wiggles around the exact solution profile and is still far from its converged state. In this example the systematic reduction of the grid step sizes and the use of higher approximation for the time derivative characterize the diminishing contribution of the respective local and global discretization errors.

Round-Off Error

These errors exist due to the difference between the machine accuracy of a computer and the true value of a variable. Every computer represents numbers that have a finite number of significant figures. The default value of the number of significant digits for many computers is 7 and this is commonly referred to as *single precision*. However, calculations can also be performed using 15 significant figures, which is referred to as *double precision*. The error due to the retaining of a limited number of computer digits available for storage of a given physical value is therefore called the round-off error. This error is naturally random and there is no easy way of predicting it. It depends on the number of calculations, rounding-off method, rounding-off type, and even sequence of calculations.

Consider the case of a simple arithmetic operation performed with a computer in single precision. Given that $a = 8888888$, $b = -8888887$, and $c = 0.3333341$, let us evaluate the operations of $D = a + b + c$ and $E = a + c + b$. The arithmetic calculation for D proceeds as:

$$D = 8888888 - 8888887 + 0.3333341$$
$$= 1 + 0.3333341$$
$$= 1.333334 (\text{Correct result})$$

while E performs the following operations:

$$E = 8888888 + 0.3333341 - 8888887$$
$$= 8888888 - 8888887$$
$$= 1.000000 (\text{In error by } 25\%)$$

In algebra we learned that $a + b + c = a + c + b$, which is reasonably accepted as a mathematically proven statement. But this is not necessarily true for the calculations that will be performed on a computer, as demonstrated above. It is noted that the sequence of calculations in single precision mode results in a solution error of 25% in just two operations. Imagine that thousands or even millions of such operations are to be performed sequentially; the rounding-off error would accumulate and lead to serious error without any prior warning signs. If computer round-off errors are suspected of being significant, one test that can be performed is to employ double precision or on a computer known to store floating point numbers at a higher precision. An attempt to continuously refine a coarse grid solution to achieve a solution of diminishing finite quantities may be feasible; however, this may not usually be possible for more complex algorithms. Nevertheless, it is noted that round-off errors are usually not the dominant contributor to the source of solution errors when compared to discretization errors. The implication of round-off errors in a CFD problem will be expounded in the test case investigated in Section 6.7.1.

Iteration or Convergence Error

These errors occur due to the difference between a fully converged solution of a finite number of grid points and a solution that has not fully achieved convergence. The majority of commercial CFD codes solve the discretized equations iteratively for steady-state solution methodologies. For procedures requiring an accurate intermediate solution at a given time step, this is solved iteratively in transient methods. It is expected that progressively better estimates of the solution are generated as the iteration step proceeds and ideally satisfies the imposed boundary conditions and equations in each local grid cell and globally over the whole domain. However, if the iterative process is terminated prematurely, then errors arise. Convergence errors therefore can occur because of either being impatient to allow the solution algorithm to complete its progress to the final converged solution or applying too large convergence tolerances to halt the iteration process when the CFD solution may still be considerably far from its converged state.

Physical Modeling Error

These errors are due to uncertainty in the formulation of the mathematical models and deliberate simplifications of the models. Here, we reinforce the definition of uncertainty from above, where the Navier-Stokes equations can be considered to be exact and solving them is impossible for most flows of engineering interest because of lack of sufficient knowledge to model them. The sources of uncertainty in physical models are that:

(1) the phenomenon is not thoroughly understood,
(2) parameters employed in the model are known to possess some degree of uncertainty,
(3) appropriate models are simplified and thus uncertainty is introduced, and
(4) experimental confirmation of the models is not possible or is incomplete.

Modeling is often required for turbulence, which places huge demands on computational resources if it is to be simulated directly. Other phenomena like combustion, multi-phase flow, chemical processes, etc., are difficult to describe exactly and they inevitably require the introduction of approximate models. Even Newton's and Fourier's laws are themselves models, though they are solidly based on experimental observations for many fluids. In addition, the underlying mathematical model is nearly exact but some fluid properties may not be exactly known. They may depend strongly on temperature, species concentration, and possibly, pressure; this dependence is ignored, thus introducing modeling errors (e.g. the use of Boussinesq approximation for natural convection). In some situations a simplified model may be adopted within the CFD code for the convenience of a more efficient computation even though a physical process is known to a high level of accuracy. Physical modeling errors are examined by performing validation studies that focus on certain models. The conceptual idea and definition of validation will be expounded in Section 6.5.3.

Human Error

There are essentially two categories of errors associated with human error. Firstly, computer programming errors involve human mistakes made in programming, which are the direct responsibility of the programmers. These errors can be removed by systematically performing verification studies of sub-programs of the computer code and the entire code, reviewing the details inserted into the code, and performing validation studies of the code. We will review and discuss the concept of verification in Section 6.5.3. Secondly, usage errors are also due to application of the code in a less-than-accurate or improper manner. Inexperience in handling CFD codes may result from either incorrect computational domains (such as improper geometry construction or grid generation) or inappropriate setting of boundary conditions. Selection of bad numerical schemes or computational models to simulate certain flow problems compounds the undesirable usage errors. The reader should note that the potential for usage errors increases with the increased level of options available in CFD. Nevertheless, usage errors can be minimized and controlled through proper training and analysis and the accumulation of experience. Some practical guidelines will be discussed in the next chapter.

6.5.2 Controlling the Solution Errors

It is good engineering practice that the CFD user examines any potential pitfalls when employing numerical methodologies. Numerical errors are primarily concerned with discretization and round-off errors. They have the tendency to accumulate through computational processes that may yield unphysical CFD solutions. The results that seem rather reasonable overall may be in considerable error at certain locations within the flow domain. Controlling the solution errors therefore represents a crucial step toward obtaining reliable and meaningful CFD solutions.

We begin by focusing on the contribution of the discretization and round-off errors obtained through numerical methods. As the mesh or time step size decreases, it is seen in Fig. 6.2 that the discretization error decreases with step size while the round-off error

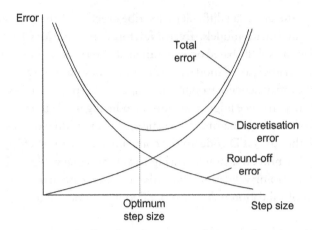

Fig. 6.2 The discretization, round-off, and total errors as a function of mesh and/or time step size.

increases. For the total error that is taken as the sum of the discretization and round-off errors, it is evident that continually decreasing the step size does not necessarily mean that more accurate results are attained. The opposite is true at small step sizes, where less accurate results are obtained because of the quicker increase in the round-off error. In order to contain this error we should therefore avoid a large number of computations with very small numbers.

In practice, we will not be able to determine the magnitude of the error involved in the numerical method. It is seen from above that the knowledge of discretization error alone is meaningless without a true estimate of the round-off error. To better assess the accuracy of the results obtained, some practical guidelines are recommended.

Firstly, the issue of *grid independence* is explored. To address this issue, we can begin by solving the flow problem with reasonable mesh sizes of Δx, Δy, and Δz (and a time step of Δt for transient problem) based on acquired experience. The computational results may look qualitatively good but let us assume that the problem is repeated with twice as many grid points, thus halving the mesh sizes in each direction by $\Delta x/2$, $\Delta y/2$, and $\Delta z/2$. If the results obtained do not differ significantly from the results obtained in the original grid layout, we can conclude that the discretization error is at an acceptable level. But if the values for the transport variables are quite different for this second calculation, then the solution is a function of the number of grid points. In all practical cases the grid needs to be refined by increasing the number of grid points until a solution is achieved where no significant changes in the results occur. This indicates that the discretization error is reduced to an acceptable error and *grid independence* is reached. This issue will be further investigated in a test case described in Section 6.7.1.

Secondly, the majority of CFD calculations are usually performed in single precision to avoid overburdening the computational resources. Nevertheless, if the round-off error is found to be significant, the flow calculations can be repeated by using double precision while holding the mesh sizes (and the size of the time step in transient problems) constant. If the results do not change considerably, we conclude that the round-off error is not a problem for the CFD solution. However, if the changes are larger than expected, we may attempt to reduce the total number of calculations by either increasing the mesh sizes or changing the order of computations such as adopting a higher approximation to evaluate the first-order spatial derivatives of the convective terms and/or the first order time derivatives in the conservation equations. As seen from Fig. 6.2, discretization error increases with increasing mesh size; the reader should therefore acknowledge this important trend and seek some reasonable compromise.

Thirdly, selection of appropriate turbulence models or other approximate models can be a daunting task especially in attempting to minimize physical modeling errors. The desired level of simplification that can be accepted to adequately model the physical flow problem is not straightforward, as it depends on the choice of models that govern and characterize the particular flow physics. Also, carelessness in setting up a feasible geometrical model and an improper choice of boundary conditions are some of the human errors that frequently arise in CFD. Succinct practical guidelines to eliminate such pitfalls will be discussed in the next chapter.

6.5.3 Verification and Validation

In addition to errors, uncertainties can also arise while performing a numerical simulation. These can be due to the improper modeling of physics such as a misunderstanding of the phenomena leading to incorrect assumptions or the incorrect computational design such as making wrong approximations and simplifications about the parameters that govern the fluid dynamics. For a CFD solution to be credible, it requires a detailed analysis to be performed, to quantify the modeling and numerical uncertainties in the simulation. Verification and validation procedures are the means by which a CFD solution can be properly assessed through quantitatively estimating the inherent errors and uncertainties.

Verification and validation have very distinct definitions. Although there is an absence of a universal agreement on the details of these definitions, there is a fairly standard and consistent agreement on their usage. In this book we will adopt definitions that are focused on numerical errors and uncertainties where they cannot be considered negligible or overlooked.

An important aspect to remember about verification and validation is that it is applied in two very distinct ways. We can differentiate verification and validation as follows. *Verification* can be defined *as a process for assessing the numerical simulation uncertainty and, when conditions permit, estimating the sign and magnitude of the numerical simulation error and the uncertainty in that estimated error.* This procedure relates primarily to the input parameters used for geometry, initial conditions, and boundary conditions. They are required to be carefully checked and systematically documented. It is also important that mesh and time step sensitivity studies are extensively performed to bound the errors, whether they may be insufficient spatial discretizations, too large temporal advancement, lack of iterative convergence, or computer programming errors, that are associated with the discrete approximations employed for the partial differential equations. On the other hand, *validation* can be defined *as a process for assessing simulation model uncertainty by using benchmark experimental data and, when conditions permit, estimating the sign and magnitude of the simulation modeling error itself.* This procedure simply means validating the calculations by establishing a range of physical conditions obtained from the calculations and performing comparisons of the results from the CFD code with experiments that span the range of conditions. This represents the final phase of the credibility process of the models applied and is interpreted as the stage to determine the degree to which a model corresponds to an accurate representation of the real physical flow problem that is solved.

We shall further demonstrate the use of the verification and validation procedures in the case studies described in Section 6.7.

6.6 EFFICIENCY

The rapid advancement and reduced costs of computer hardware and resources have revolutionized the use of CFD. Current users have nowadays the luxury of accessing many commercial or shareware computer codes. More complex engineering systems and applications can be solved because of the feasibility of constructing high-quality grids to resolve the physical flow structures. Such flow problems usually require a substantial amount of grid points to envelope the whole physical domain to achieve adequate resolution. If the mesh requires further refinement, almost all iterative solution methods suffer from slower convergence on the finer grids. The rate of convergence depends on the particular numerical method; the number of iterations for many methods is linearly proportional to the number of grid nodal points in one coordinate direction. This behavior is related to the fact that, for iterative procedures, information has to travel back and forth across the domain several times.

To overcome the convergence problem, applications of conjugate gradient and multigrid methods have received unprecedented attention to enhance the *efficiency* of a CFD procedure. The former is essentially a method that seeks the minimum of a function that belongs to the class of *steepest descent methods*. The basic method in itself converges rather slowly and is not very useful but when it is used in conjunction with some *preconditioning* of the original matrix, major enhancements in its speed of convergence have been recorded. This preconditioning technique is achieved by either applying the *incomplete Cholesky* factorization for symmetric matrices or applying *biconjugate gradients* for asymmetric matrices. The latter, the multigrid method that has been discussed in Chapter 5, employs a hierarchy of meshes. In the simplest case the coarse ones are just the subsets of the fine ones. Efficiency may be improved by the decision to switch from one grid to another on the rate of convergence through the V-cycle, W-cycle, a combination of V-W cycles, or other possible strategies. Additional details of the multigrid method can be further referred to in Appendix F. The conjugate gradient and multigrid methods are ideal for solving the Poisson-like pressure or pressure-correction equation such as those found in the SIMPLE method.

Another approach to achieving computational *efficiency* is parallel computing. The increase in the capability of single processor computers has almost reached their peak performances. It now appears that further increases in speed will require multiple processors: parallel computers. One of the advantages of parallel computers over classical vector supercomputers is scalability. Parallel computers employ standard processing chips and are therefore cheaper to produce and to attain. Commercially available parallel computers may have thousands of processors, gigabytes of memory, and computing speed measured in gigaflops. The basic idea of parallel computing involves the sub-division of the solution domain into sub-domains and assigning each sub-domain to one processor. In such a case the same computer code runs all processors. Since each processor needs data that resides in other sub-domains, exchange of data between processors and storage overlap is necessary. This demonstrates that parallel computing environments require redesign of algorithms. Good parallelization therefore needs modification of the solution algorithm.

Explicit marching methods are relatively easy to parallelize since all operators are performed on data from preceding time steps. It is only necessary to ensure that data is exchanged at the interface regions between neighboring sub-domains after each step is completed. The sequence of operations and the results are identical on one and many processors. Of course, the problem of handling the solution of the Poisson-like pressure equation remains. Implicit methods are more difficult to parallelize because the iterative solvers that are efficient in serial computations are not suited when performed in parallel. Some solvers can be parallelized and perform the same sequence of operations on multiprocessors as on a single one. However, these arithmetic operations are inefficient and the communication overhead is extremely large.

One possible way to achieve parallelization for implicit methods is by data parallelism or domain decomposition. For steady flow problems, the concept of spatial domain decomposition is to divide the solution domain into a number of sub-domains with the objective to maximize efficiency, thus distributing the same amount of work on each processor. The usual approach is to split the global matrix coefficients that comprise the central coefficient a_P and neighboring coefficients a_{nb} into a system of diagonal blocks. Each of these blocks is assigned to one processor and data is then transferred on two levels of communications. Local communication takes place between processors operating on neighboring blocks while global communication gathers information from all blocks in a "master" processor, while broadcasting some information back to the other processors. For transient flow problems, domain decomposition in space can also be equally applied in time. More details of parallel computations (see Section 9.3.3 and references given within) are left to the interested reader.

Recent advances in graphics processing units (GPUs) have also gained significant traction within the CFD community for parallel computing due to their considerably lower cost than supercomputers or workstation clusters. In essence, GPU computing has the ability to accelerate parallel computer codes for high-performance computing and has demonstrated remarkable performance and enhanced computational precisions. Typically, GPUs are employed to drive the display of a computer that requires many threads and are thus very suitable because of their inherent feature to run many threads all at once. This inherent multi-threading ability of GPU makes it ideal for general-purpose computing. In comparison to the central processing units (CPUs), GPUs have an intelligent throughput architecture that executes many concurrent threads slowly rather than executing a single thread rapidly. Massive multi-threading in hardware aims to overcome latencies that derive from the communication of data in the device. GPU programming has been greatly facilitated with the release of the CUDA device. The use of CUDA (compute unified device architecture) – a parallel computing platform developed by NVIDIA – has provided developers with direct access to virtual instruction set and memory of the parallel computational elements. For CFD, GPUs have shown the propensity of accelerating arithmetic operations with impressive speedups and a significant reduction in runtimes. More descriptions of GPUs and CUDA environment (see Section 9.3.3 and references given within) are left to the interested reader.

6.7 CASE STUDIES

A selection of case studies is presented to demonstrate the relevance of the computational solutions in *consistency*, *stability*, *convergence*, and *accuracy*. These cases will individually show two or three of the following effects: discretization error (*truncation error*), iteration or convergence error (*grid convergence*), and physical modeling error (application uncertainty of boundary conditions, geometry, and CFD models, for example, turbulence model). They will also include comparisons with an analytic solution (*verification*) and with experimental data (*validation*). It is important to note that the presence of the test case from a particular code is not intended to provide any endorsement or acceptance of the code for this particular purpose. Likewise, the absence of a test case from any particular code does not provide a statement on the unsuitability of the code for this particular application.

6.7.1 Test Case A: Channel Flow

This test case was calculated using an in-house finite volume CFD computer code.

Model description: The geometry of the test problem is a two-dimensional laminar flow between two parallel plates as used in previously worked examples. For this test problem, the channel has dimensions of height $H = 0.1$ m and length $L = 1.0$ m, with air taken as the working fluid.

Grid: The governing equations are discretized on a collocated grid arrangement. Velocity and pressure are collocated (cell-centered) as described in Fig. 5.9. A uniform mesh is generated spanning the height and length of the channel. *Grid convergence (independency)* is performed with meshes of 5×10, 10×20, and 20×40 control volumes.

Features of the simulation: The numerical technique is based on the finite volume discretization. The algorithm for the solution of the Navier-Stokes equations relies on the implicit segregated velocity-pressure formulation such as the SIMPLE scheme. This leads to a Poisson equation for the pressure correction, which is solved through a default iterative solver within the in-house CFD code. To avoid non-physical oscillations of the pressure field and the associated difficulties in obtaining a converged solution, the Rhie and Chow (1983) interpolation scheme is employed.

The fluid is incompressible and its density and viscosity have values of 1.2 kg/m^3 and 4×10^{-5} kg/m · s, respectively. With the inlet velocity specified at 0.01 m/s, the corresponding Reynolds number (see Eq. (3.26) in Chapter 3) based on this inlet velocity and height of the channel is 30.

Results: As demonstrated in Example 6.7, a stable and converged solution is evidenced by diminishing residuals as the number of iterations increases. The decreasing trends of the residuals for a coarse grid distribution of 5×10, medium grid distribution of 10×20, and a fine grid distribution of 20×40 control volumes also exhibit the desirable aspects of *stability* and *convergence*. Since the diminishing nature of the residuals is represented in all three respective grid distributions, only indicative residuals obtained from the fine grid distribution are illustrated, as shown in Fig. 6.4. The numerical calculations are terminated for all three different meshes when all velocities, pressure correction, and mass residuals fall below the convergence tolerance of $\varepsilon = 1 \times 10^{-7}$. It is noted that the pressure correction residual is the same as the mass residual; only this mass residual is presented in Fig. 6.3.

The numerical solution is verified against the analytical solution developed in Example 3.4. Recalling the relationship of Eq. (3.4A) in Chapter 3, which has been developed for $-H/2 \leq y \leq H/2$, the horizontal velocity u can be determined as:

$$u(y) = \frac{3}{2} U_m \left[1 - \frac{y^2}{(H/2)^2} \right]$$

where the average velocity U_m in this test problem corresponds to the uniform inlet velocity $u_{in} = 0.01$ m/s. Fig. 6.4 compares the numerical solutions for the three meshes against the analytical solution in the fully developed flow region. Controlling the solution error due to spatial discretization is fundamentally addressed through this parametric

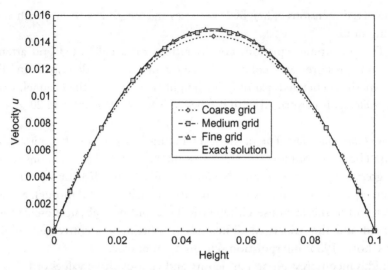

Fig. 6.3 Velocity profiles in the fully developed region for the computational results of three different grid distributions and analytical solution.

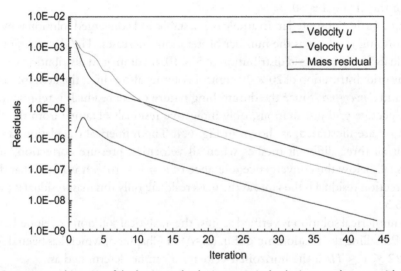

Fig. 6.4 Convergence histories of the horizontal velocity u, vertical velocity v, and mass residual for the fine grid distribution.

study. As the mesh is refined, the approximate solution approaches the exact solution. Herein, *consistency* is achieved since the truncation error diminishes with an increasing mesh resolution. Table 6.1 presents the solution error evaluated for the maximum magnitude of the horizontal velocity u between the analytical value and predicted results for four different meshes. The results demonstrate that *accuracy* is attained for the numerical

Table 6.1 Solution Errors Between the Predicted and Analytical Maximum Velocities for the Three Different Grid Distributions

Mesh	Predicted Max Velocity u	Analytical Max Velocity u	Error = $(u_{analytical} - u_{predicted})/$ $u_{analytical} \times 100\%$
5 × 10	0.01444	0.015	3.73
10 × 20	0.01472	0.015	1.87
20 × 40	0.01496	0.015	0.27
40 × 80	0.01496	0.015	0.27

solution on grid distributions of 20 × 40 and 40 × 80 control volumes. With the additional refinement made to the channel geometry consisting of 40 × 80 control volumes, *grid independency* is thus achieved, since this approximate solution is the same as the result obtained for the grid distribution of 20 × 40 control volumes.

As shown in Table 6.1, the solution error appears to be leveling off even though the mesh is further refined from a grid distribution of 20 × 40 to 40 × 80 control volumes. Here, the solution error registers below 1%, which could not be totally eliminated due to the presence of round-off error during the numerical computations. This round-off error is due to the *single precision* setting during the numerical calculations. As shown in Fig. 6.2, further refinement of the mesh may lead to an increase of the round-off error. One possible way of significantly reducing this round-off error is to extend the number of significant figures during the iterative procedure to *double precision*. Bear in mind that the *double precision* setting tends to incur additional computational burden. In practice, some trade-off of the solution *accuracy* to achieve a quicker turnaround of the numerical computations is generally required especially for fluid flows that are complex.

Conclusion: In this test case the various aspects concerning *consistency*, *stability*, *convergence*, and *accuracy* are succinctly illustrated for a two-dimensional channel flow problem. The error contribution to the numerical solution is investigated on the basis of evaluating the discretization or truncation error. The increasing mesh resolution demonstrates two important outcomes. Firstly, as the spatial discretization error becomes smaller, a grid-independent solution is achieved. Secondly, the excellent agreement between the approximate and analytical solutions verifies the numerical algorithm that is adopted and provides credibility for the computational solution.

6.7.2 Test Case B: Flow Through a 90° Bend

This test case was calculated using a commercial finite volume CFD computer code ANSYS-FLUENT, Version 6.1.

Model description: The geometry of the test problem is a three-dimensional turbulent flow through a 90° bend. The schematic view of the experimental setup comprising an open-circuit suction wind tunnel system for the 90° duct bend is shown in Fig. 6.5,

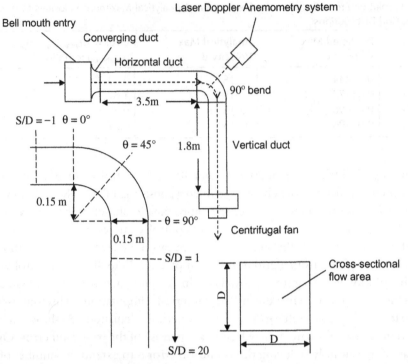

Fig. 6.5 Schematic view of the experimental rig of the 90° bend.

which comprises a 3.5 m long horizontal duct, a 90° bend with a radius ratio of 1.5, and a 1.8 m long vertical straight duct. Air flows through a 10 mm thick Perspex square test section with the bulk gas velocity U_b adjusted through the aid of a variable frequency controller. Experimental data are obtained on this experimental setup using flow visualization and laser Doppler anemometry (LDA) system.

Grid: For the 90° square-section bend, the computational domain begins at a distance of 2D upstream from the bend entrance and extends to 20D downstream from the bend exit. A structured mesh of 325 × 43 × 41 control volumes (directions along the streamwise, width, and height, respectively) is generated for the whole computational domain.

Features of the simulation: This test case illustrates the importance of evaluating the choice of the turbulence models, standard k-ε (Launder and Spalding, 1974) and Reynolds stress (Launder et al., 1975; Launder, 1989), for computing the flow separation around the 90° bend and validating the computational solutions against experimental data.

The algorithm for the solution of the Navier-Stokes equations relies on the implicit segregated velocity-pressure formulation such as the SIMPLE scheme. This leads to a Poisson equation for the pressure correction that is solved through the default iterative solver, normally the multigrid solver in the ANSYS-FLUENT computer code. Finite volume discretization is employed to approximate the governing equations. To avoid non-physical oscillations of the pressure field and the associated difficulties in obtaining a converged solution on a collocated grid arrangement, the Rhie and Chow (1983) interpolation scheme is employed.

The working fluid, air, is taken to be incompressible and the default initial conditions implemented in the computer code are used for the simulations.

Boundary conditions: At the inlet, Dirichlet conditions are used for all variables. The bulk velocity U_b was taken as constant with a value of 10 m/s. With the density and viscosity of air having values of 1.2 kg/m^3 and 2×10^{-5} kg/m · s, respectively, the corresponding Reynolds number based on this inlet velocity and height of the channel is 90,000. The turbulent kinetic energy k and dissipation ε are determined from the measured turbulence intensity I of about 1% at the center of the duct cross-section. For the Reynolds stresses, the diagonal components are taken to be equal to 2/3 k, whereas the extra-diagonal components are set to zero (assuming isotropic turbulence). At the outlet, Neumann boundary conditions are applied for all the transported variables. The non-equilibrium wall function (as will be described in the next chapter) is employed for the air flow at solid walls because of its capability to better handle complex flows where the mean flow and turbulence are subjected to severe pressure gradients and rapid change, such as flow separation, reattachment, and impingement.

Results: The comparison between the measured and calculated longitudinal mean velocities normalized by the bulk velocity U_b at the bend exit ($\theta = 90°$) and 0.5D after the bend exit is illustrated in Fig. 6.6. The prediction of the streamwise velocities using the Reynolds stress model is observed to yield better agreement in contrast to the standard k-ε model, which is due to the capability of the Reynolds stress model capturing the anisotropy behavior of the flow separation region around the 90° bend.

The predicted longitudinal mean velocity normalized by the bulk velocity U_b using the Reynolds stress model is further compared against the measured data at different locations in the duct center plane, as represented in Fig. 6.7. At the bend entrance ($\theta = 0°$), the turbulence model successfully predicts the acceleration of the air flow near the inner wall. The fluid deceleration caused by the unfavorable pressure gradient is also captured near the outer wall. More importantly, the turbulence model adequately reproduces the distorted longitudinal velocity profiles at the angles of $\theta = 30°$, $\theta = 45°$, and $\theta = 60°$ after the bend entrance.

To better understand the flow characteristics around the 90° bend, Fig. 6.8 presents the calculated velocity vectors and pressure distribution of the air flow obtained through the Reynolds stress model at different cross-sections of the duct flow. Fig. 6.8A and B show the calculated air velocity vector and pressure distribution at the cross-sectional middle plane of the duct flow. Favorable (positive) and unfavorable (negative) longitudinal pressure gradients persist near the inner and outer walls of the bend entrance. The presence of the favorable and unfavorable pressure gradients is caused by the balance of

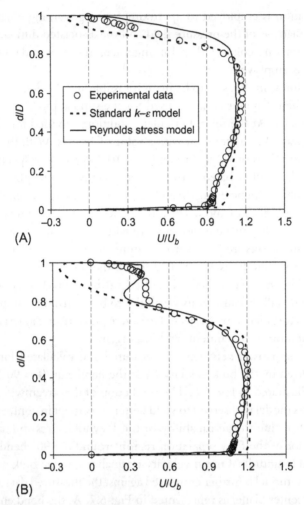

Fig. 6.6 Comparison between measured and calculated longitudinal mean velocities normalized by the bulk velocity U_b: (A) streamwise velocity at bend exit ($\theta = 90°$) and (B) streamwise velocity at 0.5D after the bend exit (see Fig. 6.6 for more description).

centrifugal force and radial pressure gradient in the bend (Humphrey et al., 1981). This physical phenomenon is typical of curved duct flows. Secondary flows are developed due to the direct consequence of the cross-stream pressure gradient. The predicted secondary flow vectors are clearly depicted by the three square cross-sectional flow areas at angles of $\theta = 45°$ and $\theta = 90°$ and distance of 3D after the bend exit (S/D = 3) in Figs. 6.8C, D, and E, respectively.

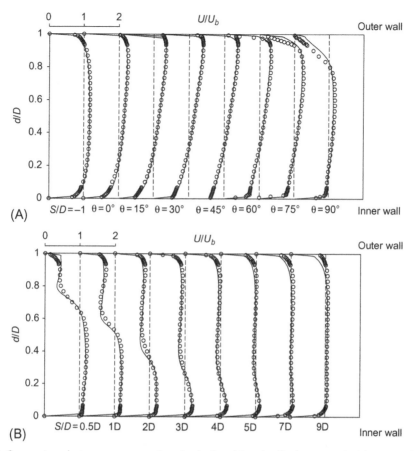

Fig. 6.7 Comparison between measured and calculated longitudinal mean velocities normalized by the bulk velocity U_b at different locations in the duct center plane: (A) between bend entrance and exit and (B) downstream of bend exit (see again Fig. 6.6 for more description).

Conclusion: This test case focuses on the use of approximate models such as the turbulence models to predict the physical characteristics of the turbulent flow around a 90° bend. Owing to the absence of analytical solutions, *validation* of the computational solutions is performed by comparing the predictions against the experimental data in order to address the simulation model uncertainty and the degree to which models correspond to an accurate representation of the real physical flow. The flow around a 90° bend, though geometrically simple, exhibits complex flow structures due to the existence of secondary flows in the vicinity of the bend region, which are generally anisotropic in nature. The results from the more sophisticated Reynolds stress model are shown to better capture the anisotropy behavior of the flow in contrast to the standard k-ε model that assumes isotropy in its original model formulation.

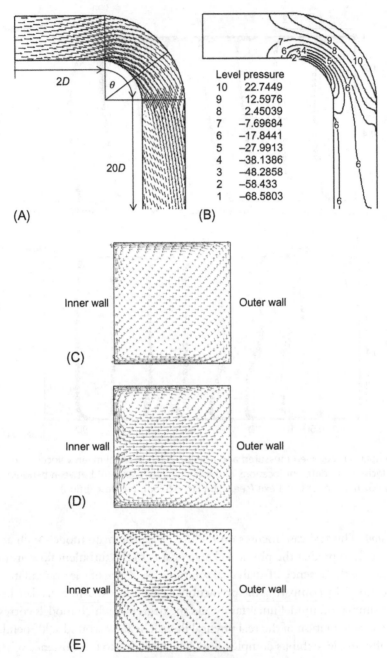

Level pressure
10	22.7449
9	12.5976
8	2.45039
7	−7.69684
6	−17.8441
5	−27.9913
4	−38.1386
3	−48.2858
2	−58.433
1	−68.5803

Fig. 6.8 Calculated flow field: (A) air flow in the duct middle plane; (B) static pressure distribution in the duct middle plane; (C) secondary flow at a cross-sectional flow area of location θ = 45°; (D) secondary flow at a cross-sectional flow area of location θ = 90° (bend exit); and (E) secondary flow at cross-sectional flow area of location S/D = 3D after the bend exit.

6.8 SUMMARY

The credibility of a computational solution is analyzed and assessed in this chapter through the consideration of the various aspects of *consistency*, *stability*, *convergence*, and *accuracy*. It is usually possible to demonstrate whether a discretized form of the governing fluid flow equations is consistent and also if the algebraic form of these equations is stable. Convergence requires, however, implications from *consistency* and *stability*; that is, *consistency + stability = convergence*. In any numerical calculations errors and uncertainties affect the accuracy of the computational solution. It is imperative that the errors and uncertainties are systematically reduced so that the computational solution better represents the real physical flow problem that is being solved. The conceptual framework linking the various aspects of *consistency*, *stability*, *convergence*, and *accuracy* beginning from the governing partial differential equations as considered in Chapter 3 and arriving at the approximate solution of the algebraic equations as described in Chapter 5 can be seen in Fig. 6.9.

The application of a Taylor series expansion to the discrete approximation of the governing equations results in an explicit expression for the truncation error (Section 6.2). This error measures the accuracy of the finite difference or finite volume approximation and determines the rate at which the error decreases as the finite quantities, time step, and/or mesh spacing diminish. It is noted that because of the close correlation between truncation error and solution error (Section 6.5.1), reducing the truncation error has beneficial consequences in the likelihood of also reducing the solution error.

Numerous worked examples and test cases presented throughout this chapter have aimed to better illustrate the conceptual properties of *consistency*, *stability*, *convergence*,

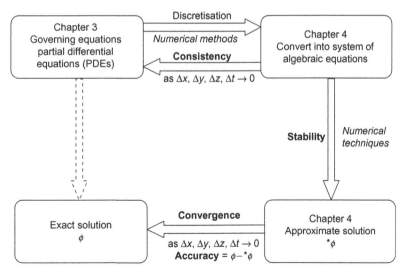

Fig. 6.9 A conceptual framework linking the various aspects of consistency, stability, convergence, and accuracy in arriving at a solution for the transport equations.

and *accuracy* while solving the discretized form of the partial differential equations. In addition, it is important that the computational solution should be subjected to the rigorous process of *verification* and *validation*, which was demonstrated through two test cases. For a simple two-dimensional channel flow problem in Test Case A, it is feasible to verify the computational solution against an analytical relationship. However, for a more complicated flow problem that employs a computational solution in a three-dimensional domain, the absence of an analytical solution requires the dependency on benchmark and/or experimental data to validate the computational solution. This is evidenced by Test Case B for a three-dimensional air flow around a 90° bend. In this example we demonstrated the use of appropriate turbulence models to better capture the physical flow behavior in a 90° bend. This aspect highlights the uncertainty that arises through the use of approximate models for solving turbulent flow, which will be further investigated in the next chapter. More practical guidelines will be further discussed in Chapter 7 to better equip the reader to handle and solve a range of CFD problems.

REVIEW QUESTIONS

6.1 Why do the results obtained through numerical methods differ from the exact solutions solved analytically? What are some of the causes for this difference?

6.2 In the analysis of CFD results what does consistency imply?

6.3 What are the key aspects of consistency?

6.4 If a system of algebraic equations is equivalent to the partial differential equation as the grid spacing tends to zero, does this also mean the solution of the system of algebraic equations will approach the exact solution of the partial differential equation? Why?

6.5 Explain why the following Taylor series expansion of the one-dimensional transient diffusion equation $\left[\frac{\partial \phi}{\partial t} - \alpha \frac{\partial^2 \phi}{\partial x^2} + \alpha \left(\frac{\Delta t}{\Delta x}\right)^2 \frac{\partial^2 \phi}{\partial t^2}\right]_i = 0$ does not show consistent properties.

6.6 Describe the concept of stability.

6.7 What are the stability criteria produced by the von Neuman analysis?

6.8 What is the Courant number and what is its function?

6.9 Consider the following discretized equation, locate the Courant number, and discuss the Courant-Friedrichs-Lewy condition for stability in this case:

$$\phi_i^{n+1} = \phi_i^n + a\frac{\Delta t}{\Delta x}\left(\phi_i^n - \phi_{i-1}^n\right)$$

6.10 Provide a definition of the concept of convergence.

6.11 State *Lax's equivalence theorem* for convergence. Does it apply to non-linear problems?

6.12 What are the three important rules when considering *iterative convergence*?

6.13 How is the concept of residual applied to describe the discretized equation of the system of transport equations?

6.14 Differentiate between a local residual and a global residual.

6.15 What is implied when the residuals become negligible with increasing iterations?

6.16 What is the usual recommended residual tolerance level?

6.17 Define under-relaxation factor. State its advantages and disadvantages when using a small value.

6.18 Discuss ways in which convergence can be accelerated.

6.19 Is a converged solution also an accurate solution? Why?

6.20 Discuss some types of errors that can cause a solution to be inaccurate.

6.21 What are discretization errors? What is the difference between a global error and a local error?

6.22 Which methods can be used to minimize discretization errors?

6.23 What are round-off errors and what kind of calculations are most affected by them?

6.24 Which method can be used to minimize round-off errors?

6.25 What does it mean to perform a grid convergence (independency) test?

6.26 What is the difference between verification and validation? Why are these two steps important in analyzing results?

6.27 Discuss briefly how multigrid methods are employed to increase the computational efficiency of solving CFD problems.

6.28 Discuss briefly how parallel or GPU computing is used to achieve computational efficiency.

6.29 Consider the following algebraic equation $a_p\Phi_p = \sum a_{nb}\Phi_{nb} + b$, as described in Chapter 5. In matrix form a_p represents the diagonal element, while a_{nb} is the neighboring element. The condition for convergence stipulates that $\sum |a_{nb}|/a_P \leq 1$, which simply means that the sum of the neighboring elements divided by the diagonal elements must be less than unity, at all grid nodal points. Analyze the condition for convergence given the system of equations below.

Case 1:	$\Phi_1 = 0.5\Phi_2 + 1.5$	(1)
	$\Phi_2 = \Phi_1 + 2$	(2)

by re-arranging the above equations, we have

Case 2:	$\Phi_1 = \Phi_2 - 2$	(2)
	$\Phi_2 = 2\Phi_1 - 3$	(1)

CHAPTER 7

Practical Guidelines for CFD Simulation and Analysis

7.1 INTRODUCTION

The necessity to understand the physics of a flow problem in hand, the basis of the numerical methods employed to solve the governing equations, and the means to obtain the most accurate and consistent results given the available computing resource are some of the common challenges faced by the CFD user. This chapter is particularly written to address these common aspects of CFD and the aim is to provide some practical guidelines for carrying out a CFD simulation, solution assessment, and analysis.

In Chapter 2 grid generation was viewed as one of the key considerations during the pre-process stage following the definition of computational domain geometry. The generation of a mesh such as has been described in Chapter 4 is an important numerical issue where the type of mesh chosen for a given flow problem can determine the *success* or *failure* in attaining a computational solution. Because of this, grid generation has become an entity by itself in CFD. As a guide to grid generation, the mesh must be sufficiently fine to provide an adequate resolution of the important flow features and geometrical structures. For flows with bounded walls, it is recommended that recirculation vortices or steep flow gradients within the viscous boundary layers are properly resolved through locally refining or clustering the mesh in the vicinity of wall boundaries. Mesh concentration may also be required for fluid flows having high shear and/or high-temperature gradients. Furthermore, the quality of the mesh has significant implications for the convergence and stability of the numerical simulation and accuracy of the computational result obtained. All of these grid generation issues will be examined in the next section.

Also in Chapter 2, special attention was given to specifying a range of boundary conditions commonly applied to a given CFD problem. The physical meanings of the various boundary conditions that were employed to close the fluid flow system were later illustrated in Chapter 3. Boundary conditions have serious implications for the CFD solution. This further reinforces the requirement to define suitable boundary conditions that appropriately mimic the real physical representation of the fluid flow. In many real applications there is always great difficulty in defining in detail some of the boundary conditions at the inlet and outlet of a flow domain that is required for an accurate solution. A typical example is the specification of turbulence properties (turbulent intensity and length scale) at the inlet flow boundary, as these are arbitrary in many CFD problems.

Computational Fluid Dynamics
https://doi.org/10.1016/B978-0-323-93938-6.00011-7

251

Nevertheless, by carrying out an uncertainty analysis, the reader can develop a good feel of the appropriateness and inappropriateness of the boundary conditions that are being imposed within the physical context of the CFD problem being solved. Some useful guidelines on the specification of inlet, outlet, wall, and other types of boundary conditions for different classes of problems will be examined and discussed.

Since most flows of engineering significance are turbulent in nature, the classical two-equation modeling approach of handling turbulence was briefly introduced in Chapter 3, where the basic formulation of the standard k-ε model was described. Nonetheless, we demonstrated in the test case of a flow over a 90° bend of Chapter 6 that the use of a more sophisticated turbulence model, the Reynolds stress model, was deemed necessary to better capture the anisotropy behavior of the flow separation around the bend region. A turbulence model is a computational procedure to instinctively close the system of mean flow governing equations. Nowadays, the two-equation and Reynolds stress models form the basis of turbulence calculations in numerous commercial, shareware, and in-house CFD codes. On the other hand, due to some inherent limitations of the standard k-ε model, the development of other dedicated models for limited categories of flows has led to the formulation of many variants of the standard model. A CFD user is therefore confronted with the pressing choice of selecting a suitable turbulence model. The provision of appropriate guidelines is therefore challenging. Although many industrial problems fall within the limited class of flow that can be resolved by the standard k-ε model, the only practical way to validate more specific flow problems is to conduct a case-by-case examination to determine the optimum turbulence model. Pertinent turbulence modeling issues are explored later in this chapter.

As we continue to unravel the mysteries of CFD, this chapter by and large assembles all the essential knowledge gathered in the previous chapters, thereby bringing to fruition the realization of CFD simulation and analysis. We begin by focusing on some practical guidelines required for handling grid generation.

7.2 GUIDELINES FOR BOUNDARY CONDITIONS

7.2.1 Overview of Setting Boundary Conditions

The setting of proper physical boundary conditions, together with the section of local refinement and solution adaptation, governs the computational stability and the numerical convergence of the CFD problem. In many real applications there is always great difficulty in prescribing some of the boundary conditions at the inlet and outlet of the computational domain. Boundary conditions for fluid flow are generally more complex due to the coupling of velocity fields with the pressure distribution. In the context of CFD defining suitable boundary conditions generally encompasses the specification of two types of boundary conditions: the *Dirichlet* and *Neumann* boundary conditions. The physical meanings of these boundary conditions have been described in Section 3.7. For brevity, the *Dirichlet* boundary condition can be simply defined for the transport property ϕ as the requirement of specifying the physical quantity over the boundary, such as:

$$\phi = f(\text{analytic}) \tag{7.1}$$

The *Neumann* boundary condition involves, however, the prescription of its derivative at the boundary given by:

$$\frac{\partial \phi}{\partial n} = \text{const} \tag{7.2}$$

A practice that is widely adopted for inflow boundaries is to set the transported quantities of either a uniform or some predetermined profile over the boundary surface (*Dirichlet*). For outflow boundaries, the convective derivative normal to the boundary face is set equal to zero; the transported quantities at the boundaries are extrapolated along the stream-wise direction of the fluid flow (*Neumann*). However, the use of such an approach is not as straightforward in some selected applications. Some difficulties may arise during the implementation of such boundary conditions. For example, *non-physical* reflection of outgoing information back into the calculation domain (Giles, 1990) such as the fluid that may inadvertently re-enter the domain through these outflow boundaries as well as in regions of possible high swirl, large curvatures, or pressure gradients may significantly affect the convergence behavior of the iterative procedure. In addressing some of these difficulties the specification of cross-stream equilibrium of a pressure field is deemed to be more preferable to the usual constant static pressure for swirling flows at an outlet. Also, when strong pressure gradients are present, special non-reflecting boundary conditions are sometimes required for the inflow and outflow boundaries (Giles, 1990).

Some other difficulties that are also important to note include the specification of suitable inlet turbulence intensity, length scales, turbulence kinetic energy and dissipation for a turbulent flow entering the flow domain, the correct description of the boundary layer velocity profile on the walls and at the inlet, and also the precise distribution of scalar or species concentrations at the inflow boundary. It is therefore not surprising that as a CFD user he/she must be fully aware of these problems and needs to develop a good feel for the certainty or uncertainty of the boundary conditions that are being imposed.

In ensuring consistency of the boundary conditions imposed with the CFD models, it is vital that the boundary conditions strongly reflect the application that is being calculated. Some general guidelines are provided. Whenever possible, it is useful to examine the potential of modifying the computational domain by moving the domain boundaries to a position where the boundary conditions are more readily identifiable and where they can be more precisely specified. It is also important to be mindful of upstream or downstream obstacles such as bends, contractions, diffusers, and blade rows, outside of the flow domain, which may significantly affect the flow distribution. Often information on these components upstream and downstream of the domain is incomplete or simply not available; it may be required to incorporate these pertinent components in the calculations so that appropriate flow predictions are obtained. In handling a wide range of different CFD applications, it may be beneficial to carry out a sensitivity analysis in which the boundary conditions are systematically altered within certain limits to detect any significant variation in the computational results. Should any of these variations prove to greatly influence the simulated results and lead to large changes in the simulation, it is then necessary to obtain more accurate data on the boundary conditions that are being specified. When commercial codes are used, default settings of the boundary conditions for the domain boundaries are given and the user is usually not required to specify any boundary conditions. Nonetheless, he/she still needs to ensure that appropriate boundary conditions are applied for the application that is being solved. We proceed to concentrate on more specific guidelines for the inlet, outlet, wall, and symmetry and periodicity boundary conditions, which are subsequently discussed below.

7.2.2 Guidelines for Inlet Boundary Conditions

Since flows inside a CFD domain are driven by boundary conditions, carrying out a sensitivity analysis by systematically changing the inlet boundary conditions within allowable limits is highly recommended. The aim of this practical guideline is to ensure that appropriate boundary conditions are applied at the inflow boundaries. Depending on the flow problem, certain key parameters can be scrutinized to ascertain the suitability of the inlet boundary conditions that are being specified. Some useful parameters that can be examined are:

- Inlet flow direction and the magnitude of the flow velocity
- A uniform distribution of a parameter or a profile specification, for example, either a uniform inlet velocity or some predetermined inlet velocity profile
- Variation of physical properties
- Variation of turbulence properties at inlet

The reader should also take note that an inlet boundary represents a potential *mass source* for the fluid flow. It is thereby imperative that an outlet boundary is imposed within the flow domain to characterize the *mass sink* of the fluid flow. Without an outlet boundary, the mass is not conserved. Not surprisingly, the CFD calculations in this circumstance will tend to "blow up" rather swiftly.

By revisiting the case for the fluid flowing between two stationary parallel plates, the various combinations of the inlet and outlet boundary conditions can be appropriately demonstrated. The various configurations that are typically adopted are illustrated in Fig. 7.1. The first configuration that involves the specification of a prescribed velocity profile at the inlet and a constant pressure at the outlet is commonly employed in many

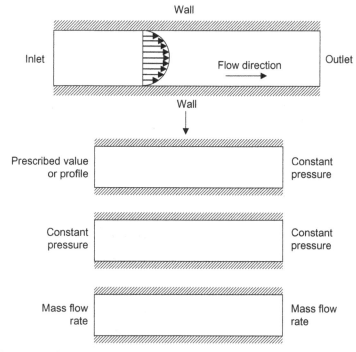

Fig. 7.1 Configurations for a simple flow between two stationary parallel plates.

CFD calculations. For rapid convergence and computational robustness, most commercial codes generally recommend the implementation of this type of configuration. Nonetheless, other highlighted configurations may be equally applicable if proper initial conditions are accommodated prior to the CFD calculations. At steady state, the CFD solution for all the configurations should yield identical velocity profiles, especially at the fully developed region situated near the channel exit.

For certain types of geometries, the specification of inlet boundary conditions for the upstream flow may require special attention. At the pre-designated inlet location, the exact distribution may be unknown. The possibility of moving the inlet boundary to a position where the fluid flow is allowed to develop through some distance inside the domain should therefore always be examined. Let us consider a typical flow problem of the backward-facing step geometry. For high accuracy, it is necessary to demonstrate that the interior solution is unaffected by choice of location of the inlet by carrying out a sensitivity study for the effect of different upstream distances from the flow expansion region as indicated in Fig. 7.2. If the inlet is too close, the velocity profile will be changing in the flow direction before the flow expansion region, which significantly affects the recirculation vortex and subsequently the re-attachment length. Such a solution, of course, cannot be fully trusted and a re-examination of the inlet location is required.

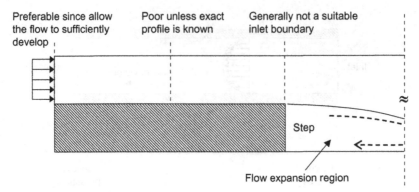

Fig. 7.2 Inlet locations for the backward-facing step flow problem.

7.2.3 Guidelines for Outlet Boundary Conditions

It is vital that the boundary condition imposed at the outlet is always selected on the basis that it has a weak influence on the upstream flow. The most suitable outflow conditions, as demonstrated in Fig. 7.1, are weak formulations involving the specification of a constant pressure such as static pressure at the outlet plane. Whenever possible, the outlet boundary must be placed as far away as possible from the region of interest and should be avoided in regions of strong geometrical changes or in wake regions with recirculation. Let us illustrate the potential hazards of the latter by revisiting the backward-facing step geometry flow problem.

Fig. 7.3 shows typical velocity profiles downstream from the step and the appropriate choice for the location of the outlet boundary. As aforementioned, a common practice that is adopted for outflow boundaries in many CFD calculations is to apply the *Neumann* boundary condition, in which the convective derivative normal to the boundary face is set to zero and the boundary value is evaluated based on the stream-wise extrapolation of the transported quantities. If the outlet is located at *Position 1*, a plane in this position cuts across the wake region of recirculation (Fig. 7.3). Not only is the assumed boundary condition not valid, but the problem is also exacerbated by the apparent area of reverse flow where the fluid enters the domain. Further downstream at *Position 2*, there may not be any reverse flow but the zero gradient condition still does not hold since the velocity profile is changing in the flow direction. As a practical guideline, the outlet boundary should be placed much further downstream by at least 10 step heights ($>10\ h$) to give accurate results – *Position 3*. It is also necessary to ensure and demonstrate that the interior solution is not affected by the location of the outlet boundary.

For some CFD problems where the outlet boundaries may not be feasibly relocated in the domain, care should be exercised so that the possibility of fluid flow entering through the outflow boundaries does not lead to progressive instability of the numerical procedure or to the extent of attaining an incorrect solution. In such an event some

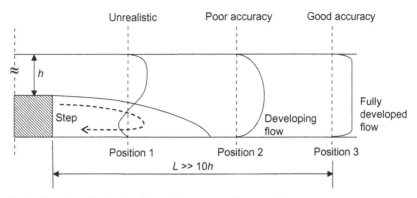

Fig. 7.3 Outlet locations for the backward-facing step flow problem.

reconfiguration of the model geometry may be required and the flow area at these outlets is restricted, provided that the outlet boundaries are situated at a location substantially far away from the region of interest. Nonetheless, if the outflow boundary condition permits flow to re-enter the domain, then appropriate *Dirichlet* boundary conditions should be imposed for all transported quantities.

7.2.4 Guidelines for Wall Boundary Conditions

Wall boundary conditions are generally employed for solid walls bounding the flow domain. Here, care should always be taken to ensure that the boundary conditions imposed on these types of walls are consistent with both the numerical model used and the actual physical features of the flow geometry. For the fluid flowing between two stationary parallel plates in Fig. 7.1, wall boundary conditions are enforced for the channel walls. Nonetheless, these boundary conditions are also often used to bound fluid and solid regions such as applied for the presence of the step within the flow domain or the two cylinders inside the open surrounding flow environment.

   For stationary walls, the default consideration is to assume that the no-slip condition applies, which simply means that the velocity component tangential to the wall is taken to be zero at the solid boundaries. This condition implies that the fluid flow comes to rest at the solid walls. We may also explore the possibility of modeling the boundary conditions on the solid walls as a free-slip condition, which assumes that the flow is parallel to the wall at this point. This condition corresponds to the absence of viscous effects in the continuum equations and is applied to the problem where the continuum approach breaks down as the fluid approaches the wall in viscous flow. For a general fluid flow case where the transport of heat takes precedence within the domain, great care must be taken to specify boundary conditions (e.g. adiabatic walls or local heat fluxes) on the solid boundaries of the numerical model that properly represent the heat transfer characteristics of the solid walls in the actual physical model.

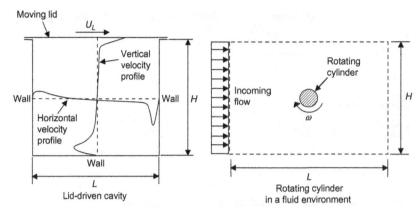

Fig. 7.4 Illustrative fluid flow examples with moving or rotating walls.

In flow cases with moving or rotating walls it is important that the boundary conditions that need to be specified are consistent with the motion of the solid walls. Lid–driven cavity and a rotating cylinder in a fluid environment, as shown in Fig. 7.4, are some typical examples exemplifying the moving boundary problems in CFD. These types of problems allow a positive or negative tangential velocity to be imposed at the top boundary of the lid–driven cavity or a clockwise or counterclockwise rotational speed to be specified on the circumferential surface of the rotating cylinder. Other more complex CFD problems may require the need to employ sliding or moving meshes to better emulate the motion of a rotating impeller stirring the fluid in a tank or the sea waves hitting the ship's hull. The use of sliding or moving meshes is beyond the scope of this introductory book. Interested readers are advised to consult other works in the literature for more information (see e.g. Ferziger and Perić, 2002).

7.2.5 Guidelines on Symmetry and Periodic Boundary Conditions

These boundary conditions are generally used when the flow geometry possesses some symmetry or periodic properties that allow the flow problem to be simplified by solving only a fraction of the domain.

For symmetry boundary conditions, they are applied when the physical geometry and the flow field have a mirror symmetry. At the symmetry plane, the following requirements must be satisfied:
- the normal velocity is zero
- the normal gradients for all transported properties are zero

Consider the flow problem where air is flowing through a square duct as described in Fig. 7.5. The application of symmetry planes can significantly reduce the computational effort since only a quarter of the flow domain needs to be modeled. Care should always be exercised especially in prescribing the mass flow rates at the inlet and outlet boundaries.

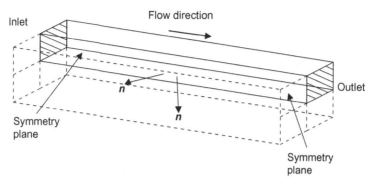

Fig. 7.5 Application of symmetry boundary conditions for the air flow through a square duct.

For this simplified flow geometry, the correct mass flow rates to be imposed at these boundaries are only a quarter of the actual mass flow rates.

To apply the periodic boundary conditions, it is imperative that the periodic planes come in pairs. The key requirement in employing these boundary conditions is that not only do they have to be physically identical but the mesh distribution in the respective planes must also be identical. In CFD applications, this condition implies that the flow leaving one of the periodic planes is equal to the flow entering in the other. For wall-bounded flows such as the fluid flowing between two stationary parallel plates or air flowing through a square duct, the inlet and outlet boundaries can otherwise be prescribed as periodic planes where now the flow field leaving the outlet boundary is taken to be the same entering the inlet boundary. Fig. 7.6 further illustrates the application of periodic boundary conditions for the fluid flow in a mixing tank and across an in-line arrangement of heat exchanger tubes. For the former, the flow field is known as rotationally or cyclically periodic, where the flow field leaving "Boundary 2" is enforced as inlet flow condition at "Boundary 1." For the latter, the flow field is, however, considered translationally periodic. As before, the flow field leaving "Boundary 2" is used as inlet flow field at "Boundary 1." It is worthwhile noting that this flow case is particularly similar to the aforementioned wall-bounded flows.

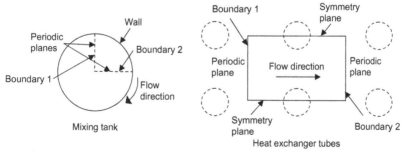

Fig. 7.6 Some illustrative examples of the application of periodic boundary conditions.

7.3 GUIDELINES FOR TURBULENCE MODELING

7.3.1 Overview of Turbulence Modeling Approaches

Some pertinent properties characterizing flows that are turbulent have been discussed and described in Chapter 3. The reader may wish to momentarily revisit Section 3.5.1 and review the various aspects detailed therein. In brief, turbulent flows can be classified as being highly unsteady and random. They are also known to contain large coherent structures responsible for the mixing or stirring processes. Nonetheless, the fluctuating property across a broad range of length and time scales particularly from the modeling standpoint makes the direct numerical simulation of these types of flows very difficult. Turbulent flows are constantly encountered in many engineering systems. They tend to demand more computational resources compared with flows that are laminar. For research and design purposes, CFD analysts and engineers are expected to understand and predict the effects produced by turbulence. In this section, we review many state-of-the-art turbulence modeling approaches that have been developed and applied to date.

Advanced Techniques: The most accurate approach to turbulence simulation is to solve directly the governing transport equations without undertaking any averaging or approximation other than the numerical discretizations performed on them. Through such simulations, all of the fluid motions contained in the flow are considered to be resolved. This approach is commonly known as the direct numerical simulation or, by its more well-established acronym, DNS. Since DNS requires all significant turbulent structures to be adequately captured (i.e. the domain on which the computation is carried out needs to accommodate for the smallest and largest turbulent eddy), it can be very expensive to be employed. Alternatively, we can consider another approach where the structure of turbulent flow can be viewed as the distinct transport of large- and small-scale motions. Fig. 7.7 presents a schematic illustration of such flow. Since the large-scale motions are generally much more energetic and by far the most effective transporters of the

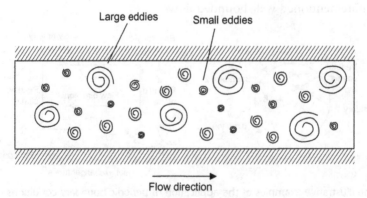

Fig. 7.7 Schematic representation of a turbulent motion.

conserved properties than the small ones, a simulation that treats the *large eddies* exactly but approximates the *small eddies* makes rather perfect sense. Such an approach is recognized as the large eddy simulation or by its more widespread acronym, LES. It is still expensive but much less costly than DNS. In general, though, DNS is the preferred method because it is more accurate. LES is, however, the preferred method for flows in which the Reynolds number is too high or the geometry is too complex to allow the application of DNS.

The results of a DNS or LES simulation contain very detailed information about the flow, producing an accurate realization of the flow while encapsulating the broad range of length and time scales. For design purposes, it is far more information than any engineer needs. DNS and LES approaches usually require high usage of computational resources and often cannot be used as a viable *design tool* because of the enormity of the numerical calculations and the large number of grid nodal points. A question therefore arises of their usefulness and what role they can play in CFD. Because of the wealth of information, DNS and LES taken as a *research tool* can provide a qualitative understanding of the flow physics and also can construct a quantitative model allowing other, similar, flows to be computed. More importantly, they can assist in some cases to improve the performance of currently applied turbulence models in practice. We shall demonstrate the use of DNS as a research tool in some sample aerodynamic investigations in Section 8.4.5.

Practical Techniques: For most engineering purposes, it is unnecessary to resolve the details of the turbulent fluctuations such as the DNS and LES approaches. In general, the effects of the turbulence on the mean flow are usually sufficient to quantify the turbulent flow characteristics. For a turbulence model to be practical and useful in a general-purpose CFD code, it must have wide applicability and be simple, accurate, robust, and economical.

In Chapter 3 the Reynolds-averaged approaches to turbulence results in the formulation of the two-equation turbulence model – the *standard k-ε model* proposed by Launder and Spalding (1974). This model is well established and widely validated and gives rather sensible solutions to most industrially relevant flows. Nonetheless, numerous limitations are also identified especially for flows with large, rapid, extra strains (e.g. highly curved boundary layers and diverging passages) since the model generally fails to fully describe the subtle effects of the streamline curvature on turbulence. This weakness has been exemplified in Chapter 6 through the worked example in Section 6.7.2. The test case results revealed that the *Reynolds Stress Model* works better because of the ability of the model to accommodate the *anisotropic* turbulent stresses occurring around the 90° bend. The *standard k-ε model*, which is a consequence of the eddy viscosity concept, assumes that the turbulent stresses are linearly related to the rate of strain by a scalar turbulent viscosity, and the principal strain directions are aligned to the principal stress directions – *isotropic*. Owing to the deficiencies of the treatment of the normal stresses, secondary flows that exist in the 90° bend that are driven by *anisotropic* normal Reynolds stresses could not be properly predicted.

The *Reynolds stress model*, also called the second-moment closure model, dispenses the notion of turbulent viscosity and determines the turbulent stresses directly by solving a transport equation for each stress component. This requires the solution of six additional equations ($\overline{u_1'^2}$, $\overline{u_2'^2}$, $\overline{u_3'^2}$, $\overline{u_1'u_2'}$, $\overline{u_1'u_3'}$, and $\overline{u_2'u_3'}$), which are solved accordingly in the form of Eq. (3.53) in Chapter 3 in order to account for the directional effects of the Reynolds stress field. An additional equation for ε is also solved to provide a length scale determining quantity. For a comprehensive explanation of this particular type of turbulence model, interested readers can consult relevant reference texts by Launder (1989) and Rodi (1993). In a similar manner, the turbulent heat fluxes can be determined directly by solving three additional equations, one for each flux component, thereby removing the notion of a turbulent Prandtl number. There is no doubt that the *Reynolds stress model* has a greater potential to represent the turbulent flow phenomena more correctly than the *standard k-ε model*. This type of model can handle rather complex strain and, in principle, can cope with non-equilibrium flows. The shortcoming of this model is the very high computing costs that may be incurred because of the extra governing equations. Also, the model success thus far has been rather moderate, specifically for axisymmetric and unconfined recirculating flows, where they have been shown to perform as poorly as the *standard k-ε model*.

A lot of research is still being performed in this field, and new models are constantly being proposed. The turbulent states that can be encountered across the whole range of industrially relevant flows are rich, complex, and varied. It is now accepted that no single turbulence model can span these states since none is expected to be universally valid for all flows. Accordingly, Bradshaw (1994) refers to turbulence as *the invention of the Devil on the seventh day of creation, when the Good Lord wasn't looking*. Because of its difficult nature, we will attempt to propose some useful strategies for selecting appropriate turbulence models in handling turbulent fluid engineering problems in the proceeding section.

7.3.2 Strategy for Selecting Turbulence Models

In CFD different types of turbulent flows require different applications of turbulence models. In the event that insufficient knowledge precludes the selection of an appropriate model, we strongly encourage the use of the two-equation model such as the *standard k-ε model* as a starting point for turbulent analysis. This model offers the simplest level of closure since it has no dependence on the geometry or flow regime input. As a first step to turbulence model selection, the *standard k-ε model* is robust and stable and is as good as any other more sophisticated turbulence models in some applications. The majority of in-house and commercial codes generally set this model as the default modeling option for handling flows that are turbulent. It is not entirely surprising because it has been a *de facto* standard in industrial applications and still remains the work-horse of industrial computations.

Nevertheless, the *standard k-ε model* as aforementioned is not without any weaknesses. It is therefore imperative that the major weaknesses associated with this model are cataloged in some fashion so as to instigate palliative actions that might be fruitfully considered for improving the numerical predictions. These advisory actions, which will be given shortly, should not be viewed as definitive cures but rather recommendations whereby possible alternatives to the *standard k-ε model* can be systematically investigated. There is also no guarantee that the specific advice will yield significantly improved results. The necessary task of carrying out careful *validation* and *verification* remains the defining step in order to fully justify the application of turbulence models for the particular CFD problem being solved.

Guideline on a Particular Weakness of the Standard k-ε Model: Historically, the five adjustable constants C_μ, σ_k, σ_ε, $C_{1\varepsilon}$, and $C_{2\varepsilon}$ in the *standard k-ε model* have been calibrated against comprehensive data for a wide range of turbulent flows but of simple geometrical flow origins. This has been evidenced by its remarkable successes in handling thin shear layer and recirculating and confined flows. Deviation from such flow behaviors has, however, resulted in the *standard k-ε model* performing rather poorly especially in important flow cases having:

- flow separation (Baldwin and Lomax, 1978)
- flow re-attachment (Kato and Launder, 1993)
- flow recovery (Ince and Launder, 1995)
- some unconfined flows (e.g. free shear jet) (Apsley et al., 1997)
- secondary flows in complex geometrical configurations (Flow around a poppet valve in Ferziger and Perić, 1999)

Let us demonstrate a specific palliative technique for the weakness under consideration for the *standard k-ε model* of a worked example represented by the backward-facing step flow problem described earlier. It is well known that the re-attachment length l_r (Fig. 7.8 below) is generally poorly predicted by this model under turbulent flow because of the over-prediction of the turbulent kinetic energy. The high turbulence levels predicted upstream following the flow expansion at the step are transported downstream and the real boundary layer development is subsequently swamped by this effect. For illustration purposes, we investigate the use of other more sophisticated turbulence models such as the *RNG k-ε model* and *realizable k-ε model*, respectively proposed by Yakhot et al. (1992) and Shih et al. (1995), to exemplify the improvements that can be achieved in the numerical predictions.)

For this flow problem, the CFD commercial code, ANSYS-FLUENT, Version 6.1, is utilized to predict the continuum gas phase under steady-state conditions through solutions of the conservation of mass and momentum in two dimensions. A computational domain with a size of 12 h (length) × 1 h (height) before the step and 50 h (length) × 2 h (width) after the step is considered. The Reynolds number based on the free stream velocity u_∞ of 40 m/s and step height h for this investigation is evaluated as 64,000.

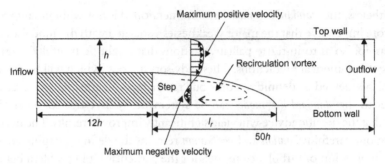

Fig. 7.8 A schematic illustration of the re-attachment length location and respective dimensions for the regions before and after the step for the backward-facing step geometry.

Some pertinent differences are worthwhile mentioning with regard to each of the turbulence models before proceeding to discuss the numerical results. The *RNG k-ε model* includes a modification to the transport ε-equation where the source term is solved as:

$$S_\varepsilon = \frac{\varepsilon}{k}(C_{\varepsilon 1}P - C_{\varepsilon 2}D) - R \tag{7.3}$$

In the *standard k-ε model* the rate of strain term R in the above equation is absent (compare Eq. (3.57) in Chapter 3). The presence of this R term is formulated in the form of:

$$R = \frac{C_\mu \eta^3 (1 - \eta/\eta_o)}{1 + \beta \eta^3} \frac{\varepsilon^2}{k} \tag{7.4}$$

where β and η_o are constants with values of 0.015 and 4.38, respectively. The significance of the inclusion of this term is its responsiveness toward the effects of rapid rate of strain and streamline curvature, which cannot be properly represented by the *standard k-ε model*. According to the renormalization group theory (Yakhot and Orzag, 1986), the constants in the turbulent transport equations are given by:

$$C_\mu = 0.0845, \quad \sigma_k = 0.718, \quad \sigma_\varepsilon = 0.718, \quad C_{\varepsilon 1} = 1.42, \quad C_{\varepsilon 2} = 1.68$$

For the *realizable k-ε model*, the term "*realizable*" means that the model satisfies certain mathematical constraints on the normal stresses, consistent with the physics of turbulent flows. In this model the development involved the formulation of a new eddy-viscosity formula involving the variable C_μ in the turbulent viscosity relationship. It also differs in the changes imposed to the transport ε-equation (based on the dynamic equation of the mean–square vorticity fluctuation) where the source term is now solved according to:

$$S_\varepsilon = C_1 \rho \left(2S_{ij}^2\right)^{1/2} \varepsilon - C_2 \rho \frac{\varepsilon^2}{k + \sqrt{\upsilon_T \varepsilon}}, \quad S_{ij} = \frac{1}{2}\left(\frac{\partial u_i}{\partial x_j} + \frac{\partial u_j}{\partial x_i}\right) \tag{7.5}$$

and the variable constant C_1 is expressed as:

$$C_1 = \max \left[\frac{\eta}{\eta + 5} \right], \quad \eta = \frac{k}{\varepsilon} \left(2S_{ij}^2 \right)^{1/2}$$

The variable C_μ, no longer a constant, is evaluated from:

$$C_\mu = \frac{1}{A_o + A_s \frac{kU^*}{\varepsilon}} \tag{7.6}$$

Consequently, model constants A_o and A_s are determined as:

$$A_o = 4.04, \quad A_s = \sqrt{6} \cos \varphi, \quad \varphi = \frac{1}{3} \cos^{-1} \left(\sqrt{6} W \right), \quad W = \frac{S_{ij} S_{jk} S_{ki}}{\tilde{S}^3}, \quad \tilde{S} = \sqrt{S_{ij}^2}$$

while the parameter U^* is given by:

$$U^* \equiv \sqrt{S_{ij}^2 + \tilde{\Omega}_{ij}^2}, \quad \tilde{\Omega}_{ij} = \Omega_{ij} - 2\varepsilon_{ijk}\omega_k, \quad \Omega_{ij} = \tilde{\Omega}_{ij} - \varepsilon_{ijk}\omega_k$$

Other constants in the turbulent transport equations for this model are: $C_2 = 1.9$, $\sigma_\kappa = 1.0$, and $\sigma_\varepsilon = 1.2$. The k-equation in both the *RNG k-ε model* and *realizable k-ε model* is the same as that in the *standard k-ε model* except for the model constants.

The computed velocity profiles normalized by the free stream velocity u_∞ at locations of $x/H = 0, 1, 3, 5, 7$, and 9 behind the step, predicted through each turbulence model for the backward-facing step geometry, are illustrated in Fig. 7.9. In order to validate the numerical predictions, the results are compared against measurements of Ruck and Makiola (1988). We can observe that the flow velocities are better predicted by the *RNG k-ε model* and *realizable k-ε model* and that significant deviations from the experimental data are evident for the *standard k-ε model* at downstream locations of $x/H = 5, 7$, and 9.

Under the same flow condition, the maximum negative velocity profiles of the flow in the recirculation zone are shown in Fig. 7.10. Much lower values of the maximum negative velocities are predicted by all the three turbulence models in comparison to the experimental value. Among these three turbulence models, the re-attachment length of the *realizable k-ε model* comes closest to the measurement (a predicted value of $x/H = 8.2$ compared with the measured value of $x/H = 8.1$). Note the re-attachment length l_r, as described in Fig. 7.8. The *RNG k-ε model* marginally over-predicts the re-attachment length with a value of $x/H = 8.5$. The *standard k-ε model* claims the last spot as it severely under-predicts the re-attachment length by a substantial margin yielding a value of $x/H = 6.9$.

As aforementioned, the *standard k-ε model* generally over-predicts the gas turbulence kinetic energy k in the recirculation region, which leads to the evaluation of a higher turbulent viscosity ν_T ($= C_\mu k^2 / \varepsilon$). Fig. 7.11A demonstrates the normalized profiles of

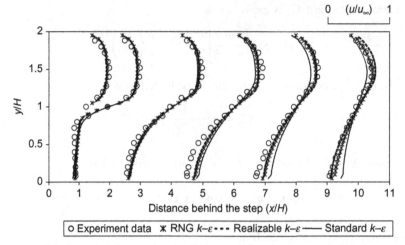

Fig. 7.9 Measured and predicted normalized velocity profiles at locations of $x/H = 0$, 1, 3, 5, 7, and 9 behind the step for the backward-facing step geometry.

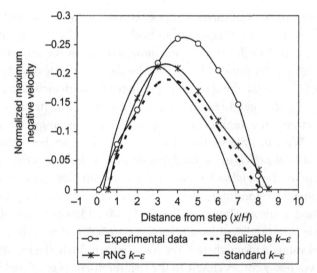

Fig. 7.10 Maximum measured and predicted negative velocity profiles of the flow in the recirculation zone.

the turbulent kinetic energy k at the location of $x/H = 5$, while Fig. 7.11B provides the normalized profiles of the turbulent dissipation rate ε at the same location determined by the respective three turbulence models. It is evident that the *standard k-ε model* predicts higher turbulent kinetic energy in the recirculation zone. This therefore results in the production of excessive mixing in the *standard k-ε model*, which significantly reduces the intensity of this zone (as also confirmed by Murakami, 1993). Another possible cause

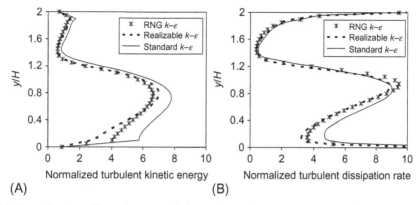

Fig. 7.11 Normalized profiles at location of $x/H = 5$ (recirculation region): (A) turbulent kinetic energy and (B) turbulent dissipation rate.

is the ε-transport equation of the *standard k-ε model*. The modifications in the ε-transport equation of the *RNG k-ε model* and *realizable k-ε model* accommodate the possibility of handling large rates of flow deformation. Both the rate of production of k and rate of destruction of ε can be reduced to yield smaller eddy viscosity. We note that even though the turbulence dissipation rate ε is predicted lower by the *RNG k-ε model* and *realizable k-ε model* in Fig. 7.11B, the turbulent viscosities predicted through these two models are still much lower because of the apparent k^2 evaluation in the turbulent viscosity relationship, which is dominated by the higher turbulent kinetic energy k predicted by the *standard k-ε model*.

Other Useful Guidelines: It is noted that for wall attached boundary layers such as that found within a simple flow between two stationary parallel plates or the backward-facing step geometry, turbulent fluctuations are suppressed adjacent to the wall and the viscous effects become prominent in this region known as the *viscous sub-layer*. This modified turbulent structure generally precludes the application of the two-equation models such as *standard k-ε model*, *RNG k-ε model*, and *realizable k-ε model* or even the *Reynolds stress model* at the near-wall region, which thereby requires special near-wall modeling procedures. Selecting an appropriate near-wall model represents another important strategy in the context of turbulence modeling. Here, the reader has to decide whether he/she adopts the so-called wall-function method, in which the near-wall region is bridged with *wall functions*, or a *low Reynolds number turbulence model*, in which the flow structure in the *viscous sub-layer* is resolved. This decision will certainly depend on the availability of computational resources and the accuracy requirements for resolution of the boundary layer. Some useful guidance on the application of relevant low Reynolds number models that can be employed all the way through the wall is provided herein. More discussions and practical guidelines will also be given for near-wall treatments using the wall-function method in the next section below.

The *k-ω model* developed by Wilcox (1998), where *ω* is a frequency of the large eddies, has been shown to perform splendidly close to walls in boundary layer flows. Such a model is common in the majority of commercial codes and it works exceptionally well particularly under strong adverse pressure gradients, hence its popularity in aerospace applications. Like the *standard k-ε model*, a modeled transport equation is solved for *ω* to determine its local distribution within the fluid flow. Nonetheless, the model is very sensitive to the free-stream value of *ω* and unless great care is taken in prescribing this value, spurious results are obtained in both boundary layer flows and free shear flows. In general, the *standard k-ε model* is less sensitive to the free-stream values but is often inadequate under adverse pressure gradients. To overcome such problems, Menter (1994a, 1994b) proposed to combine both the *standard k-ε model* and *k-ω model,* which retains the properties of *k-ω* close to the wall and gradually blends into the *standard k-ε model* away from the wall. This *Menter's model* has been shown to eliminate the free-stream sensitivity problem without sacrificing the *k-ω* near-wall performance.

To account for strong non-equilibrium effects, the SST (shear stress transport) variation of *Menter's model* (1993,1996) leads to a significant improvement in handling non-equilibrium boundary layer regions such as those found behind shocks and close to separation. It is therefore highly recommended for flow separation since the real flow is more likely to be much closer to separation (or more separated) than the calculations from the *standard k-ε model* suggest. Bear in mind that SST should not be viewed as a universal cure for turbulence modeling because it still inherits noticeable weaknesses. SST, for example, is less able to cope with flow recovery following flow re-attachment. For this, a promising possibility is the use of a length-scale limiting device, as proposed by Ince and Launder (1995). Interested readers may also wish to refer to Patel et al. (1985) for the applications of other various low Reynolds number versions of the *standard k-ε model* and the *Reynolds stress model* where modifications to the governing transport equations are used to deal with near-wall effects allowing these models to be deployed directly through to the wall. Alternatively, the *standard k-ε model* and the *Reynolds stress model* can be employed in the interior of the flow and coupled to the one-equation *k-L model* (Wolfshtein, 1969) that is dedicated to resolving mainly the wall region (see the review by Rodi, 1991), a so-called two-layer model. It is imperative that whatever low Reynolds number models are adopted, a sufficient number of grid nodal points must be placed into a very narrow region adjacent to the wall to adequately capture the rapid variation in the flow variables.

7.3.3 Near-Wall Treatments

It is inevitable that appropriate near-wall models are required for handling wall-bounded turbulent flow problems. In addition to turbulent models that can be applied all the way through the wall, another modeling procedure commonly adopted is *wall functions*. The use of these functions is prevalent in industrial practice and can be found practically in every CFD commercial and in-house computer code. Through this approach, the difficult near-wall region is not explicitly resolved within the numerical model but rather is bridged using such functions (Launder and Spalding, 1974; Wilcox, 1998). To illustrate the modeling procedure, attention will be primarily directed toward the consideration of the flow domain having smooth walls; the alternative treatment of rough walls will be subsequently mentioned as a special case.

To construct these functions, it is usual that the region close to the wall is characterized in terms of dimensionless variables with respect to the local conditions at the wall. If we let y be the normal distance from the wall and U be the time-averaged velocity parallel to the wall, then the dimensionless velocity U^+ and wall distance y^+ can be appropriately described in the form as U/u_τ and $y\rho u_\tau/\mu$, respectively. Within these dimensionless parameters, the wall friction velocity u_τ is defined with respect to the wall shear stress τ_w, as $\sqrt{\tau_w/\rho}$. If the flow close to the wall is solely determined by the conditions at the wall, then to some limiting value of dimensionless wall distance y^+, the dimensionless velocity U^+ can be expected to be a *universal* (wall) function as:

$$U^+ = f(y^+) \tag{7.7}$$

For a wall distance of $y^+ < 5$, the layer is dominated by viscous forces that produce the no-slip condition and is subsequently called the *viscous sub-layer*. We may assume that the shear stress is approximately *constant* and equivalent to the wall shear stress τ_w. A linear relationship between the time-averaged velocity and the distance from the wall can thus be obtained and making use of the definitions of U^+ and y^+ leads to:

$$U^+ = y^+ \tag{7.8}$$

Outside the *viscous sub-layer*, turbulent diffusion effects are felt and a logarithmic relationship is usually employed to account for this. The profile is:

$$U^+ = \frac{1}{\kappa} \ln(E y^+) \tag{7.9}$$

The above relationship is often called the *log-law* and the layer where the wall distance y^+ lies in the range of $30 < y^+ < 500$ is known as the *log-law layer*. Fig. 7.12 illustrates the validity of Eqs (7.8) and (7.9) inside the turbulent boundary layer. The values for κ (~ 0.4) and E (~ 9.8) are universal constants valid for all turbulent flows past smooth walls at high Reynolds numbers. For rough surfaces, the constant E in Eq. (7.9) is usually reduced. Additionally, the law of the wall must also be modified by scaling the normal wall distance y on the equivalent roughness height, h_o (i.e. y^+ is replaced by y/h_o), and appropriate values must be selected from data or literature. A similar universal, non-dimensional function can also be constructed to the heat and scalar fluxes. This can be used to bridge the near-wall region when solving the energy and scalar equations.

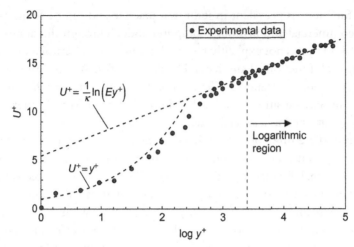

Fig. 7.12 The turbulent boundary layer: respective dimensionless velocity profile as a function of the wall distance in comparison to experimental data.

The reader should be aware that the universal profiles presented above have been derived based on an attached two-dimensional Couette flow configuration with the assumptions of *small pressure gradients*, *local equilibrium of turbulence* (production rate of k equals its dissipation rate), and *a constant near-wall stress layer*. Therefore it is imperative that care is always exercised to check the validity of wall functions to the CFD problem that is being solved. The calculated flow must be consistent or nearly consistent with the specific assumptions made, in arriving at the wall function relationships. Applying the wall functions outside this application range will lead to significant inaccuracies attained in the CFD solution.

To remove some of the limitations imposed by the above standard wall functions, a two-layer-based, non-equilibrium wall function is also available. A more in-depth discussion regarding the formulation of such a wall function can be found in Kim and Choudhury (1995). Briefly, the key elements are: (1) the *log-law*, which is now sensitized to pressure gradient effects, and (2) the two-layer-based concept, which is adopted to calculate the cell-averaged turbulence kinetic energy production and dissipation in wall-adjacent cells. Based on the latter, the turbulence kinetic energy budget for the wall-adjacent cells is sensitized to the proportions of the *viscous sub-layer* and the *fully turbulent layer*, which can significantly vary from cell to cell in highly non-equilibrium flows. This effectively relaxes the *local equilibrium of turbulence* that is adopted by the standard wall functions. Because of its capability to partly account for the effects of pressure gradients and departure from equilibrium, the non-equilibrium wall functions are recommended for complex flows that may involve flow separation, flow re-attachment, and flow impingement. In such flows improvements are obtained in the CFD solution, particularly in the prediction of wall shear and heat transfer. We note that the non-equilibrium wall functions were employed for the results obtained through the worked example of the backward-facing step turbulent flow problem considered in the preceding Section 7.3.2.

Another useful near-wall modeling method worth considering is the enhanced wall treatment that combines a two-layer model with enhanced wall functions. The main thrust of this approach is to achieve the goal of implementing the standard two-layer approach for fine meshes while not significantly reducing the accuracy for coarse meshes. By formulating the law of the wall as a single wall law for the entire wall region, the enhanced wall function extends its applicability throughout the near-wall region. This can be achieved by blending the linear and logarithmic laws of the wall as:

$$U^+ = e^{\Gamma} U_{lam}^+ + e^{1/\Gamma} U_{turb}^+ \tag{7.10}$$

where Γ is a blending function that allows the two different models to be smoothly blended.

$$\Gamma = -\frac{0.01(y^+)^4}{1 + 5y^+}$$

Similarly, the general equation for $\partial U^+/\partial y^+$ can also be expressed by:

$$\frac{\partial U^+}{\partial y^+} = e^{\Gamma} \frac{\partial U_{lam}^+}{\partial y^+} + e^{1/\Gamma} \frac{\partial U_{turb}^+}{\partial y^+} \tag{7.11}$$

The above equation allows the turbulent law to be easily modified and extended to account for effects such as pressure gradients or variable properties. It also guarantees the correct asymptotic behavior for large and small values of the wall distance y^+ and a reasonable representation of the velocity profiles in cases where y^+ lies inside the wall buffer region ($5 < y^+ < 30$). More details of this approach can be referred to in Kader (1993).

Near-Wall Meshing Guidelines on Wall Functions: Since the use of wall functions is to relate the flow variables to the first computational mesh point, thereby removing the requirement to resolve the structure in between, the lower limit of y^+ at this point must be carefully placed so that it does not fall into the *viscous sub-layer*. In such a case the meshing should be arranged so that the values of y^+ at all the wall-adjacent integration points are considered only slightly above the recommended limit, typically between 20 and 30. This procedure offers the best opportunity to resolve the turbulent portion of the boundary layer. Besides checking the lower limit of y^+, it is important that the upper limit of y^+ is also investigated during the computational calculation. For example, a flow with a moderate Reynolds number has a boundary layer that extends up to y^+ between 300 and 500. If the first integration point placed at a value of $y^+ = 100$, then this will certainly yield an impaired solution due to insufficient resolution for the region. Adequate boundary layer resolution generally requires at least 8 to 10 grid nodal points in the layer, and it is recommended that a post-analysis of the CFD solution is undertaken to determine whether the degree of resolution is achieved or the flow calculation is subsequently performed with a finer mesh. This can be achieved by plotting the ratio between the turbulent diffusion and the molecular diffusion (due to molecular viscosity) that is generally high inside the fully turbulent part of the boundary layer.

Near-Wall Meshing Guidelines with Low-Reynolds-Number Turbulence Models: A universal near-wall behavior over a practical range of y^+ may not be realizable everywhere in a flow such as that found for low Reynolds number flows. Under such circumstances, the wall function concept breaks down and its use will lead to significant errors. The alternative is to fully resolve the flow through the wall, which can be achieved by using low-Reynolds-number turbulence models as aforementioned, but it should be noted that the cost of the solution is around an order of magnitude greater than when wall functions are used because of the additional grid nodal points involved. With the intention of resolving the viscous sub-layer inside the turbulent boundary layer, y^+ at the first node adjacent to the wall should be set preferably close to unity (i.e. $y^+ = 1$). Nevertheless, a higher y^+ is acceptable so long as it is still well within the *viscous sub-layer* ($y^+ = 4$ or 5). Depending on the Reynolds number, the reader should ensure that there are between 5 and 10 grid nodal points between the wall and the location where y^+ equals 20, which is within the viscosity-affected near-wall region in order to resolve the mean velocity and turbulent quantities. This most likely will result in 30 to 60 grid nodal points inside the boundary layer to achieve adequate boundary layer resolution.

7.3.4 Setting Boundary Conditions

Specifying appropriate boundary conditions is particularly important in turbulence modeling. In this section we survey the various approaches to handling the various types of boundaries within the flow domain and provide some useful guidelines for setting proper boundary conditions for turbulence.

In many practical CFD problems specification of the turbulence quantities at the inlet can be rather difficult, and some sensible engineering judgment usually needs to be exercised. This is because the magnitude of turbulent kinetic energy k and dissipation ε can have a significant influence on the CFD solution. In most cases readily accessible measurements of k and ε are rare. In exploratory design computations the problem is compounded by the non-existence of any available boundary condition information to operate the turbulence models. Preferably, experimentally verified quantities should always be applied as inlet boundary conditions for k and ε. Nonetheless, if they are unavailable, then the values need to be prescribed using sensible engineering assumptions, and the influence of the choice taken must be examined against sensitivity tests with different simulations. For the specification of the turbulent kinetic energy k, appropriate values can be specified through a turbulence intensity I that is defined by the ratio of the fluctuating component of the velocity to the mean velocity. In general, the inlet turbulence is a function of the upstream flow conditions. Approximate values for k can be determined according to the following relationship:

$$k_{inlet} = \frac{3}{2}\left(U_{inlet}I\right)^2 \tag{7.12}$$

In external aerodynamic flows over airfoils the turbulence intensity level is typically 0.3%. For atmospheric boundary layer flows, the level can be as high as two orders of magnitude, that is, 30%. In internal flows the turbulence level between 5% and 10% is deemed to be appropriate. Similarly, the specification of the dissipation ε can be approximated by the following assumed form:

$$\varepsilon_{inlet} = C_\mu^{3/4} \frac{k^{3/2}}{L} \tag{7.13}$$

where L appearing in Eq. (7.13) is the characteristic length scale. If the $k\text{-}\omega$ *model* is employed, ω can be approximated by:

$$\omega_{inlet} = \frac{k^{1/2}}{C_\mu^{1/4}L} \tag{7.14}$$

For external flows remote from the boundary layers, a value determined from the assumption that the ratio of turbulent and molecular viscosity between 1 and 10 is a reasonable guess. For internal flows, a constant value of length scale derived from a characteristic geometrical feature can be employed such as 1% to 10% of the inlet hydraulic diameter. If the *Reynolds stress model* is applied, each stress component ($\overline{u_1'^2}$, $\overline{u_2'^2}$, $\overline{u_3'^2}$, $\overline{u_1'u_2'}$, $\overline{u_1'u_3'}$, and $\overline{u_2'u_2'}$) is required to be properly specified. If these are unavailable, as is often the case, the diagonal components ($\overline{u_1'^2}$, $\overline{u_2'^2}$, and $\overline{u_3'^2}$) are taken to be equal to 2/3 k, whereas the extra–diagonal components ($\overline{u_1'u_2'}$, $\overline{u_1'u_3'}$, and $\overline{u_2'u_3'}$) are set to zero (assuming isotropic turbulence). In cases where problems arise in specifying appropriate turbulence quantities, the inflow boundary should be moved sufficiently far away from the region of interest so that the inlet boundary layer and subsequently the turbulence are allowed to be developed naturally.

For solid walls, boundary conditions for k and ε or ω are substantially different depending on whether low Reynolds number turbulence models or the wall function method is employed. For the former, it is appropriate to set $k = 0$ at the wall but the dissipation ε is determined through either

$$\frac{\partial \varepsilon}{\partial n} = 0 \ \text{ or } \ \varepsilon = \left(\frac{\partial v_t}{\partial n}\right)^2 \tag{7.15}$$

where v_t is the velocity component tangential to the wall. For the k-ω model, the rough-wall method by Wilcox (1993) can be adopted. The surface value for ω can be written as follows:

$$\omega_{wall} = \frac{u_\tau}{\nu_{wall}} S_R \qquad (7.16)$$

The variable ν_{wall} is the kinematic viscosity on the wall, while S_R is a non-dimensional function determining the degree of surface roughness of the wall. However, when the law-of-the-wall type boundary conditions are employed instead, the diffusive flux of k through the wall is usually taken to be zero, thus yielding:

$$\frac{\partial k}{\partial n} = 0 \qquad (7.17)$$

The dissipation ε is, however, derived from the *local equilibrium of turbulence* assumption. It is noted that ε is not applied at the wall but rather is calculated at the first computational mesh point. For the finite volume method, ε_P is evaluated at the control volume center and is given by:

$$\varepsilon_P = \frac{C_\mu^{3/4} k_P^{3/2}}{\kappa \, n_P} \qquad (7.18)$$

where n_P denotes the normal distance from the wall to the first computational point. In cases where the non-equilibrium wall function is used instead, ε_P is now calculated according to:

$$\varepsilon_P = \frac{1}{2 n_P} \left[\frac{2\mu}{\rho y_\nu} + \frac{C_\mu^{3/4} k_P}{\kappa} \ln\left(\frac{2 n_P}{y_\nu} \right) \right] k_P \qquad (7.19)$$

where y_ν is the physical viscous sub-layer thickness computed from $y^* \mu / \rho C_\mu^{1/4} k_P^{1/2}$ and y^* is set at 11.225. More details considering the formulation of Eq. (7.19) can be referred to in Kim and Choudhury (1995).

At the outlet or symmetry boundaries, the *Neumann* boundary conditions are applicable, namely,

$$\frac{\partial k}{\partial n} = 0; \quad \frac{\partial \varepsilon}{\partial n} = 0; \quad \frac{\partial \overline{u_i' u_j'}}{\partial n} = 0 \qquad (7.20)$$

In the free-stream flow where the computational boundaries are far away from the region of interest, the following boundary conditions can be used:

$$k \approx 0; \quad \varepsilon \approx 0; \quad \overline{u_i' u_j'} \approx 0 \qquad (7.21)$$

which results in the turbulent viscosity $\mu_T \approx 0$.

7.3.5 Test Case: Assessment of Two-Equation Turbulence Modeling for Hydrofoil Flows

This test case was calculated using a commercial finite volume CFD computer code ANSYS-FLUENT, Version 6.1.

Model Description: The geometry is a two-dimensional hydrofoil, spanning the test section with a width of 3.05 m and a chord length (C) of 2.134 m. The cross-section profile is represented by a generic naval propeller of moderate thickness (t) and camber (f), utilizing a NACA-16 airfoil profile ($t/C = 8\%$ and $f/C = 3.2\%$) with two modifications. A detailed diagram of the hydrofoil geometry is illustrated in Fig. 7.13 while the anti-singing trailing edge geometry is detailed in Bourgoyne et al. (2000). Experiments performed on this test hydrofoil were conducted in the world's largest water tunnel, the William B. Morgan Large Cavitation Channel (LCC) in Memphis, USA.

Grid: For the hydrofoil geometry, the computational domain extends $1.5 \times C$ upstream of the leading edge, $1.5 \times C$ above and below the pressure surface, and $3 \times C$ downstream from the trailing edge. A mesh overlay of quadrilateral elements is constructed for the flow domain. Particular attention to the mesh generation is directed toward an offset *inner region* encompassing the hydrofoil. Within this region, a considerable fine O-type mesh is applied to sufficiently resolve the hydrofoil surface and the boundary layer region. For the *wake region* (downstream from the trailing edge of the inner region), a considerably fine H-type mesh is, however, applied to accurately resolve the near- and far-wake flow behavior. The remaining *outer region* of the domain is subsequently filled with a coarser H-type mesh. A total number of 208,416 grid nodal points is generated for the whole computational domain with an average distribution of y^+ at 2.31 and a minimum and maximum y^+ at 0.09 and 4.06, respectively. The computational grid is shown in Fig. 7.14 below.

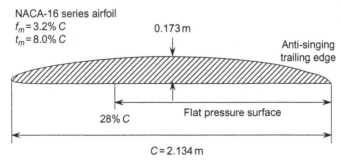

Fig. 7.13 Schematic view of the two-dimensional hydrofoil geometry.

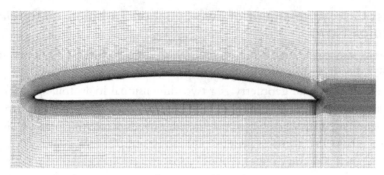

Fig. 7.14 Close-up view of the computational grid around the foil.

Features of the Simulation: This test case illustrates the importance of not only evaluating the choice of various turbulence models described above but also more importantly assessing the wall functions employed to model the near-wall region of the hydrofoil at high Reynolds numbers. Three wall treatments, namely, standard logarithmic wall function, non-equilibrium wall function, and enhanced wall treatment, are investigated.

The algorithm for the solution of the Navier-Stokes equations utilized an implicit segregated velocity-pressure formulation such as the SIMPLE scheme. This led to a Poisson equation for the pressure correction and was solved through the default iterative solver, normally the multigrid solver, of the ANSYS-FLUENT computer code. Finite volume discretization was employed to approximate the governing equations. To avoid non-physical oscillations of the pressure field and the associated difficulties in obtaining a converged solution on a collocated grid arrangement, the Rhie and Chow (1983) interpolation scheme was employed. A second-order upwind scheme was used for the convection, while the central-differencing scheme was used for the diffusion terms.

The working fluid, water, was taken to be incompressible and default initial conditions implemented in the computer code were used for the simulations.

Boundary Conditions: At the inlet, Dirichlet conditions were used for all variables. The inlet velocity based on the free-stream velocity U_{ref} was taken as constant with values of 3 m/s and 6 m/s. With the density and viscosity of water having values of 995.1 kg/m^3 and 7.69×10^{-4} kg/m $\cdot$ s, the corresponding Reynolds numbers based on the inlet velocities and the chord length are 8.284×10^6 and 1.657×10^7, respectively. At the free stream field, the inlet velocity was also applied to the computational domain walls above and below the hydrofoil. The turbulent kinetic energy k and dissipation ε were determined from the measured turbulence intensity I of about 0.1%. At the outlet, zero gradient conditions were applied for all the transported variables. No-slip wall boundary conditions were applied to the pressure and suction surfaces of the hydrofoil.

Results: The comparison between the measured and calculated coefficients of pressure distribution at the surface of the hydrofoil using different wall treatments is illustrated in Fig. 7.15. In this figure the realizable k-ε model is adopted in conjunction with the three wall treatments to predict the coefficients of pressure distribution. It is apparent that the use of different wall treatment approaches impacts the solution behavior. For strong adverse pressure gradients and boundary layer separation flows, the enhanced wall treatment produces the most accurate distribution. The prediction by the standard logarithmic wall function is slightly less accurate but still performs rather well at the leading and trailing edges. Nevertheless, the use of the non-equilibrium wall function produces questionable results, particularly at the leading edge where the coefficient of pressure distributions at the suction and pressure surfaces cross each other.

Fig. 7.16 represents the experimental and predicted pressure surface boundary layer velocity profiles at 93% C at a free-stream velocity of 3 m/s. The realizable k-ε model is also employed here in conjunction with the three wall treatments to predict the pressure surface boundary layer velocity profiles. As expected in the near-wall region, the use of different wall treatment procedures results in different predictions of the velocity profiles. Among these three different wall treatments, good agreement is achieved between the measurements and the enhanced wall treatment approach. The use of the standard logarithmic wall function produces a boundary layer velocity profile similar to that produced by the enhanced wall treatment, except having a larger boundary layer thickness. An even greater boundary layer thickness is predicted for the non-equilibrium wall function.

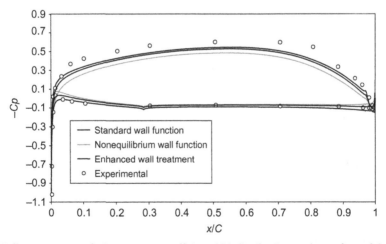

Fig. 7.15 Wall treatment analysis: pressure coefficient (C_p) distribution at the surface of the hydrofoil ($U_b = 3$ m/s).

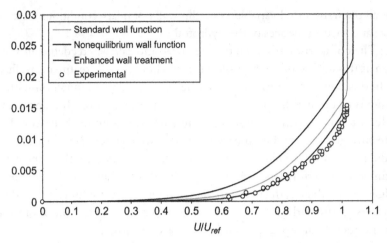

Fig. 7.16 Wall treatment analysis: pressure surface boundary layer normalized mean velocity profile at 93% C (U_{ref} = 3 m/s).

Figs. 7.17 and 7.18 demonstrate the measured and predicted pressure surface boundary layer velocity profiles applying different turbulent models at 93% C for free-stream velocities of 3 m/s and 6 m/s, respectively. It is observed that the three turbulent models of standard k-ε, standard k-ω (Wilcox's), and SST k-ω (Menter's) generally over-predict the boundary layer thickness when compared with the experimental result. The realizable k-ε appears, however, to be the only turbulence model

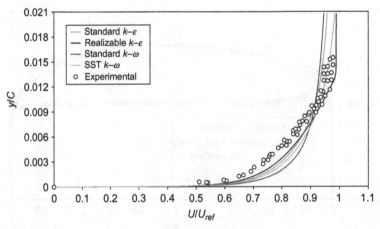

Fig. 7.17 Turbulence model performance: pressure surface boundary layer normalized mean velocity profile at 93% C (U_{ref} = 3 m/s).

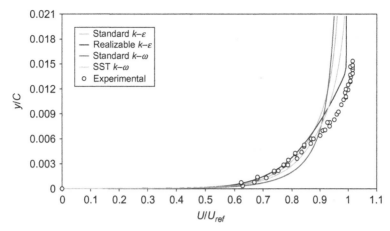

Fig. 7.18 Turbulence model performance: pressure surface boundary layer normalized mean velocity profile at 93% C (U_{ref} = 6 m/s).

to accurately predict the boundary layer velocity profile. Close to the surface ($y/C <$ 0.5%), this model predicts the boundary layer exceptionally well and continues to maintain a high degree of correlation further from the surface. Unlike the other models, the realizable k-ε model produces a definable gradient change where the velocity profile becomes rather blunt.

Conclusion: This test case focused on the evaluation of turbulence model applications based on the standard k-ε, standard k-ω (Wilcox's), SST k-ω (Menter's), and realizable k-ε in conjunction with three wall treatment approaches, namely, standard logarithmic wall function, non-equilibrium wall function, and enhanced wall treatment, for a turbulent boundary layer flow over a hydrofoil at high Reynolds numbers. As for the test case presented in Chapter 6, *validation* of the computational solutions is performed by comparing the predictions against experimental data to address the simulation model uncertainty and the degree to which models correspond to an accurate representation of the real physical flow in the absence of analytical solutions. The realizable k-ε model is found to accurately predict the pressure coefficient distribution at the surface of the hydrofoil leading to good overall predictions of the pressure-derived lift and drag coefficients. It also resolves the velocity profile and correctly predicts the thinning of the boundary layer, the commencement of boundary layer separation, and the full separation point of the turbulent boundary layer moving rearward with increasing Reynolds number. For flows with adverse pressure gradients and boundary layer separation, the enhanced wall treatment is considered to be a good candidate to handle such near-wall flow complexities. It is nevertheless noted that the numerical results obtained pertained to only the above test case and they may well be different for other types of flow problems considered.

7.4 OTHER PRACTICAL GUIDELINES FOR CFD USERS

7.4.1 Guidelines on Problem Definition

CFD users should always give careful thought at the outset of defining the CFD problem to be solved. The following questions and issues that are worth considering especially for problem definition include:

- What are the local or global information required for the intended CFD simulation?
- What are the important flow physics to be solved? For example, laminar, turbulent, transitional, steady, unsteady, compressible single phase or multi-phase, heat transfer, internal flow, external flow, etc.
- What is the area of primary interest or domain of the flow calculation? Any simplifications that can be imposed through the specifications of the boundary conditions at inlets or outlets or the use of symmetry or periodic boundaries?
- What level of validation is necessary for the CFD problem? For a routine application where validation and calibration of models has been performed on similar flow fields, only relatively small changes can be expected from earlier similar simulations. For a non-routine application where little earlier validation work has been carried out, the assessment should be based in assessing the flow physics and conditions of the applied models.
- What sensitivity tests are to be performed to assess the simulation accuracy? For example, mesh refinement on the prediction of the coefficient of drag, lift, or pressure or the local distribution of flow quantities.
- What results are meaningful for reporting for the CFD problem?

7.4.2 Guidelines on Solution Strategy and Global Solution Algorithm

Translating the problem definition into adopting a suitable solution strategy and global solution algorithm forms another important part of attaining the intended results of the CFD problem. Some recommendations are as follows:

- The adequacy of the solution procedure must always be checked with respect to the physical properties of the fluid flow. For example, compressible or incompressible flow.
- For CFD users employing existing commercial computer codes, it is always preferable as a first step to employ the default mathematical and physical models, numerical schemes, and pressure- or density-based solution method.
- If it is necessary to alter the parameters within the existing commercial computer codes to aid convergence, a step-by-step change of the relaxation factors or time step sizes would allow a CFD user to keep track of the parameters that will influence the convergence of the flow calculations. Rectifying the specification of boundary conditions, decreasing relaxation factors or time step sizes, or performing more number of iterations could alleviate the divergence problem.

- Convergence of the numerical solution should not be assessed purely in terms of the achievement of a particular level of residual error. The numerical solution should be evaluated on the convergence of global balances (conservation of mass, momentum and energy) where possible.
- A systematic one-by-one testing of the models, schemes, and solution method should always be performed in order to keep track of the models, schemes, and solution method that have been employed for the CFD problem. For example, the use of more sophisticated turbulence models with different mesh densities should be carried out to assess the accuracy of the numerical solutions.
- Time development of the flow quantities in the locations of interest will allow CFD users to examine whether the flow has reached its steady-state condition even though a solution has been computed. Using the steady solution as the initial condition, an unsteady simulation, preferably with a small perturbation, can be carried out to determine whether the fluid flow has achieved its steady state condition.

7.4.3 Guidelines on Validation

In assessing the CFD results the following guidelines on the validation of models may be adopted. They include:

- The experimental data sets to be employed should be independent of any data set for the construction of the CFD model. They should be carefully checked for quality and accuracy and should include measurements targeted for the purpose of CFD validation. These intended data sets should also be checked in order for them to be consistent with initial and boundary conditions for CFD simulation.
- The CFD model should be validated against the data set of an application with similar flow structures and flow physics. For a design application where accurate performance and flow-field data is required, the CFD model would require more detailed validation than just the preliminary investigation of a fluid flow. The CFD results attained should always be checked against the prominent and relevant flow features.
- For complex fluid flows, the taxonomy of the flow should be considered and it is recommended that the CFD model is validated against relevant simpler flows. Simple test cases would allow the assessment of different mesh densities so that the results of calculations at typical mesh resolution expected in the application can be investigated.
- The fluid flow should always be examined against a well-understood and well-tested baseline design application comprising the same important flow features as means of calibrating the expected accuracy. Reliability of a new design can subsequently be assessed by taking into account the calibration with the baseline design.

7.5 SUMMARY

Practical guidelines specifying appropriate boundary conditions within the CFD context and turbulence modeling as well as other useful guidelines on problem definition, solution strategy, and global solution algorithm and validation have been discussed in this chapter. The guidelines provided thus far are by no means exhaustive but they should put the reader in an advantageous position to understand how an appropriate CFD solution can be attained that is physically valid and meaningful through the above guidelines. It was also shown in this chapter that CFD simulation and analysis are more than a mechanically driven oriented exercise. They actually encapsulate, on the one hand, an understanding of the application of the essential fundamental theory of conservation equations and numerical methods to solve the CFD problem while, on the other hand, an appreciation of setting a viable mesh, defining suitable boundary conditions and models to arrive at a proper CFD solution.

Although not expounded within this chapter, the reader may well benefit from considering the use of either the steady state or transient approach to achieve a steady state solution. Generally, a steady-state calculation yields a solution in a shorter execution time in comparison to a transient calculation. Nonetheless, there may be some underlying circumstances where convergence and/or stability cannot be guaranteed through the steady-state approach and a transient calculation may be required to march the numerical procedure toward the steady-state condition. Other issues concerning the aspects of *computational accuracy* and *efficiency* can also have a strong influence on the CFD solution. Subject to the availability of computational resources, it is nearly always inevitable that some compromise has to be reached for solving complex CFD problems. By increasing the number of cells, that is, with decreasing mesh spacing, in the computational domain geometry, the *accuracy* of the computational solution is usually enhanced. There is, however, a trade-off that needs to be considered between increased computer storage and running time. One possible way that comparable *accuracies* can be obtained on a coarser mesh while maintaining *computational efficiency* is to employ higher-order discretization schemes to solve such problems. More sophisticated CFD approaches such as DNS and LES techniques for pertinent applications can yield solutions of high *computational accuracy* due to the application of fine meshes but are still subject to stringent *computational efficiency* due to computer hardware requirements. In practice, DNS and LES techniques should only be employed as a last resort when nothing else succeeds or to check the validity of a particular turbulence model being applied. Hence it is still beneficial to explore other practical turbulence models to solve a number of real engineering flows. As demonstrated in the above test case, a more advanced turbulence model based on the realizable k-ε with enhanced wall treatment can be applied to successfully resolve flows with adverse pressure gradients and boundary layer separation especially such as those experienced on hydrofoil or even aerofoil geometries.

In the next chapter we will apply the CFD techniques to a variety of practical problems of various degrees of complexity. The culmination of theory and practice will be demonstrated through these worked examples.

REVIEW QUESTIONS

7.1 What are some problems/difficulties in setting up correct boundary conditions?

7.2 What types of conditions may be applied for an inlet boundary and why do you need a corresponding outlet condition?

7.3 What is the *Neumann* boundary condition? Explain how it is used as an outlet boundary condition.

7.4 The geometry for an air-conditioning problem in two rooms separated by a partitioned wall is shown below. Label the boundaries that have to be defined and discuss what types of conditions may be applied.

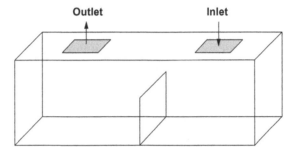

7.5 What requirements must be satisfied for a symmetry boundary condition, and what are the benefits of using this condition?

7.6 Discuss the main difference between a symmetry boundary condition and a periodic boundary condition.

7.7 The geometry of staggered tube-bank heat exchanger in two dimensions is shown below. Show, by a sketch, how you would define a computational domain for this geometry, by making use of the periodicity or the symmetry boundary condition.

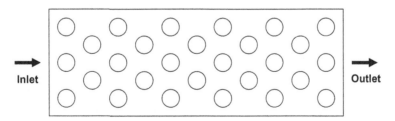

7.8 In general, describe how LES (large eddy simulation) solves a turbulent flow. How does this differ from DNS (direct number simulation)?

7.9 Why do engineers prefer the Reynolds-averaged-based turbulence models such as the k-ε model over the complex LES model?

7.10 What are some of the major problems experienced with the k-ε model?

7.11 The standard k-ε turbulence model over-predicts the turbulent kinetic energy k when compared with the *realizable* and *RNG* turbulence models in the recirculation zones. For the backward-facing step case given below, provide an explanation of your final predictions of the re-attachment length L_R for these turbulence models in comparison with the experimental result (L_E is the actual length). In modeling turbulent flow we effectively solve the following equations:

$$ u\frac{\partial \phi}{\partial x} + v\frac{\partial \phi}{\partial y} = \mu_t \left[\frac{\partial^2 \phi}{\partial x^2} + \frac{\partial^2 \phi}{\partial y^2} \right] + S_\phi $$

(a) Use the relationship between $\mu_t = \rho C_\mu k^2/\varepsilon$ and k to discuss the flow prediction.

(b) Which area within the domain requires fine mesh and where can coarse mesh be used?

(c) What kinds of boundary conditions do we impose for the inlet and outlet and at the wall?

(d) How should the flow pattern at the outlet be described?

(e) What are the values of the horizontal u velocity (positive or negative) before, after, and at the re-attachment point?

(f) Sketch to show the development of the velocity profile throughout the flow from the inlet to the outlet at X_1, X_2, and X_3.

(g) Keeping the inlet velocity the same, what would be the L_E if the working fluid is changed from air to water?

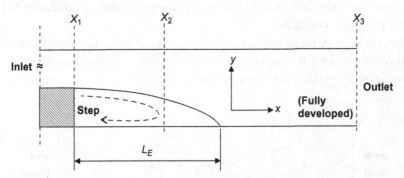

7.12 What is y^+ and why is it important in the context of near-wall turbulent modeling?

7.13 What should a typically recommended y^+ value be for the first adjacent node next to the wall in order to account for the viscous sub-layer?

7.14 Without experimental data for turbulent inlet profiles, what is the recommended method to consider turbulence effects?

7.15 (a) For the modeling of flow around a simplified car model given below, state and justify the values of y^+ to be used around the car, if the "enhanced wall function" model is employed.

(b) If C_d (coefficient of drag) is used for testing mesh independency, on what basis would you think of achieving it? Substantiate your answer with an X–Y plot.

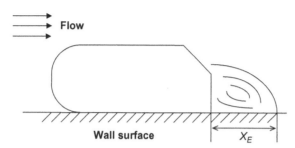

CHAPTER 8

Some Applications of CFD With Examples

8.1 INTRODUCTION

The increasing demand for CFD in resolving numerous fluid flow problems is growing within the scientific community. Needless to say, such an accomplishment could only have been made possible through the meticulous developments by persistent researchers and, more recently, the extension to industrial usage by dedicated code developers that have resulted in the availability of a number of commercial CFD packages. The cornerstone of any CFD analysis lies in the heart of transport equations and the building blocks of efficient numerical techniques. Throughout this book, the authors have continually stressed the importance of grasping the essential conservation equations as described in Chapter 3 and the basic understanding of numerical approximations considered in Chapters 5 and 6. In Chapter 7 the authors further provided useful guidelines for handling practical flow problems such as the requirement of suitable turbulence models to resolve real fluid flow processes, which incidentally exemplifies the range of computed results presented in Chapter 1.

We should be thoroughly aware that CFD is not confined to predicting only the fluid flow behavior. Consideration of adequate physical models that may appropriately handle chemical reactions (e.g. combustion), multi-phase flows (e.g. transport of gas-liquid, gas-solid, liquid-solid or even gas-liquid-solid mixtures), or phase changes (e.g. solidification or boiling) is increasingly being incorporated within the CFD framework to tackle some of these complex and challenging industrial processes. Obtaining real engineering solutions through CFD is now a very realizable prospect due to not only the evolution of computer hardware but also the mature development in numerical methods.

Having worked through the many important aspects of CFD in previous chapters, it is therefore fitting that this chapter culminates the knowledge gathered by aptly describing the application of theory into practice. From a practical viewpoint, it is envisaged that the selected examples that are illustrated in the subsequent sections, ranging from rudimentary to complex flow physics and simple to complicated geometrical domains, will assist in establishing some concrete steps toward undertaking any CFD problems that are exemplified through a wide range of engineering disciplines. The important aims of this chapter are thus:

- To illustrate the basic steps required for the reader to solve a practical flow problem

Computational Fluid Dynamics
https://doi.org/10.1016/B978-0-323-93938-6.00017-8

287

- To guide the reader on how to appropriately apply the knowledge gained within this book and any additional knowledge that may be required to solve other complex flow problems
- To demonstrate how CFD can be adopted as a research tool for better comprehending particular flow behaviors
- To demonstrate how CFD can be employed as a design tool in enhancing performance through a better understanding of the flow systems

8.2 TO ASSIST IN DESIGN PROCESS – AS A DESIGN TOOL

As previously described in Chapter 1, CFD is progressively being adopted to better optimize existing equipment and/or predict the performance of new designs even while in their conceptual stage of development. To illustrate how CFD can function as a design tool, a specific flow system is solved in the subsequent section, for example, a three-dimensional airflow in an office room layout. In practice, this flow problem is very prevalent in the design of ventilation systems.

The flow process within these systems is generally turbulent in nature and the suitability of which turbulence models (Chapters 3 and 6) to be applied characterized by low-Reynolds-number (LRN) turbulence will be assessed and discussed in the first half of this example. CFD is also increasingly being considered the preferred approach in the ventilation design process since scale-up experimental room measurements can be rather difficult to perform due to the expensive costs of instrumentation. The use of CFD as a design tool in enhancing the diffuser outlet design of a room ventilation system will be demonstrated in the second half of this example.

8.2.1 Indoor Airflow Distribution

Understanding the airflow distribution in enclosed environments is integral to indoor air quality control. It is well known that this is usually a function of the building's ventilation systems. With the availability of CFD computer codes, building engineers are increasingly embracing this methodology as an attractive alternative tool to predict the airflow distribution instead of employing scale modeling methods.

Despite many encouraging successes, some uncertainties still remain particularly in the application of turbulence models for ventilation design. An important aspect with regard to modeling indoor airflows is the characterization of LRN turbulence. The improper handling of LRN turbulence can contribute to inaccurate calculations since the airflow is strongly affected by the air phase velocity and turbulent fluctuations. Before CFD can be confidently applied for assisting engineers in ventilation system design, it is imperative to evaluate and validate the range of available turbulence models.

In this specific example (Tian et al., 2006) the application of three turbulence models: standard k-ε, RNG k-ε, and RNG-based LES models are investigated for an indoor

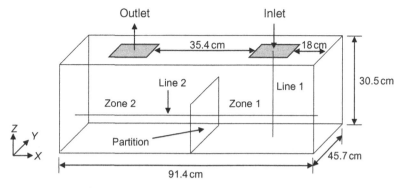

Fig. 8.1 Schematic view of the ventilation inside model room geometry.

airflow environment. The calculated airflow velocities are evaluated and validated against experimental data obtained by Posner et al. (2003) in a ventilated model room.

Model description: The geometrical structure of the model room for the purpose of this investigative study is illustrated in Fig. 8.1. The width, depth, and height of the room are respectively 91.4 cm, 45.7 cm, and 30.5 cm. Within this model room, a partition with a height of 15 cm is located in the middle of the room. Air is allowed to enter the room through one ceiling vent and leave through the other, as indicated by the outlet vent shown in Fig. 8.1.

Grid: For the room geometry, a structured mesh with rectilinear elements, distributed uniformly, is allocated for the whole physical domain, yielding an elemental volume size of 0.8 cm × 0.8 cm × 0.8 cm. The computational mesh is shown in Fig. 8.2, which results in a total number of 246,924 finite volumes generated for the whole computational domain. In any CFD calculation it is recommended that grid-independency

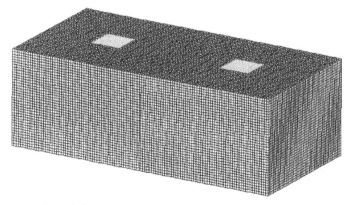

Fig. 8.2 Computational mesh for the model room geometry.

(Chapter 6) is performed to assess the numerical errors generated by the computed results. For this flow problem, the mesh is further refined to an elemental volume size of 0.5 cm × 0.5 cm × 0.5 cm. The difference in the air phase velocity predicted by the turbulence RNG k-ε model for the original mesh and the refined mesh is found to be less than 1%. For computational efficiency, the following results presented below are obtained with a mesh with an elemental volume size of 0.8 cm × 0.8 cm × 0.8 cm.

Features of the simulation: The vertical inlet velocity (U_{inlet}) of 0.235 m/s and a characteristic length of 0.1 m provides a flow Reynolds number of 1500. All computations are performed in transient state in which the time-dependent terms appearing in the transport equations are handled through an implicit second-order backward differencing in time. A non-dimensional time step of 0.035 is used. This time step has been defined by $t' = U_{inlet}\, t/H$, where U_{inlet} is the inlet air velocity as given above, t is the physical time step with a value of 0.05 s, and H is the room height. The transport equations are discretized using the finite-volume method (Chapter 5). For the RANS approach employing the standard k-ε and RNG k-ε models (see Chapters 3 and 6 again for a more detailed description), a third-order interpolation scheme such as the *QUICK* scheme (see Appendix B) is used to approximate the convective terms at the faces of the control volumes. In LES energetic eddies that exist near the cut-off wave number can significantly influence the spatial discretization errors (Park et al., 2004). The contribution of the LES subgrid-scale force may be overwhelmed by the use of *Upwind* and *Upwind-biased* schemes (Mittal and Moin, 1997) and hence the use of a central difference scheme is thus adopted for the LES calculations. For the pressure-velocity coupling, the SIMPLE algorithm is employed (Chapter 5). As the airflow in near-wall meshes can be at a very low Reynolds number ($y^+ \approx 1$), this study employs an enhanced wall treatment, a near-wall modeling method that combines a two-layer model with enhanced wall functions, for the k-ε models (Chapter 7). For the LES model, a very fine mesh is required to resolve the wall layer, which is very computationally expensive especially for engineering applications. Therefore, a wall model, like the RANS approach, is used to bridge the wall with the adjacent turbulent airflow. Convergence for the airflow governing variables (velocities, pressure, k, and ε) is assumed to have been reached when the iteration residuals are reduced by five orders of magnitude (e.g. 1×10^{-5}).

Model validation against experimental data: The case of the ventilated model room as investigated by Posner et al. (2003) is used to evaluate the turbulent indoor airflow through the standard k-ε, RNG k-ε, and RNG-based LES models. The initial condition of the flow field in a room is assumed to have a randomly perturbed velocity about the magnitude of the mean velocity U_{inlet}. To ensure that the solution for the LES computations achieves sufficient statistical independence from the initial state, time-averaged results are obtained from the instantaneous transient values after the airflow simulation is marched for 2000 non-dimensional time steps, which represents 100 s in physical time. After this time, the instantaneous values such as the airflow velocities are averaged over

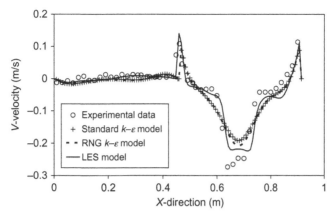

Fig. 8.3 Comparison between predicted and measured results of the vertical velocity component along the horizontal line at mid-partition height.

10,000 non-dimensional time steps (500 s in physical time). The simulated vertical air velocity component along the vertical inlet jet axis (line 1 in Fig. 8.1) and the vertical air velocity component along the horizontal line at mid-partition height (line 2 in Fig. 8.1) are validated against the measured results.

Fig. 8.3 shows the comparison between the predicted and measured vertical air velocity component along the horizontal line at the mid-partition height. Between the location of $x = 0$ m and the partition position, all three turbulence models yield almost similar results. Here again, the RNG-based LES model provides a slightly better prediction in the region between $x = 0.2$ m and the partition position. In the near-wall regions about the locations $x = 0.46$ m and $x = 0.9$ m, the RNG-based LES model successfully captures the highest positive vertical velocities while the two k-ε models significantly under-predict the velocities. From these results, it is apparent that better prediction is achieved through the RNG-based LES model, as demonstrated through the excellent agreement with the experimental results in the region from the partition position to the horizontal location of $x = 0.6$ m. Considerable under-prediction of the negative vertical velocity by the k-ε models, found in the region immediately beneath the inlet, may well be attributed to the excessive diffusion caused by the eddy-viscosity modeling. The marginal discrepancy between the measured data and the simulation results found in the region about the location $x = 0.85$ m shows that the k-ε models' results are slightly better than the RNG-based LES model. Overall, all three turbulence models perform rather well; a good agreement is achieved between the predictions and measured data and the flow trends are successfully captured. One significant aspect of this study is that the RNG-based LES model has been shown to provide significantly better results, especially in zone 2, because of the model's inherent ability to better capture the fluid flow characteristic within a confined space. Despite this important discovery, we observe that the predicted velocities by the two k-ε models are still within reasonable limits of the measurements. Practically, these models can still be applied with some degree of confidence and are generally quite sufficient for the majority of engineering applications.

Design of a room ventilation system: The practical use of CFD is illustrated herein. As a design tool, CFD can be systematically employed to parametrically investigate a number of ventilation design system features. Let us investigate the scenario of a ventilated compartment as depicted in Fig. 8.4. In this example cold air is supplied through the top plenum duct. A portion of this main air supply diverts into the diffuser grills at the ceiling, and consequently exhausts at the bottom of the diffuser outlet into an open space inside the enclosure. The diffuser grill is connected to the top plenum duct by an adjoining duct at a length H. Two design features – Design A and Design B – in the vicinity of the diffuser outlet are shown in Fig. 8.5. From a ventilation perspective, Design A generally results in a distortion of the air distribution as it exits through the diffuser grill. In order to achieve a more symmetrical diffusion pattern, guiding vanes near the adjoining duct are introduced as part of the improved design feature to better redistribute the air flow before reaching the diffuser outlet.

CFD simulations of Design A and Design B with the adjoining duct of length H are presented in Fig. 8.6. The color spectrum of the velocity vectors exemplifies the velocity magnitude whereby the lowest speed is designated by the color blue while the highest speed is indicated by red. As observed, the proposed design change with the guiding vanes being positioned upstream of the adjoining duct in Design B has a dramatic impact on the air distribution at the diffuser exit. The air flow appears to be more uniformly distributed before reaching the diffuser grill. A symmetrical flow pattern is obtained. This is not, however, the case that is seen in Design A. A stagnant flow region persists and the air flow beyond the diffuser grill behaves more like a jet stream being injected into the ventilated compartment.

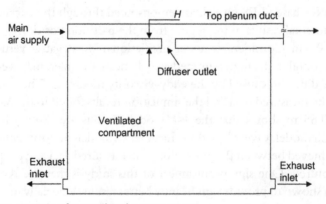

Fig. 8.4 Schematic scenario of a ventilated compartment.

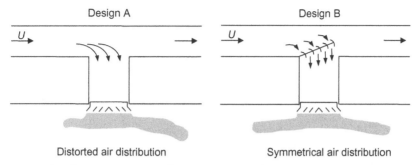

Fig. 8.5 Distorted and symmetrical air diffusion through diffuser grill.

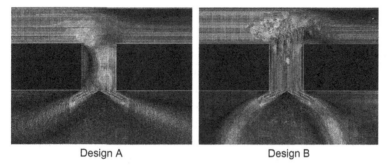

Fig. 8.6 Comparison between Design A and Design B for the predicted air distribution in the vicinity of the diffuser outlet with an adjoining duct of length H.

Conclusion: Two key points are demonstrated through this example. Firstly, the mature development achieved in turbulence modeling demonstrates the feasibility of employing such models for engineering applications with some degree of confidence. Secondly, it exemplifies the possible use of CFD in assisting the design process where multiple parametric investigations can be performed at low costs and hence providing engineers with the opportunity to better assess numerous design options toward the eventual selection of a viable system.

Before we proceed toward how CFD can be used nowadays as a research tool to enhance understanding of some challenging complex flows, the authors would like to allude to an important aspect regarding the application of turbulence models. Despite the significant advances achieved in the area of turbulence, even to the point where LES models nowadays can be feasibly applied with current computational resources, the standard k-ε still remains the workhorse model that can provide reliable predictions for many engineering flow problems. In practice, the standard k-ε model should be advocated for carrying out initial exploratory design calculations. The results obtained may well be adequate within the acceptable limits of engineering solutions but in the event that the model fails to capture the essential flow physics or crystallize any viable solutions, more advanced turbulence models can be applied as the next step of the design process to improve the CFD results.

8.3 TO ENHANCE UNDERSTANDING – AS A RESEARCH TOOL

In retrospect, CFD had been established more as a research tool rather than a design tool to comprehend the many important physical aspects of a flow field. We have briefly observed in Chapter 1 the ways whereby CFD can be applied nowadays to elucidate some interesting flow structures over cylindrical obstacles. In spite of the many significant advances and the maturity achieved in CFD research of single-phase flow problems, the use of this methodology is currently being realized in two-phase flow investigations. To further demonstrate the versatility of this tool, we further concentrate on another challenging area in CFD; that is, the numerical study of gas-particle flow.

Gas-particle flows are frequently encountered in numerous important processes such as in the minerals, petroleum, chemical, metallurgical and energy industries, and environmental engineering industries. Scaled-up experiments involving such flows are usually rather difficult to be performed mainly due to the inherent complexity of the flow phenomena (e.g. clusters appearing within the gas-particle flows) that require high-precision instrumentation. Fig. 8.7 illustrates an experimental flow visualization of solid particles (glass beads with a mean diameter of 66 μm) suspended in the gas flow through a 90° bend.

During the last decade, significant research efforts have been made in both academic and industrial research to better understand the dispersed two-phase flows using CFD techniques. Note that this next CFD example should only be construed as merely

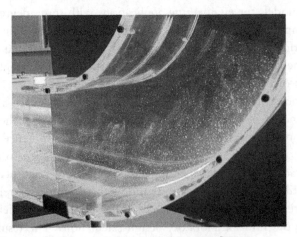

Fig. 8.7 Flow visualization of solid particles suspended in a gas flow.

providing a brief overview of the additional modeling effort that is required in modeling dispersed two-phase flows. Handling such flows generally requires the consideration of additional conservation equations and appropriate models where the gas phase is usually regarded as a carrier and the particle phase is taken to be dispersed within the gas flow. Firstly, the Lagrangian approach can be adopted by treating the gas phase as a continuum fluid with discrete particles in the fluid space. Secondly, both the gas and particle phases can alternatively be assumed to be continuum fluids, which is the essence of the Eulerian approach. Transport equations (mathematical models) describing the gas-particle flow according to the two different approaches will be discussed in the next section.

8.3.1 Gas-Particle Flow in a 90° Bend

Two different CFD computer codes, an in-house research code (Tu, 1997) and a commercial CFD code ANSYS-FLUENT, Version 6.1, are employed to simulate the gas-particle flow. The discrete approach described by the Lagrangian model is handled by the ANSYS-FLUENT code, while the continuum approach, which is essentially the Eulerian model, is solved through the in-house research code. The simulated results are validated against the experimental data of Kliafas and Holt (1987).

Model description: The geometry of this example is a three-dimensional turbulent airflow with particles traveling around a 90° duct bend as exemplified in Fig. 8.8. The schematic view of the experimental setup for this geometry is similar to Fig. 6.5 of the Test Case B in Chapter 6, which comprises a 1.2 m long horizontal duct, a 90° bend with a radius ratio of 1.76 (bend radius to bend hydraulic diameter), and a 1 m long vertical straight duct. Air enters the square test section with a bulk gas velocity U_b adjusted through the aid of a variable frequency controller. Experimental data for both the gas and particle fields are obtained through the laser Doppler anemometry (LDA) system.

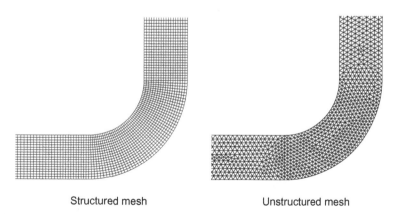

Structured mesh   Unstructured mesh

Fig. 8.8 Computational meshes for the 90° square-section bend.

Grid: For the 90° square-section bend, the computational domain begins at 1 m (10D) upstream from the bend entrance and extends to 1.2 m (12D) downstream from the bend exit. A structured mesh overlaying with hexahedral elements is generated for the in-house computer code. On the other hand, an unstructured mesh comprising tetrahedral elements fills the whole 90° bend geometry as illustrated in Fig. 8.8 for the ANSYS-FLUENT computer code.

Physics and mathematical models: For both approaches, the governing equations for the gas phase are the same as described by Eq. 3.53 (Chapter 3), which can be re-written as

$$\frac{\partial \phi}{\partial t} + \frac{\partial (u\phi)}{\partial x} + \frac{\partial (v\phi)}{\partial y} + \frac{\partial (w\phi)}{\partial z} = \frac{\partial}{\partial x}\left[\Gamma_\phi \frac{\partial \phi}{\partial x}\right] + \frac{\partial}{\partial y}\left[\Gamma_\phi \frac{\partial \phi}{\partial y}\right] + \frac{\partial}{\partial z}\left[\Gamma_\phi \frac{\partial \phi}{\partial z}\right] + S_\phi$$

(8.1)

By setting the transport property ϕ equivalent to 1, u_g, v_g, w_g, k_g, ε_g and selecting appropriate values for the diffusion coefficient Γ_ϕ and source terms S_ϕ, we obtain the set of transport equations as presented in Table 3.2 (see Chapter 3) for each partial differential equation describing the conservation of mass, momentum, and the turbulent quantities of the gas phase.

8.3.1.1 Governing Equations of Particle Phase Using the Lagrangian Approach

This approach calculates the trajectory of each individual discrete particle by integrating Newton's second law, written in a Lagrangian reference frame (Chiesa et al., 2005). Appropriate forces such as the drag and gravitational forces can be incorporated into the equation of motion. The equation can thus be written as:

$$\frac{du_p}{dt} = F_D\left(u_g - u_p\right) + \frac{g\left(\rho_p - \rho_g\right)}{\rho_p}$$

(8.2)

where $F_D\left(u_g - u_p\right)$ is the drag force per unit particle mass and F_D is given by:

$$F_D = \frac{18\mu_g}{\rho_p d_p^2} \frac{C_D \text{Re}_p}{24}$$

(8.3)

The relative Reynolds number Re_p is defined as:

$$\text{Re}_p = \frac{\rho_p d_p \left|u_p - u_g\right|}{\mu_g}$$

(8.4)

while the drag coefficient C_D is determined from:

$$C_D = a_1 + \frac{a_2}{\text{Re}_p} + \frac{a_3}{\text{Re}_p^2}$$

(8.5)

In Eq. (8.5) the coefficients denoted by a_1, a_2, and a_3 are empirical constants for smooth spherical particles over several ranges of particle Reynolds number (Morsi and Alexander, 1972). The *eddy lifetime* model is used to account for the effect of gas-phase turbulence on the particle phase. More details regarding this model can be found in Tian et al. (2005).

8.3.1.2 Governing Equations of Particle Phase Using the Eulerian Approach

For this approach, the governing equations for the particle phase can also be expressed in the form of Eq. (8.1) described above. The set of transport equations can be similarly obtained by setting the transport property ϕ_p equivalent to 1, u_p, v_p, w_p, k_{gp}, k_p, ε_{gp} and selecting appropriate values for the diffusion coefficient $\Gamma_{\phi p}$ and source terms $S_{\phi p}$; the special forms tabulated in Table 8.1 describe the partial differential equations for the conservation of mass, momentum, and the turbulent quantities of the particle phase. Details on the derivations of these equations can be found in Tu (1997).

Features of the simulation: The governing transport equations are discretized using a finite-volume approach with a non-staggered grid system (see Chapter 5 for more explanation). A third-order *QUICK* scheme (Appendix B) is used to approximate the convective terms, while a second-order accurate central difference scheme is adopted for the diffusion terms. The velocity-pressure linkage is realized through the SIMPLE algorithm (Chapter 5). In order to mimic the experimental conditions, a uniform velocity ($U_b =$ 52.19 m/s) for both gas and particle phases is imposed at the top inlet 1 m away from the bend entrance, which corresponds to a Reynolds number of 3.47×10^5. The inlet

Table 8.1 General Form of Governing Equations for the Particle Flow in an Eulerian Reference Frame.

$\phi_\mathbf{p}$	$\Gamma_{\phi p}$	$S_{\phi p}$
ρ_p	0	0
u	ν_{pT}	$F_D^u + F_G^u + F_{WM}^u$
v	ν_{pT}	$F_D^v + F_G^v + F_{WM}^v$
w	ν_{pT}	$F_D^w + F_G^w + F_{WM}^w$
k_{gp}	$\dfrac{\nu_{pT}}{\mathrm{Pr}_T}$	$P_{kgp} - \overline{\rho}_p \varepsilon_{gp} - II_{gp}$
k_p	$\dfrac{\nu_{pT}}{\sigma_k}$	$P_{kp} - I_{gp}$
ε_{gp}	$\dfrac{\nu_{pT}}{\sigma_\varepsilon}$	$\dfrac{\varepsilon_g}{k_g}\left(C_{\varepsilon 1} P_g - C_{\varepsilon 2} D_g\right)$

turbulence intensity is prescribed at 1%, whereas the particles are taken to be glass spheres of density $\rho_s = 2990$ kg/m^3 and size 50 μm. The corresponding particulate loading and volumetric ratios are 1.5×10^{-4} and 6×10^{-8}, respectively, for which the particle suspension is considered to be very dilute. At the outflow, the normal gradient for all dependent variables is set to zero. A non-slip boundary condition is employed along the wall for the gas phase as well as the particulate phase at the wall. Appropriate boundary conditions also need to be specified to represent the particle-wall momentum transfer for the Eulerian approach, more details of which can be found in Tu and Fletcher (1995).

For the Lagrangian model, the particle transport using a discrete random walk (DRW) model is computed from the converged solution of the gas flow. For the DRW model, 20,000 individually tracked particles are released from 10 uniformly distributed points across the inlet. The independence of statistical particle phase predictions is tested using 10,000, 20,000, and 50,000 particles. The difference in the maximum positive velocities of 20,000 and 50,000 particles is found to be less than 1 %.

In the Eulerian approach the numerical method for the solution of governing equations of the particle phase is similar to the method for scalar transport variables such as temperature for the gas phase. This is due to the fact that there is no pressure term in the particle momentum equations based on an assumption of no collisions among particles in a diluted gas-particle flow. All the governing equations for both gas and particle phases are solved sequentially at each iteration step. They are iteratively solved via the strongly implicit procedure (SIP) solver. The above solution process is marched toward steady state until convergence is attained.

The CPU time for the Lagrangian approach is generally much greater than the Eulerian approach. It is thus not surprising since significantly more computations are required to determine each individual particle trajectory in the Lagrangian approach when compared with the much reduced computational effort in only calculating the particle phase as one entity in the Eulerian approach.

Model validation against experimental data: The mean quantities of both gas and particle phases employing the two different numerical approaches are compared against the well-established experimental results of Kliafas and Holt (1987). In this example an important dimensionless parameter particularly for gas-particle flow is the Stokes number, which is defined as the ratio between the particle relaxation time to a characteristic time of the fluid motion; in other words, $St = t_p/t_s$. This dimensionless number determines the kinetic equilibrium of the particles with the surrounding gas. The system relaxation time, t_s, in the Stokes number can usually be derived from the characteristic length (L_s) and the characteristic velocity of the system under investigation. In this example it is the free stream velocity (U_b); hence $t_s = L_s/U_b$. A small Stokes number ($St \ll 1$) signifies that the particles are in near velocity equilibrium with the carrier fluid. For larger Stokes number ($St \gg 1$), particles are no longer in equilibrium with the surrounding fluid phase; they divert rather substantially from the fluid stream path.

Mean quantities are of utmost interest in engineering applications. For the gas and particle phases, the mean velocity, concentration, and fluctuation distributions along the bend are compared against measurements at the mid-plane along the spanwise direction of the duct geometry. All the values reported here (unless otherwise stated) are normalized using the inlet bulk velocity (U_b). Fig. 8.9 shows the comparison of the numerical results against the experimental data for the mean streamwise gas velocity (the case for $St < 0.1$) along various sections of the bend. There appears to be a qualitative agreement for both of the numerical models against the data reported by Kliafas and Holt (1987).

Enhance understanding through research: As a research tool, CFD can further enhance our understanding of particle behavior around the carrier gas phase by carrying out *numerical experiments* for the gas-particle flow. The following results demonstrate the use of this important tool.

To better understand the particle paths along the bend of different Stokes numbers, Fig. 8.10 depicts the paths using Lagrangian tracking for each respective case. For a Stokes number of 0.01, it is clearly evident that the particles have a tendency to act as "gas tracers." In such a flow scenario they are generally considered to be in equilibrium with the carrier phase. However, this phenomenon is less pronounced as the Stokes number is

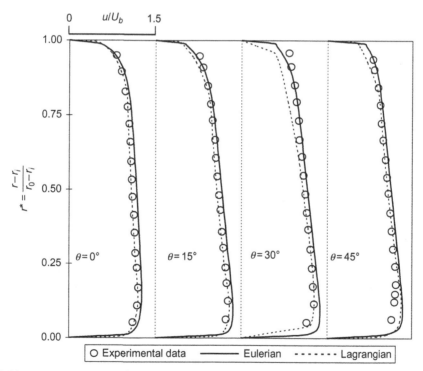

Fig. 8.9 Mean streamwise particle velocities along the bend for Stokes number $St = 0.01$.

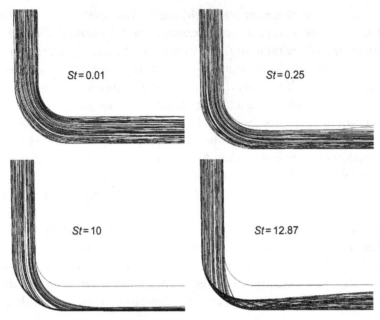

Fig. 8.10 Lagrangian particle paths for varying Stokes number.

increased. A positive slip velocity between the particulate and gas phase at the outer walls exists alongside the gas velocities at the inner walls due to the presence of a favorable pressure gradient. This "gas tracing" phenomenon of the particles becomes less pronounced as they approach the bend exit since the flow regains the energy it lost due to slip. It is observed that for flows with $St \geq 1$, the positive slip velocity between the particle and gas velocity decreases along with the bend radius and reverts to negative slip velocity at the bend exit, where the particles now lag the gas. They are unable to keep pace with the gas due to its own inertia as well as the energy lost through particle wall collisions. As the Stokes number increases, the particles show a greater tendency in migrating toward the outer bend.

Conclusion: With the increasing computing power and availability of a viable discrete approach described by the Lagrangian model or a continuum approach by the Eulerian framework, which entails the consideration of additional conservation equations, many complex multi-phase flow problems are increasingly being tackled through CFD. The feasibility of attaining qualitative and quantitative numerical results in this example represents not only the many rich physical insights that can be gained through the CFD methodology but also a testimonial feat of the rigorous advancements that have been made to the state-of-the-art multi-phase flow research.

8.4 OTHER IMPORTANT APPLICATIONS

8.4.1 Heat Transfer Coupled With Fluid Flow

There are other applications within the CFD framework whereby heat transfer significantly influences the fluid flow behavior. To demonstrate how to properly handle such flows, two examples are presented below to further elucidate pertinent features associated with modeling and procedures for obtaining meaningful practical solutions associated with these types of CFD problems.

8.4.1.1 Heat Exchanger

Heat exchangers are employed in numerous industries. Steam generation in a boiler, air cooling within the coil of an air conditioner, and automotive radiators represent just some of the conventional applications of this mechanical system. Of particular importance in the design of heat exchangers is the pivotal understanding of heat transfer in flow across a bank of tubes. Tube banks, as used in many heat exchangers, can be systematically arranged in an in-line or staggered manner as described in Fig. 8.11. Here, the cross-flow exchanger of cooler fluid removing the heat from the warmer fluid within the tubes flowing perpendicular to the schematic drawing is shown. For design purposes, tube banks are usually characterized by a number of important dimensionless parameters such as the transverse, longitudinal, and diagonal pitches. These parameters allow engineers to assess various heat transfer augmentation methods to specifically design special types of heat exchangers such as compact heat exchangers. According to Žukauskas and Ulinskas (1988), these types of heat exchangers (the main focus in this CFD example) are categorically considered by the dimensionless transverse and longitudinal pitches being less than 1.25. They have been shown to provide a higher heat transfer coefficient in *laminar flow* than that offered by a highly *turbulent flow*. Typically, the heat transfer surface areas are substantially increased while augmenting the flow, creating various secondary flows and simultaneously destroying the hydrodynamic and thermal boundary layers created by the repetitive nature of the flow.

Clearly, the long-standing practice in designing and manufacturing heat exchangers has been dominated by experimental methods and semi-empirical integral approaches based on simple criteria relations for the heat transfer coefficient. As early as the 1930s, Colburn (1933) had proposed a simple correlation for the heat transfer having flow across banks of staggered tubes, while others such as Aiba et al. (1982a, 1982b) and Žukauskas and Ulinskas (1988) have reported extensive experimental heat transfer and fluid friction during viscous flow across in-line and staggered banks of tubes covering both isothermal and isoflux boundary conditions. Lately, Khan et al. (2006) have provided analytical correlations that could be used for either in-line or staggered arrangement for a wider range of parameters. With the advent of commercial CFD packages, engineers are nonetheless warming up to the immense possibility of gaining insight into processes that are not easily amenable to analysis by measurements or simple overall

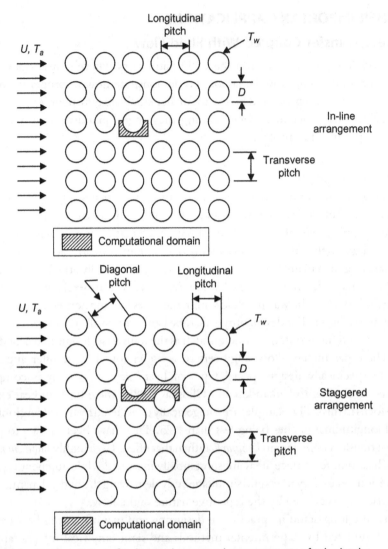

Fig. 8.11 Schematic description of in-line and staggered arrangements of tube banks.

computations, assessing both qualitatively and quantitatively their performance, and optimizing design and operation parameters through CFD. There are many advantages to the use of CFD for the design of heat exchangers. Recent pieces of literature by Witry et al. (2005), Hájek et al. (2005), and Wang et al. (2006) exemplify some of the current trends in the application of CFD to a variety of heat exchanger systems. With the increasing usage of this methodology, it is therefore appropriate to demonstrate and provide some useful practical guidelines of how CFD can be applied in solving the flow and heat transfer characteristics for compact heat exchangers.

Problem considered: For the in-line and staggered arrangements of tube banks illustrated in Fig. 8.11, fluid at a prescribed mass flow rate or velocity U and an inlet ambient temperature T_a much lower than the wall temperature T_w enters the tube banks from the left and exits at the right. By taking advantage of special geometrical features such as the inherent repetitive nature of the flow behavior, the computational fluid domain allows the possible exploitation of *symmetric* and *cyclic (periodic)* boundary conditions (as described in Chapters 2 and 6) in speeding up the computations and in turn enhancing the computational accuracy of the simplified geometries. The use of these boundary conditions, which are important features in practical CFD applications, will be demonstrated through this example.

Cyclic (periodic) boundary conditions can be suitably prescribed to ensure that the characteristics of the fluid leaving and entering the domain are identical (except the pressure). They are usually imposed on walls *perpendicular* to the flow direction. The rationale behind *symmetric* boundary conditions is, however, to replicate the fluid flow within the solution region to adjacent regions containing the same flow structures. They are commonly applied on walls *along* the flow direction. The solution regions of the two configurations are depicted by the shaded areas, as shown in Fig. 8.11 above. Specification of the relevant boundary conditions and the computational meshes that have been generated are described in Fig. 8.12 below. A hybrid mesh (structured and unstructured elements) is generated for each of the computational domains. For both cases, a boundary-layer stretched mesh of rectangular elements (structured) is concentrated near the tube walls, while the rest of the internal flow domain is filled by triangular elements (unstructured).

CFD simulation: The governing equations for the two-dimensional laminar flow within the simplified computational domains can be written as

$$\frac{\partial(\rho u \phi)}{\partial x} + \frac{\partial(\rho v \phi)}{\partial y} = \frac{\partial}{\partial x}\left[\Gamma_\phi \frac{\partial \phi}{\partial x}\right] + \frac{\partial}{\partial y}\left[\Gamma_\phi \frac{\partial \phi}{\partial y}\right] + S_\phi \tag{8.6}$$

By setting the transport property ϕ equal to 1, u, v, $H \, (= C_p T$, where C_p is the specific heat of constant pressure) and selecting appropriate values for the diffusion coefficient Γ_ϕ and source terms S_ϕ, we obtain the special forms presented in Table 8.2 for each of the partial differential equations for the conservation of mass, momentum, and energy. The solution is marched toward steady state where convergence is deemed to have been reached when the iteration residual of enthalpy is reduced by six orders of magnitude (e.g. 1×10^{-6}).

CFD results: This example is calculated using a commercial finite volume CFD computer code ANSYS-FLUENT, Version 6.1.22. The relevant data employed for the CFD calculations of the in-line and staggered arrangements are provided in Table 8.3. The working fluid is taken to be water. Fluid properties of water at a reference temperature

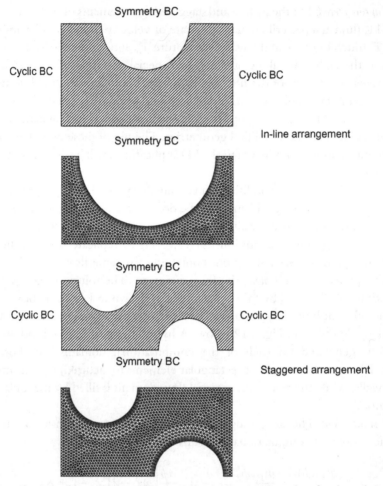

Fig. 8.12 Specification of boundary conditions and computational meshes for the solution regions associated with the in-line and staggered arrangements.

of 25°C are also given in the table below. Based on these properties, the dimensionless Prandtl number Pr can be evaluated; a value of 7.0 is thus obtained. For the computed results presented henceforth, it shall be assumed that there is a negligible effect of change in temperature on the fluid properties.

In Chapter 2 the Reynolds number has been shown to be a useful parameter in characterizing the fluid flow. Another important dimensionless parameter that is commonly associated with heat transfer is the Nusselt number (Nu), named after Wilhelm Nusselt, which may be obtained by multiplying the convective heat transfer coefficient h with the ratio L/k, where L is the characteristic length scale and k is the thermal conductivity, *viz.*,

$$Nu = \frac{hL}{k} \tag{8.7}$$

Table 8.2 Conservation Equations Governing the Fluid Flow Within the In-Line and Staggered Arrangements.

ϕ	Γ_ϕ	S_ϕ
ρ	0	0
u	μ	$-\dfrac{\partial p}{\partial x} + \dfrac{\partial}{\partial x}\left[\mu\dfrac{\partial u}{\partial x}\right] + \dfrac{\partial}{\partial y}\left[\mu\dfrac{\partial v}{\partial x}\right]$
v	μ	$-\dfrac{\partial p}{\partial y} + \dfrac{\partial}{\partial x}\left[\mu\dfrac{\partial u}{\partial y}\right] + \dfrac{\partial}{\partial y}\left[\mu\dfrac{\partial v}{\partial y}\right]$
H	$\dfrac{\mu}{Pr}$	0

Table 8.3 Data Used for the In-Line and Staggered Tube Bank Arrangements.

Quantity	Dimension
Tube diameter D, mm	10.0
Longitudinal pitch, mm	12.5
Transverse pitch, mm	12.5
Tube wall temperature T_w, °C	70
Mass flow rate, kg/s	0.01–0.4
Water properties:	
Thermal conductivity k, W/m · K	0.6
Density ρ, kg/m^3	998.2
Specific heat C_p, J/kg · K	4182
Kinematic viscosity ν, m^2/s	1×10^{-6}
Prandtl number Pr	7.0

A large Nu essentially signifies the *convection heat transfer* superseding the *conduction heat transfer* in the bulk fluid flow. For the design of heat exchangers, the diameter of the tube is usually taken as the characteristic length in the definition of Nusselt as well as Reynolds numbers. These numbers can be defined as: $Nu_D = D\,h/k$ and $Re_D = D\,U_{max}/\nu$. Instead of the usual inlet velocity being used as the reference velocity, the definition of the Reynolds number here differs by employing U_{max}, which is the maximum velocity in minimum flow area to be determined directly from the CFD solution. It is noted that the dimensions given for the longitudinal and transverse pitches in Table 8.3 correspond to the configuration of compact heat exchangers where the dimensionless longitudinal and transverse pitches are respectively 1.25 (12.5 mm/10 mm).

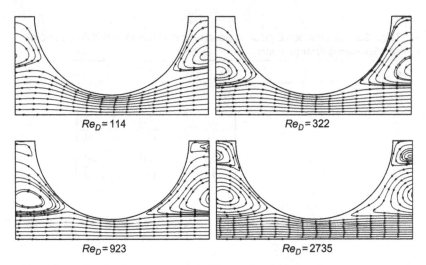

Fig. 8.13 Streamlines of in-line arrangement at different Reynolds numbers.

Fig. 8.13 illustrates the streamlines at different Reynolds numbers for the in-line arrangement. There appear to be two distinct flow regimes within the spaces separating the tubes along the longitudinal and transverse distances at low Reynolds number flows. In between the spacing of the top and bottom tubes lies the primary fluid flow of the system, which represents the multiple horizontal layers of fluid flowing along the longitudinal direction of the entire in-line tube bank. Sandwiched between these layers is the presence of two secondary recirculation vortices in between the spacing of the front and rear tubes. At sufficiently high Reynolds number flows, the breakdown of these larger secondary cells results in the formation of two smaller cells, further adding to the complexity of the fluid flow and heat transfer processes.

Fig. 8.14 shows the streamlines predicted for the staggered arrangement with boundary layer separation prevailing downstream of the fluid flow (indicated by points "S" in the figure). Here, the flow patterns resemble more like a snake weaving around the tubes. Further increase of Reynolds numbers tends to elongate and intensify the fixed recirculation eddies, which clearly show the flow breaking away from the tube at high Reynolds numbers. Such observation shares many similarities with the classical vortex shedding phenomenon, which represents an important aspect from the viewpoint of ensuring structural integrity. The shedding of these vortices gives rise to a lateral force acting on the tube. Since these forces are generally periodic following the frequency of vortex shedding, the tubes may be subjected to forced vibration, sometimes called self-induced vibration. As also observed in Fig. 8.14, the mainstream of these wavy flow layers in between the staggered spacing of the top and bottom tubes envelope the secondary elongated recirculation eddies residing in between the spacing of the front and rear tubes.

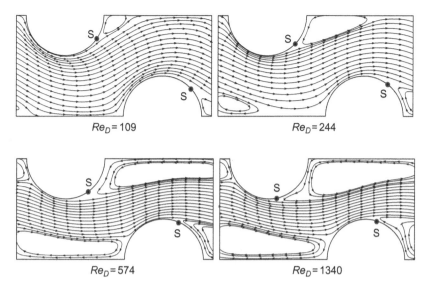

Fig. 8.14 Streamlines of staggered arrangement at different Reynolds numbers.

Higher heat transfer rates being experienced in staggered arrangement over in-line arrangement of tube banks in compact heat exchangers is further confirmed by the plot of the average Nusselt number against Reynolds number for the same dimensionless pitches in Fig. 8.15. From thermal-hydraulic considerations, one plausible explanation is that in a staggered bank the path of the main flow is deemed to be more tortuous and extends a greater coverage over the surface areas of downstream tubes, thereby resulting in more efficient removing of the heat that is contained within the internal tubes of

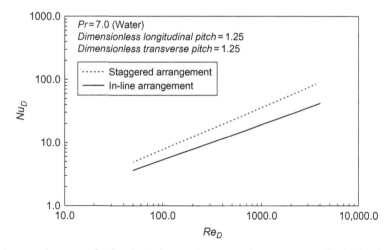

Fig. 8.15 Average heat transfer for the in-line and staggered arrangement of tube banks.

this compact heat exchanger system. This is quite the opposite for the in-line arrangement, where the flow structures observed above the main flow path appear to only glide over the tips of the top and bottom tube surfaces, which may explain the reduced effectiveness experienced in the in-line arrangement.

8.4.1.2 Solar-Induced Natural Ventilation

Transparent envelopes of buildings account for a huge portion of energy consumption due to large heat gain or losses. A common technique to improve the performance of transparent envelopes is the use of double skin facades (DSF), which have become indispensable in modern buildings owing to not only their aesthetic and constructional advantages but also providing better thermal comfort and consuming less energy. A naturally ventilated double-skin façade (NVDSF) is a type of DSF that utilizes absorbed solar radiation to enhance natural ventilation economically. The schematic representation of the configuration of a typical exhaust NVDSF is shown in Fig. 8.16. The inflow originates from the indoor space, while the outflow is connected to the outdoor environment. When solar radiation strikes transparent facades, the glazing temperature rises due to the absorption of solar radiation, hence inducing natural convection on the façade surfaces as a result of temperature differences between the façade and the ambient air. The

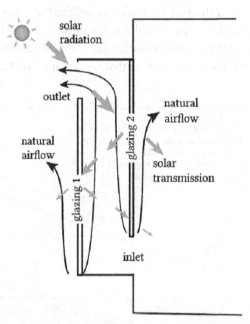

Fig. 8.16 Schematic representation of an exhaust NVDSF receiving solar radiation and inducing natural ventilation throughout the cavity.

NVDSF channel promotes natural ventilation that exhausts air from the inlet to the outlet to ventilate the connected space.

From a modeling viewpoint, this CFD example exposes a number of challenging aspects. Firstly, heat being absorbed by transparent facades is radiated to the surroundings as well as taken away by the natural convective air currents. This highlights the importance of the need to consider the heat transfer processes between both solid and gas regions, which require conservation of energy equations to be solved for each region, in order to accurately predict the solid-gas interface and temperatures of the solid façade. Secondly, the modeling of buoyant flow in air generally requires additional considerations particularly in the governing equations for momentum and turbulence. Thirdly, this example demonstrates the requirement of employing suitable turbulence and radiation models as well as imposing appropriate boundary conditions in order to sensibly attain numerical results that predict the actual ventilation performance of such a design.

Problem considered: In this specific example (Tao et al., 2021) the geometric domain consists of three parts: the NVDSF, the indoor domain, and the outdoor domain (see Fig. 8.17A). The NVDSF channel has a width of 1.2 m and a height of 2 m. Different channel gap depths (D) between about 0.05 and 0.4 m are chosen to examine the natural ventilation performance. The ambient domain has open boundaries with pressure set to 0 Pa at the in- and out-boundaries. Solar radiation strikes the NVDSF's outer glazing after passing through the semi-transparent surfaces at the front border of the ambient domain,

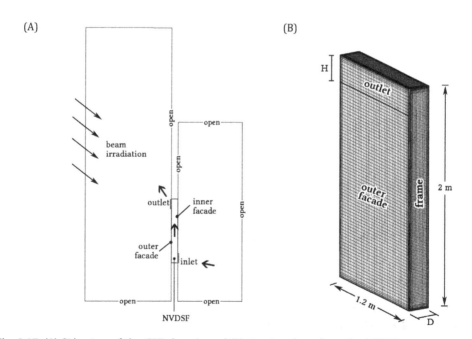

Fig. 8.17 (A) Side view of the CFD domain and (B) structured mesh on the NVDSF.

which acts as incoming solar beams with set radiation intensity and direction. The NVDSF's inner and outer glazing are both solid domains with a 6 mm thickness, and all of their surfaces are semi-transparent. The entire area is filled by air, and the density is evaluated from the consideration of the ideal gas law.

Grid: The domain is discretized with a structured mesh. As an example, only the mesh on the NVDSF is illustrated in Fig. 8.17B. Since the ventilation is purely driven by buoyancy force in the channel, it is important to resolve the thermal boundary layer on both facades by a fine mesh of prism elements. The prism layers on the glazing façades have fine mesh sizes of $y^+ < 1$. Five mesh sizes are used to evaluate the grid's sensitivity to the numerical solutions, and each mesh size is successively refined by around 1.4 at each dimension (2.7 times finer in total for the three-dimensional case). The number of cells tested have grid numbers of 144,880 (#1), 481,440 (#2), 1,166,319 (#3), 2,734,955 (#4), and 6,334,923 (#5).

The velocity magnitude and temperature distribution on the cavity's midline are depicted in Fig. 8.18. It can be seen that mesh #4 performs the best when accuracy and computational efficiency are both taken into account. For the remainder of the validation cases and numerical scenarios, mesh size #4 (NVDSF: 2 mm for the first layer grid thickness and 16 mm maximum within the channel; Room: 3 mm for the first layer grid thickness and 50 mm maximum in the domain) is utilized.

CFD simulation: The three-dimensional NVDSF system is modeled according to the computational domain as described above. Incident radiation through a semi-transparent media will be handled via the discrete ordinates (DO) model, which solves the RTE for a

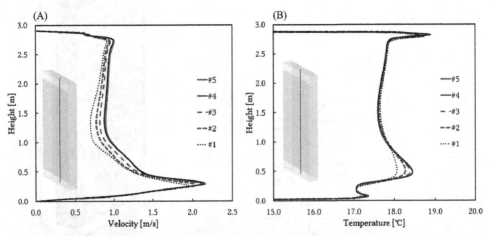

Fig. 8.18 (A) Velocity magnitude and (B) temperature over the vertical middle line of the cavity (line location as shown in the figure) for five mesh sizes.

set of discrete directions by substituting the radiation intensity in spatial coordinates into no solid angle formulation. The equation employed in the DO model is written as

$$\nabla \cdot \left(I_\lambda\left(\vec{r}, \vec{s}\right)\vec{s}\right) + \left(a_\lambda + \sigma_s\right)I_\lambda\left(\vec{r}, \vec{s}\right) = \frac{\sigma_s}{4\pi}\int_{4\pi} I_{b\lambda}\left(\vec{r}, \vec{s}'\right)\Phi\left(\vec{s} \cdot \vec{s}'\right)d\Omega' + a_\lambda n^2 \frac{\sigma T^4}{\pi}$$

(8.8)

where I, α, n, Φ, and Ω′ denote the radiation intensity, absorption coefficient, refractive index, phase function, and solid angle, respectively. $\vec{r}$ is the position vector and $\vec{s}$ is the direction vector. $\vec{s}'$ and σ_s are the scattering direction vector and scattering coefficient, respectively. σ is the Stefan-Boltzmann constant and T is the local temperature.

For the semi-transparent materials, the interface reflectivity on side a is calculated from ANSYS Inc. (2009):

$$\rho_a\left(\vec{s}\right) = \frac{1}{2}\left(\frac{n_a\cos\theta_b - n_b\cos\theta_a}{n_a\cos\theta_b + n_b\cos\theta_a}\right)^2 + \frac{1}{2}\left(\frac{n_a\cos\theta_a - n_b\cos\theta_b}{n_a\cos\theta_a + n_b\cos\theta_b}\right)^2$$

(8.9)

where ρ_a is the interface reflectivity on the side a; n_a and n_b are refractive index for media a and b; and θ_a and θ_b are the angle of incidence and angle of transmission, respectively.

All glazing surfaces are assumed to be purely specular. The spectral absorption coefficient a_λ can be found from the following relation:

$$\tau_i = e^{-a_2 x_g}\left(1 - \rho_i\right)$$

(8.10)

where τ_i is the transmissivity of glass material i; a_λ is the spectral absorption coefficient, x_g is the thickness of the glass, and ρ_i is the reflectivity of glass material i.

Model validation against experimental data: To validate the setting of solar radiation and DO model, a CFD model is developed according to the experiment conducted by Alvarez et al. (2000) (Fig. 8.19A). Their experiment evaluated the thermal properties of several glasses using a calorimeter device. A halogen–tungsten lamp was utilized in the experiment as a solar simulator to replicate the incident radiant energy. The incident radiant power that struck a plane surface of 0.50 m × 0.50 m at a distance of 0.5 m from

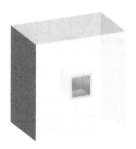

(A)   (B)

Fig. 8.19 (A) Experimental setup by Alvarez et al. (2000). (B) CFD model of the validation case – hot box and the ambient domain.

the plane was 900 ± 5.6 W/m^2. A heat exchanger, made out of an absorber copper plate with a thickness of 0.012 m, was positioned at the back wall in addition to the glass and walls.

The CFD model replicates the experimental setting as depicted in Fig. 8.19B. An ambient domain is built around the thermal box to mimic the lab environment. The heat exchange by the copper wall (the back wall) in the CFD model is set as a negative heat flux according to experimental data. Three of the glass materials tested in the experiment are chosen for validation: the 6 mm clear glass (CG-6), the 6 mm Filtrasol glass (FG-6), and the 6 mm Reflectasol glass (RG-6) (optical properties in Table 8.4).

Fig. 8.20 shows the transient variation of temperature on three glass materials over the 10 h duration. Comparison of CFD predictions and measured experimental data reveals a maximum 5.5% difference between results at the 10th hour (when temperatures become steady) for the three different glazing materials: clear glass (CG-6), Filtrasol glass (FG-6) with medium visible light transmissivity, and Reflectasol glass (RG-6) with low visible light transmissivity. Therefore it is believed that this CFD model, which incorporates the DO model and spectral optical properties, is sufficient to predict the future thermal performance of the NVDSF.

Table 8.4 Optical Properties and Results Comparison Between CFD and Experimental Data by Alvarez et al. (2000).

Glass Type	τ		α		T_Glass Exp. (°C)	T_Glass CFD (°C)	Error
	VIS	Solar	VIS	Solar			
CG-6	0.89	0.81	0.03	0.12	59.78	60.60	5.5%
FG-6	0.43	0.48	0.52	0.47	70.22	70.99	3.1%
RG-6	0.09	0.18	0.53	0.48	72.11	73.05	−2.6%

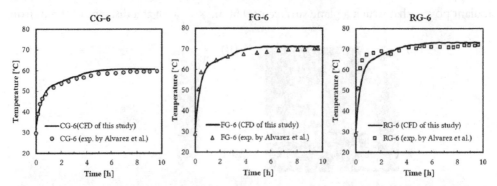

Fig. 8.20 Comparisons between transient CFD results and experimental measurements of temperatures on three glass materials (CG-6, FG-6, and RG-6) by Alvarez et al. (2000).

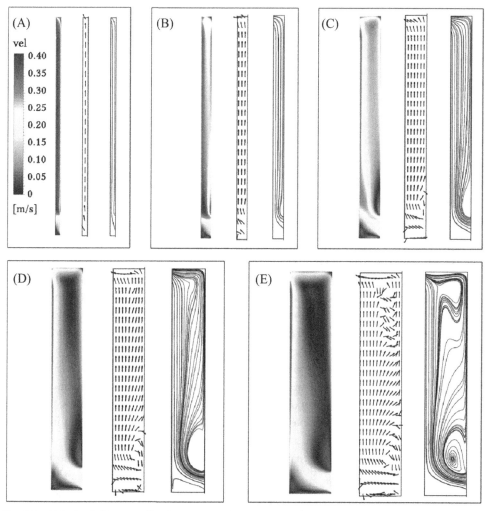

Fig. 8.21 Velocity field, normalized vectors and streamlines of gap depth (D) at (A) 0.05 m, (B) 0.1 m, (C) 0.2 m, (D) 0.3 m, and (E) 0.4 m.

CFD results: Fig. 8.21 illustrates the flow velocity contour, normalized vectors, and streamlines on the center plane for five channel gap sizes: 0.05m, 0.1 m, 0.2 m, 0.3 m, and 0.4 m. The interaction between two streams of natural currents from both facades plays a significant role in the influence of the NVDSF gap. Fig. 8.21A–C at gap sizes between 0.05 and 0.2 m show the cooperation of air currents, indicating smooth and directed airflow from the inlet to the outlet. However, when gap sizes are greater than 0.3 m, the trend changes to counteractions, which are depicted in Fig. 8.21D and E as a small region of counterflow at the top and a reversed direction. This is due to the feature of natural

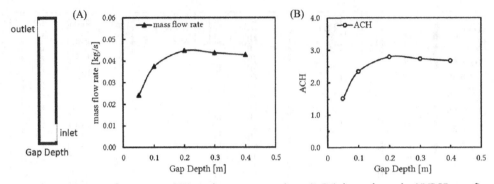

Fig. 8.22 (A) The mass flow rate and (B) air change rate per hour (ACH) throughout the NVDSF over five gap depths (D): 0.05 m, 0.1 m, 0.2 m, 0.3 m, and 0.4 m.

convection, which exhibits stronger airflow within the boundary layer while weaker at the domain boundary. Flow velocities are higher in narrower gaps due to the close spacing between boundary layers, but the insufficient path greatly hinders the flow. Contrarily, larger gaps allow more air to enter the channel, but heat exchange is less efficient; cold air reverses direction at the cooler inner façade and obstructs the outflow.

Fig. 8.22 depicts the ventilation rates. The analysis demonstrates that a gap deeper than 0.3 m is less conducive to ventilation due to the presence of reversed flow. However, it highlights a greater detrimental impact of smaller gap sizes on ventilation. The maximum ACH (air change per hour) of 2.8 among the five gap depths is found at a medium gap size of 0.2 m, which improves the overall ventilation by allowing a balanced airflow from both facades. Gaps larger than 0.3 m have a flow rate drop of about 2.2% per 0.1 m increment; nevertheless, smaller gaps provide a much more substantial reduction.

In terms of the vent sizes the main discrepancies in the flow field are in the air stream and initial flow intensity (Fig. 8.23). The small vents severely restrict the air's ability to exit, which causes circulations in the flow direction. The worst-case scenario happens at an extremely low h_v of 0.05 m, where the absence of an outflow renders the entire channel resembling an enclosure. Airflow from the inlet to the outlet becomes more directed by gradually expanding the vent's size. The airflow that blows on the exterior façade is another observable distinction, as the vent size affects the angle of the incoming airflow and causes different regions of stagnation. Along the exterior façade, $h_v = 0.2$ m (Fig. 8.23C) generates airflow velocity of around 0.4 m/s, whereas larger vents produce a smaller region that reaches 0.4 m/s due to the weaker entrance airflow.

Similar to the gap depth, Fig. 8.24 shows that a vent size neither too small nor large yields the best ventilation. A suitable size of 0.2 m results in a notable improvement in the ACH of 2.80. Analogously, smaller vent dimensions produce more serious deficiencies than larger ones. Ventilation fell significantly compared to optimal ventilation since small and sharp vents constrict airflow and circulations and increase the minor loss from the entrance.

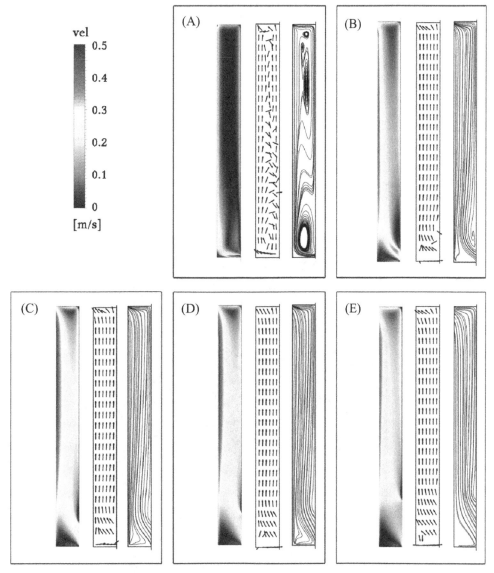

Fig. 8.23 Velocity field, normalized vectors and streamlines of NVDSF vent height (h_v) at (A) 0.05 m, (B) 0.1 m, (C) 0.2 m, (D) 0.3 m, and (E) 0.4 m.

8.4.2 A Buoyant Free-Standing Fire

As evidenced through experiments performed by McCaffrey (1979) and Cox and Chitty (1980), buoyant fires can be distinguished into three distinct regions: *a persistent flame*, *an intermittent flame*, and *a buoyant plume*. Fig. 8.25 exemplifies a typical three-zone flame structure for a buoyant fire (Cheung et al., 2007).

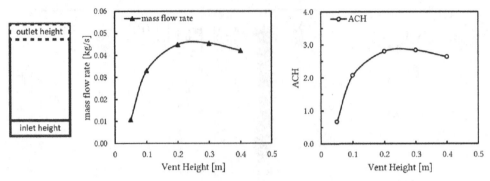

Fig. 8.24 Mass flow rate and air change rate per hour (ACH) over five vent heights (h_v): 0.05 m, 0.1 m, 0.2 m, 0.3 m, and 0.4 m.

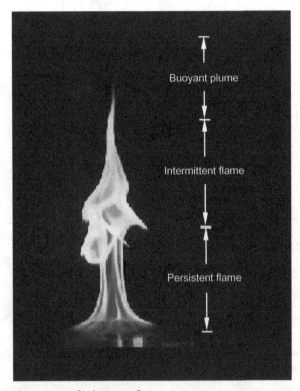

Fig. 8.25 A photographic image of a buoyant fire.

Buoyant fires are generally characterized by a very low initial momentum and they are strongly governed by buoyancy effects. It is also recognized that such fires exhibit an oscillatory behavior. The occurrence of this oscillation, generally known as the "puffing" effect, stems from the presence of coherent structures above a fire plume. These structures are a consequence of the developing buoyancy-driven instabilities, which subsequently leads to vortex shedding especially through the formation of large flaming vortices that rise up until they burn out at the top of the flame. The pulsating characteristics of such fires are strongly governed by the rate of air entrainment into the flame, flame height, combustion efficiency, and radiation heat output of the flames. Simulating such fires can be rather challenging.

According to McCaffrey (1983), the "puffing" mechanism can be described as follows. The initial momentum and buoyancy of combustion gases in stagnant air sets a large toroidal vortical structure. The entrainment of air is assisted into the reaction zone that causes the combustion gases to accelerate, creating the characteristic "necking." As the vortex propagates upward, it leaves an area of low pressure that is immediately filled by the combustion gases, creating a "bulge." This bulge rises, resulting in the formation of another toroidal vortex at the base of the fuel source. The vortex below the bulge shifts the plume surface outward in the radial direction, while the bulge above drags the plume surface inward. Hence the bulge structure is maintained. The rotational motion in the upper vortex causes the plume to be stretched in the axial direction. This periodic oscillation is usually referred to in the literature as the pulsation frequency of the flame-flickering phenomenon. Pulsation frequencies for buoyant diffusion flames have been reported by many researchers (Portscht, 1975; Zukoski et al., 1984), with Malalasekera et al. (1996) providing an excellent review.

CFD simulation: The turbulent buoyant fire is solved using an in-house large eddy simulation (LES) computer code to examine the coupled turbulence, combustion, soot chemistry, and radiation effects. The three-dimensional, Favre-filtered, compressible mass, momentum, energy, and mixture fraction and its scalar variant conservation equations are closed using the LES Smagorinsky subgrid-scale (SGS) turbulence model. The numerical method is based on a two-stage predictor-corrector approach for low Mach number compressible flows to account for the strong coupling between the density and fluid flow equations. The infinitely fast chemistry approach is adopted as the combustion model. A combination of a presumed beta-filtered density function and a conservation equation for the scalar variance is used to account for the SGS mixture fraction and scalar dissipation fluctuations on the filtered composition and local heat release rate. A soot model incorporating nucleation and surface growth agglomeration is considered. The radiation heat transfer has been accommodated through the discrete ordinates model. More details regarding this model can be found in Cheung et al. (2007).

A three-dimensional computation is performed on a designated 3 m $\times$ 3 m $\times$ 3 m domain. The gaseous fuel is injected through a porous square burner on the floor level with dimensions of 0.3 m $\times$ 0.3 m. A structured mesh totaling 96 $\times$ 96 $\times$ 96 cells is generated for the computational domain. Transient analysis is performed with about 27,500 time steps. To ensure that the analysis achieves stability and a pseudo-steady-state status, a simulation of 35 s in real time is used, for which, by employing a Pentium IV PC with a speed of 3.0 GHz

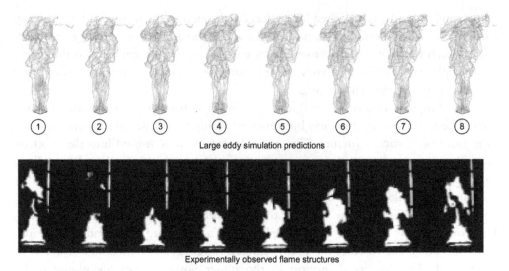

Large eddy simulation predictions

Experimentally observed flame structures

Fig. 8.26 Demonstration of the puffing effect during one flickering cycle.

and 2.0 Gbytes RAM, the CPU time for a real 35 s simulation is of the order of 400 h in order to ensure that it achieves stable and pseudo-steady-state status.

CFD results: Fig. 8.26 illustrates a series of frames capturing a flickering period comparing the macroscopic predictions using the LES approach against the photographic images obtained from McCaffrey's (1979) experiments. The naturally flickering behavior demonstrates the flow undergoing a phase shift along the axial direction. Numerically predicted images in Fig. 8.26 correspond to three isometric surfaces of temperatures at 800 K (visible flame), 450 K, and 310 K (cold smoke). Two distinct regions of the flame that comprises the upper flame separated from the lower persistent flame depicted in frames 1 and 8 by the numerical model show remarkable resemblance to the associated frames of the photographic images. The surging upward flame structures in frames 6 and 7 before flame separation in frame 8 are also adequately captured by the model when compared against the corresponding images of the developing fire observed during experiments.

The radiative cooling due to soot is demonstrated through the instantaneous soot distribution accompanied by the temperature contours and velocity vectors shown in Fig. 8.27. Buoyant diffusion flame behaves rather differently from jet diffusion flame. At the beginning of the puffing period, soot is formed at the outer fringes of the fire bed, where the meeting of the fuel and air is ideal. Owing to the self-excited toroidal vortex motion above these fringes, the soot being formed is found to be entrained in the continuous flame region, which is then carried upward during the middle and end of the puffing periods. The peak soot content is located predominantly in the middle portion of the continuous flame region for the buoyant fire; the temperature is thus subsequently lowered due to soot radiation.

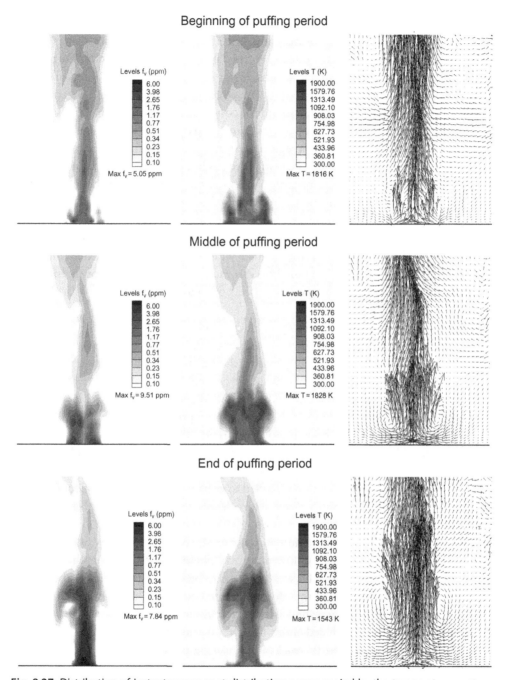

Fig. 8.27 Distribution of instantaneous soot distribution accompanied by the temperature contours and velocity vectors during one puffing period.

8.4.3 Dynamic Mesh Modeling of Human Motion

Indoor airflow patterns are significantly influenced by occupant activity that produces distinct wake flow regions and unsteady vortex shedding over the body. Particularly for indoor environments like aircraft cabins, hospital wards, operating rooms, and clean rooms where occupants' exposure to airborne contaminants is of major concern, it has attracted a lot of interest for its effect on indoor contaminant transmission and dispersion.

The occupant-induced wake displays the same fluid dynamics as separated flows over bluff bodies when air is drawn into a region of reverse flow by a mixing zone of vortices. Previous studies on wake flow characteristics such as boundary layer separation and vortex shedding usually adopt incoming airflow over stationary bluff bodies, which is simpler to achieve experimentally or numerically compared to setting up an absolute motion of an object. However, in some special scenarios which are difficult to achieve through relative motion, such as the one discussed in this section where human motion leads to the dispersion of pollutants existing in the stagnant indoor environment, introducing absolute body motion is essential. CFD method can examine how human movement affects wake dynamics by capturing the transient effects of a moving mannequin with dynamic meshing techniques.

Problem considered: An anthropomorphic mannequin with realistic motion can be simulated with the walking and swinging movement defined by the dynamic mesh technique, which can therefore be integrated with the discrete phase model (DPM) to help identify the influence of human walking on indoor particle re-suspension from floors. Compared to many studies that take simplified geometries such as blocks to represent the human body, anthropomorphic mannequins provide a better representation of the flow behavior in the near body regions. To reveal the spatial and temporal characteristics of the wake flow induced by walking, three walking speeds (0.8 m/s, 1.2 m/s, and 1.8 m/s) are adopted to represent slow, medium, and fast walking, respectively. In addition, the swinging of limbs is also included to enhance the evaluation of contaminant exposure influenced by occupant activity. More details regarding this model can be found in Tao et al. (2017).

It is anticipated that the effect of human wake on the surrounding flow field caused by walking is of small spatial range and low intensity. Therefore other factors that may also exert impacts on the flow field in this scenario need to be included as well. For the human body, the intensity of thermal plumes due to natural convection may generate flow disturbances at an approximate level; hence the heat transfer process should also be included in the analysis of motion-induced wake flow. Consequently, the solved flow domain with both human motion and thermal plume can be used to determine the effect of walking-induced indoor pollutant transport.

CFD simulation: The computational domain is 2.6 m (x-coordinate), 10.0 m (y-coordinate), and 2.7 m (z-coordinate). The mannequin model is 1.7 m in height

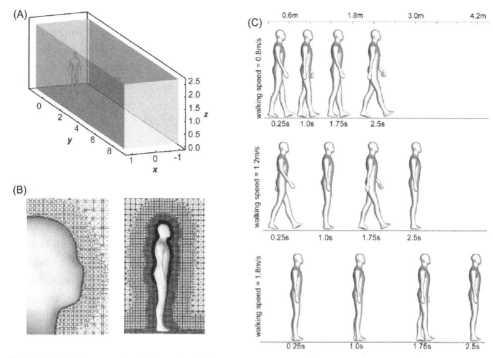

Fig. 8.28 (A) Computational domain of the room, x, y, z coordinates are in meters; (B) Prism layers and tetrahedral mesh around the mannequin body; (C) gait phases at the same time steps under three walking speeds.

and 0.58 m in width. The mannequin stands initially at x = 0 m, y = 0 m, with 1 m space from the back wall (Fig. 8.28A). Prism layers and unstructured tetrahedral cells are created for the domain (Fig. 8.28B). Ten layers of prism cells with $y^+ < 5$ are used to form the near-wall mesh surfaces. The moving mannequin's location and stance are updated for the dynamic meshing at each time step. Fig. 8.28C compares the distances covered in parallel steps when walking at three different walking speeds. The mannequin walks to 4.2 m in each of the models, and the simulations proceed until the flow field vanishes.

Walking postures are simplified in a way that arms and legs are regarded as rigid swinging pendulums with pivot points located at the shoulder and hips rather than a more complicated movement to lessen the computing burden of dynamic meshing. The arms and legs have a total flexion angle of 60° and 40°, and the rotating limb angular velocities are calculated using the length of the limbs, the walking speed, and the swinging period. A baseline walking speed of 1.2 m/s with a cadence of 2 steps per second is selected to represent the normal walking pace. A cadence of 1.33 steps per second for 0.8 m/s walking speed and 3.0 steps per second for 1.8 m/s walking speed are adjusted accordingly. The mannequin's surface temperature is maintained at 32°C, and the room temperature (air) is 22°C. The thermal plume of the body forms due to natural convection, while radiation, evaporation, respiration, and clothing are not taken into account.

The mannequin motion in the computational domain is achieved by layering and re-meshing dynamic mesh methods. For a general scalar ϕ on an arbitrary control volume V with a moving boundary, the integral form of the conservation equation is given as:

$$\frac{\partial}{\partial t}\int_V \rho\phi\,\mathrm{d}V + \int_A \rho\phi\left(\vec{u} - \vec{u}_g\right)\cdot\mathrm{d}A = \int_A \Gamma\nabla\phi\cdot\mathrm{d}A + \int_V S_\phi\,\mathrm{d}V \qquad (8.11)$$

where ϕ represents a scalar, ρ is the fluid density, $\vec{u}$ is the flow velocity vector, $\vec{u}_g$ is the grid velocity of the moving mesh, A is the boundary of the control volume V, Γ is the diffusion coefficient, and $S\phi$ is the source term. The discretization schemes used are the QUICK scheme for momentum, second-order upwind for the convection and diffusion terms, and the SIMPLE algorithm for pressure-velocity coupling. A time step of 0.001 s is adopted and the mesh is updated at each time step to progress the mannequin motion. In the commercial CFD program ANSYS Fluent v16, the gait cycle describing the angular limb velocities is achieved through a user-defined function (UDF), in which four macros (DEFINE_CG_MOTION) are defined for the left arm, right arm, left leg, and right leg, respectively, to achieve a coordinated walking stance. With the UDF defining the variation of the human geometry, the grid around the mannequin is re-meshed in each time step accordingly.

The particle re-suspension is modeled using the discrete phase model (DPM). In this Euler-Lagrange approach the discrete phase (particles) is added after the continuous phase (air) has been obtained as the initial flow field state. Particle parcels with a number of 256,557 and a total mass of 200 mg are distributed uniformly across the floor area and are introduced at the first time step (t = 0.001s). The particle tracking terminates once it reaches the surface since the boundary conditions at the floor, walls, and body surfaces are set to trap.

CFD results: The human wake can be visualized through velocity contour and vector plots on the x plane after the mannequin has come to a stop, as shown in Fig. 8.29. The total walking time varies according to the walking speeds. The velocity magnitude (U) and vertical velocity (w) contours are non-dimensionalized by the walking speed, and these are capped for velocities below U/Um = 0.10 to provide distinct variation between the wake flow and the ambient surroundings.

Airflow is driven from the front to the rear as a result of pressure build-up at the mannequin frontal body (stagnation points) and the pressure difference at the rear. Near the mannequin backline and buttocks, the disturbed air in the center plane created a triangular zone of flow behind the solid body with peak velocities of about U/U$_\mathrm{m}$ = 1.2. A zone of weak airflow is created behind the lower leg region as a result of the gap between the legs speeding up the air through and counteracting the wake downwash from the upper torso. Accelerated flow is visible through the space between the arm

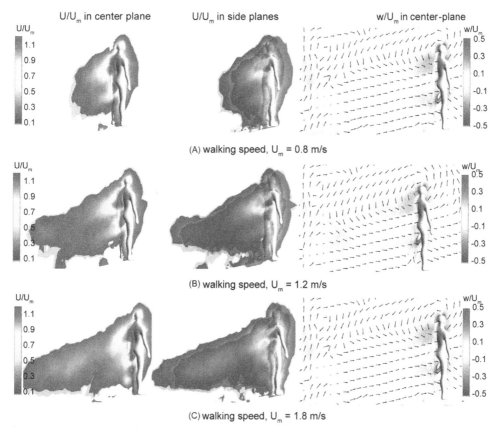

Fig. 8.29 Non-dimensionalized velocity field (U/Um) and non-dimensionalized vertical velocity field (w/U) on the vertical middle plane (x = 0 m) under walking speeds of (A) 0.8 m/s, (B) 1.2 m/s, and (C) 1.8 m/s.

and body in two side planes that are offset from the body's center and are placed at either side's shoulders. The leg that is finishing its step and getting ready to produce the flat foot pivot posture has high velocity along the calf.

The local flow behavior around the mannequin is demonstrated with streamlines originating from the body for a walking speed of 1.2 m/s, as shown in Fig. 8.30. Streamlines coming from the mannequin's front right side loop around before curving back in. This shows that the moving mannequin, which pushes the air in front of it to make room for the mannequin to occupy the space, causes considerable flow entrainment. Following the mannequin, the displaced air wraps back around the body and fills the empty area. The streamlines are seen to remain near the body for the first step before becoming wider for the second step. In general, the streamlines don't exhibit much updraft, which could potentially re-disperse particles from the ground into the breathing area. While it goes backward, the streamlines at knee height will rise while those closer to the feet stay close to the floor.

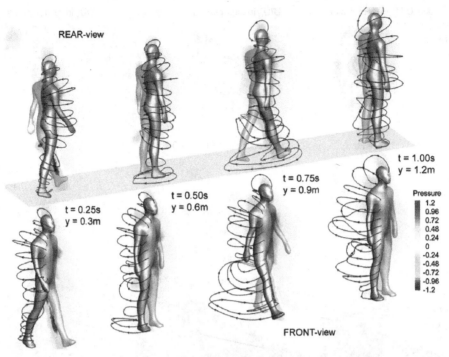

Fig. 8.30 3D streamlines generated from the right side of the body under a walking speed of 1.2 m/s with overlayed pressure contour during the first gait cycle, which occurs between $t = 0.0$ and $t = 1.0$ s.

The particle dispersion between 0.15 and 0.15 m laterally from the body's midplane (x-coordinate) is depicted in Fig. 8.31. It is demonstrated that the largest particle concentrations are formed when walking at a speed of 0.8 m/s. The deepest color denotes a place that is closest to the body (x = 0 m) and is determined by the lateral distance of the particles from the x-coordinates (lateral direction). The floor-distributed particles are vigorously agitated and entrained into the flow during the early stages of walking. Particles suspended in the air are moved from the front to the back due to the velocity at knee height brought on by airflow passing through the space between the legs. When particles enter the wake, the airflow created by the legs moves upward into the back, where it eventually spreads to higher regions. Under a strong upward airflow caused by the upper body, the particles in front of the body move both toward the body and upward into the breathing zone.

The mannequin has ceased moving at time T = 1.00, but the forward motion of the particles is still being carried out by the residual airflow. The particles advance and separate from the body from T = 1.00 to T = 3.80. The particles are accelerated by the jet-like flow through the leg gap, which creates a parabolic profile (T = 1.80). The thermal plume from natural convection becomes large as the airflow continues to advance and

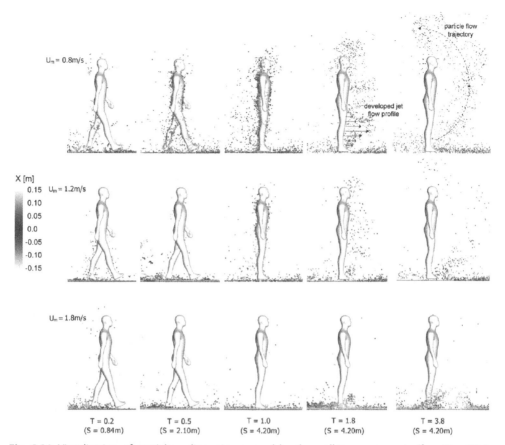

Fig. 8.31 Visualization of particle redispersion caused by the walking mannequin (from T = 0.2 to T = 1.0) and after the mannequin has stopped walking (from T = 1.0 to T = 3.8).

dissipate, and the particles travel ahead and upward in a circular profile (T = 3.80). Particles continued to travel after the mannequin stopped moving, first forward due to the remaining wake momentum and then upward as a result of the thermal plume effect.

8.4.4 Air/Particle Flow in the Human Respiratory System

8.4.4.1 Drug Delivery in the Human Nasal Cavity

An alternative to oral and injection routes of delivering systemic drugs for a variety of diseases is nasal drug delivery. The many advantages associated with such delivery are well documented and one of which is a viable treatment for respiratory ailments such as congestion and allergies. One aspect that still concerns the prescription of such treatment is the availability of useful particle deposition information within the nasal cavity that can significantly improve the effectiveness of a nasal sprayer device to deliver the drug to specifically targeted sites within the human nasal cavity.

Current in vivo and nasal cavity replica methods are usually rather limited in the scope of more intensive studies because of their intrusiveness as well as their being very time consuming and expensive in implementation. Nevertheless, the accessibility of rapid computers to perform numerous numerical analyses such as the repeatability and accuracy of a nasal spray injection released from the same location with quick turnaround times, and possibly extending a wider range of investigations, has certainly revolutionized how medical research can be carried out nowadays. Advanced application of CFD is certainly gaining enormous interest, as evidenced in Keyhani et al. (1995), Yu et al. (1998), and Hörschler et al. (2003). With the aid of graphical representations of the local particle deposition sites, particle and air flow paths, velocity contours, and vectors at any location, detailed critical assessments can be obtained on the effectiveness of delivery for a particular nasal sprayer device being tested.

Spray particle deposition can be ascertained through several parameters:

(1) Gas phase flow field such as velocity and turbulence effects.

(2) Deposition mechanisms involving the interaction between particles and its continuum.

(3) Material properties of the particle and initial spray conditions such as particle density, size, and spray cone angle.

Knowledge of the fluid flow field allows for prediction of the particle dispersion and deposition. The construction of the complex nasal cavity geometry needs to be appropriately managed for CFD simulations before calculations can proceed. This has been addressed in detail by Inthavong et al. (2006).

Problem considered: The schematic illustration of the human nasal cavity model from Tong et al. (2016) is shown in Fig. 8.32. The geometry was obtained through a CT scan of a healthy 25-year-old Asian male who is 170 cm in height and 75 kg in weight. To better capture the airflow patterns within the breathing zone especially near the nostril inlets, facial characteristics are reflected in the current model. When taking the driving fingers into account, the spray nozzle's maximum insertion depth L_d is 11 mm (Fig. 8.32).

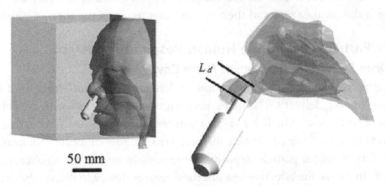

Fig. 8.32 Computational model including facial features and nasal spray device.

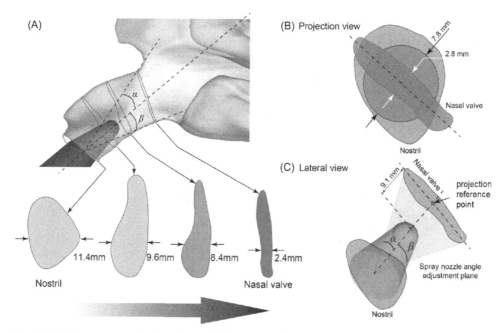

Fig. 8.33 (A)The space variation of vestibule passage for the nasal spray plume development and (B and C) the alignment of the spray nozzle and the nasal valve.

The anatomical characteristics of the vestibule region are described in cross-sectional planes in Fig. 8.33A. The nasal valve serves as the primary restriction for the development of the spray plume because the vestibule channel continuously narrows from the nostril to the most narrow nasal valve (width 2.4 mm). A poor spray nozzle-nasal valve alignment will filter the majority of the sprayed droplets due to this limited width in the vicinity of the nasal valve, which will greatly reduce the effectiveness of the drug delivery.

A spray nozzle adjustment plane is suggested to optimize the spray nozzle-nasal valve alignment. Fig. 8.33B shows how the spray nozzle is pointed at the nasal valve's center line, and Fig. 8.33C shows how the nozzle's orientation can be adjusted. Angles α and β denote the allowance for adjusting the angle upward and downward, respectively. To compare their effects on the sprayed particle deposition performance, three representative spray orientations – the center direction (OA), the tilted up direction (OB, halfway of the upward adjustment allowance), and the tilted down direction (OC, halfway of the downward adjustment allowance) – are taken into consideration.

CFD simulation: The reader should take note that the mesh generation step presents the most labor-intensive component of the entire simulation process because of the requirement to appropriately mesh the intricate surface topology of the airway geometry. Prism layers were imposed at the near wall region to better capture the airflow and particle deposition patterns. The optimum mesh size for the current CAD model, according

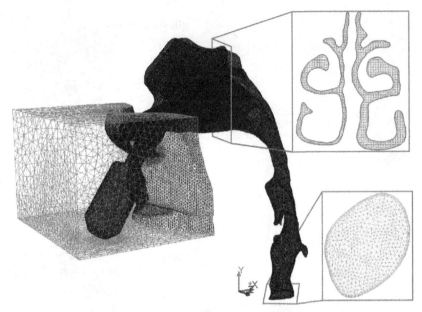

Fig. 8.34 Computational mesh results including the presence of a nasal spray device.

to the grid independence test, is 1,921,484 cells. Fig. 8.34 shows the computational mesh results, in which a coronal plane and the outlet of the computational domain display the internal mesh with a mesh-refined region close to the walls.

ANSYS-Fluent v14.5 is utilized to solve the full Navier-Stokes equations for the flow field of the continuum gas phase under steady-state conditions. The momentum equation is approximated by the QUICK scheme while the pressure–velocity coupling is achieved through the SIMPLE method. There is only one flow rate taken into account, which is 15 L/min and represents an adult's resting breathing condition. Since instructions of some commercial nasal spray devices recommend that the nostril on the side not receiving the medication to be closed by a finger, the right nostril is closed in this study.

According to Naftali et al. (2005) and Schroeter et al. (2006), the nasal main passage experiences a laminar flow regime with flow rates of 15 L/min. However, significant velocity acceleration and turbulent flow are anticipated in the anterior vestibule region of the left nasal cavity with the closure of the right nostril and the insertion of the spray nozzle. Hence the right nostril closed mode adopts the transitional SST model for flow field simulation.

CFD results: Fig. 8.35 illustrates the primary flow characteristics in the drug-administering half nasal cavity (left chamber). Three key characteristics are vertical flow, separation, and flow acceleration. The acceleration found in the vicinity of the nasal valve is primarily caused by the reduced cross-sectional area. The flow separation found at the superior main passage is primarily caused by the sudden vertical passage expansion, while

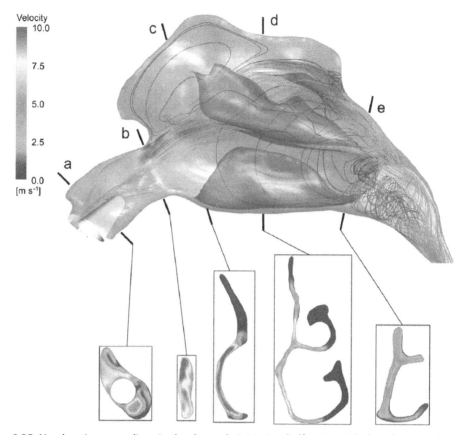

Fig. 8.35 Nasal cavity streamlines in the drug-administering half cavity with the other nostril closed.

the vertical flow observed at the posterior of the nasal cavity is attributed to the guiding effect of the middle and inferior turbinates.

Despite a rest inhalation flow condition of 15 L/min being adopted in this study, significant airflow accelerations are discovered at the vestibule and the anterior of the main passage, with airflow velocity peaking at 11.7 m/s at the nasal valve (Plane b). In contrast, prior work (Inthavong et al., 2011) provided the streamlines for a plain nasal cavity model, and the maximum velocity is about 5 m/s in the vicinity of the nasal valve. This is due to the closure of the right nostril, as recommended by the instructions, which causes a twofold increase in airflow rates in the right cavity compared to that in the previous case.

The presence of the spray device is highlighted in "Plane a," and the effective area of the right nostril to the open air is significantly reduced with apparent airflow acceleration. Additionally, the airflow speed at the spray head periphery is doubled from less than 3.5 m/s to 7.5 m/s due to the closure of the left nostril.

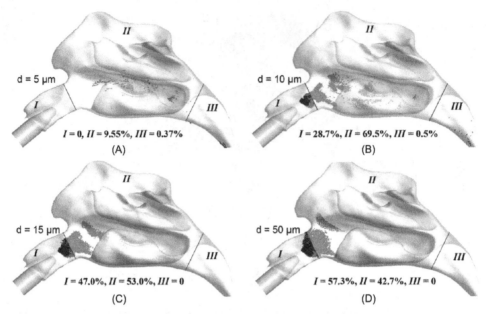

Fig 8.36 Sprayed particle deposition comparison among four selected micron sizes.

Fig 8.36 compares the depositions of selected micron particles. All other factors, including the spray direction, spray plume external features, and the beginning velocity, are held identical in order to isolate the effect of particle size. Based on anatomical features and epithelium types, the left side of the nasal chamber is classified into three main regions: the vestibule (I), the main route (II), and the pharynx (III), for drug delivery performance assessment. The main passage is selected as the targeted area for drug delivery performance assessment. The deposited particles are colored red or green, to distinguish the deposition along the septum or the outer lateral wall of the main route.

For the 5 μm case (Fig 8.36A), only 9.6% of sprayed particles are deposited in the main passage, and the nasal septum captures most of them (in red color). The fact that only a tiny amount of particles (0.4%) are deposited in the pharynx region shows that the majority of sprayed particles (90.1%) are transported by the airflow and travel further downstream of the respiratory system. The particle deposition percentage significantly increased for the 10 μm particles (Fig 8.36B), with a total deposition of 98.7% throughout the entire chamber. In detail, after being released from the spray orifice, 28.7% of particles are deposited at the vestibule region shortly, and 69.5% of particles are captured by the inferior main passage with 25.2% for the lateral wall and 44.30% for the nasal septum. While the particle distribution at the lateral wall is highly condensed at the anterior, the deposition at the nasal septum displays a widely dispersed pattern along the inferior chamber. Fig 8.36C and D demonstrate a total deposition in the nasal chamber with similar patterns for 15 μm and 50 μm cases. Fewer particles are trapped at the anterior of the lower passage because of more early deposition in the vestibule region (47.0% for the 15 μm case and 57.3% for the 50 μm case).

8.4.4.2 Deposition of Viral Droplets in the Human Upper Airway

The spread of airborne/aerosol viral particles such as SARS and COVID-19 had posed huge health and economic challenges to the global community. Having an in-depth understanding of the underlying virus transmission patterns between people is essential for limiting the disease's spread and preventing outbreaks. As the issue is closely related to the human respiratory tract, CFD shows advantages in offering a thorough analysis of viral particles' transmission features, particularly for aerosol viruses that are largely spread through respiratory droplets produced by coughing, sneezing, breathing, and talking. The deposition characteristics of such viral particles in the respiratory tract determine how the inhaled viral droplets may infect people or contaminate the environment to cause an escalating secondary transmission, hence revealing that the deposition details from an aerodynamic perspective can greatly contribute to the outlining of virus re-transmission mechanisms.

Depending on the size of the particles and the intensity of the respiratory effort, inhaled droplets have been shown to settle in either the upper or lower airways (Shang et al., 2015). An investigation by Morawska et al. (2009) suggested that particles released from close contact during coughing and talking have typical starting concentrations 10 times higher than those generated during breathing, hence being primary vehicles for respiratory virus transmission. As a result, it is reasonable to deduce that inhaled respiratory viruses trapped in upper airways are more likely to be expelled into the air than those deposited in lower airways, which can cause direct and indirect contact transmissions. Improved knowledge of the virus spreading procedure in the transmission chains, particularly its role of the secondary contacts in highly populated areas, can therefore be established by analyzing the region-specific deposition fractions of inhalable droplets in the respiratory system.

Problem considered: The example in this section simulates the particle movement in respiratory airways (Shang et al., 2021). The geometry includes the upper airway (the nasal cavity, mouth cavity, nasopharynx, oropharynx, and larynx) and the lower airway (the trachea, major bronchi, and segmental bronchi up to 15 generations), as shown in Fig. 8.37. They are realistic respiratory airway models based on current CT images. Readers can refer to Dong et al. (2019) for the specifics of mesh generation and airflow simulation methodologies, which used the same realistic airway model to examine airflow dynamics and nanoparticle deposition characteristics.

The size of the droplets that are expelled can vary widely from 1 to 1000 μm (Chao et al., 2009). The droplet would immediately fall to the ground within 1 m if its size exceeded the historical threshold of 5–10 μm for droplet transmission (World Health Organization, 2014). The parameters to discriminate between aerosol and droplet

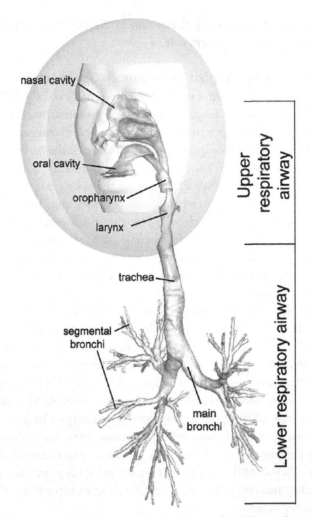

Fig. 8.37 The geometrical model of human respiratory airway for CFD simulations.

transmissions, however, should be greatly extended to 100 μm, larger than which the droplet would descend to the ground within 2 m (Prather et al., 2020). On the other hand, the inhalability of the droplet sharply declines as the droplet size exceeds 50 μm. Therefore a cut-off diameter of 50 μm is determined, and the size range from 1 to 50 μm is considered inhalable droplets for the study of particle deposition in airways.

CFD simulation: The oral–nasal inhalation scenario that inhales air through noses and mouth opening simultaneously is used in this model to more effectively disclose the normal inhaling conditions. An external sphere is configured as a pressure inlet with a

Table 8.5 Velocity Boundary Conditions at Outlets of Segmental Bronchi According to the Measurement From Horsfield et al. (1971).

Lobar Section	Percentage (%)	Volume Flow Rate (L/min)	Outlet Total Area (cm²)	Outlet Velocity (m/s)
Right Upper Lobe	20	3.6	1.40	0.43
Right Middle Lobe	10	1.8	0.57	0.52
Right Lower Lobe	25	4.5	1.44	0.52
Left Upper Lobe	20	3.6	1.24	0.49
Left Lower Lobe	25	4.2	1.32	0.57
total	100	18	5.97	

pressure of zero (Fig. 8.37). The flow rate weights measured by Horsfield et al. (1971) are used to set each group of lobar bronchi outlets as velocity outlets that sucked the airflow. Table 8.5 contains a list of the boundary condition specifics. The laminar-turbulent transitional flow is modeled using the k-ω SST turbulence equation. For particle transport and deposition simulations, micron particles with specific sizes of 1, 2, 3, 5, 7, 10, 15, 20, 30, and 50 μm are utilized.

The equations and numerical techniques of particle tracking simulation are explained in the previous section (Section 8.3.1). To more accurately predict inhaled particle depositions, the concept of inhalability, defined as the ratio of inhaled particles to discharged particles for each particle size, is used; 100,000 particles of each size are passively released from a sphere with a radius of 3 cm to precisely calculate the inhalability and regional deposition efficiencies. This radius is large enough to cover both the nasal and oral openings and small enough to guarantee a compact particle source being continuously released.

CFD results: The influence of particle sizes on airway deposition fractions and locations are shown in Fig. 8.38. The areas where large and small droplets tend to deposit are significantly diverse, despite the fact that large droplets have a far greater deposition percentage throughout the entire respiratory system (12.1%, 81.9%, and 100% for 1, 10, and 50 μm droplets, respectively). Small droplets have a lower propensity to remain in the upper airway. According to deposition fractions of 9.5% and 76.7% in the upper airway compared to 2.4% and 5.1% in the lower airway, the upper airway receives more droplet deposition than the lower airway overall, with ratios of 4:1 and 15:1 for 1 μm and 10 μm, respectively (Fig. 8.38).

Another obvious variation is found in the upper airway, specifically in the nasal cavity and the combined oral/laryngeal region. In comparison to 6.2%, 58.2%, and 45.5% in the

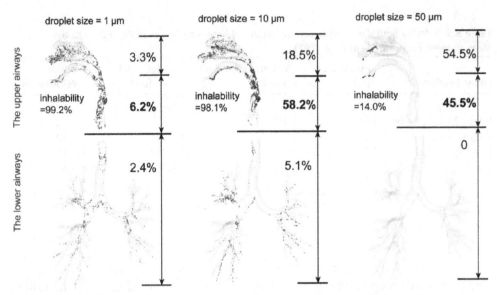

Fig. 8.38 1 μm, 10 μm, and 50 μm particles deposition patterns in the respiratory system.

oral/laryngeal combined region, virus droplets at 1, 10, and 50 μm deposit 3.3%, 18.5%, and 54.5% in the nasal cavity. Much lower deposition fractions in the nasal cavity for droplets at 1 μm and 10 μm suggest reduced infectivity by regular breathing as opposed to coughing. However, 50 μm droplets cause a noticeably high deposition fraction of 54.5% in the nasal cavities and a comparable rate of 45.5% in the combined oral and pharyngeal region. This finding suggests a high risk of rapid expelling through casual respiration and talking, especially given the fact that larger particles typically carry higher virus loads.

8.4.5 High-Speed Flows

All the examples demonstrated above have thus far concentrated where the application of CFD pertains only to *subsonic* flows, flows below the velocity of sound. The velocity of sound or sonic velocity is an important consideration in fluid mechanics. At *subsonic* velocities, small pressure waves can be propagated from both upstream and downstream. Nevertheless, when the velocity of fluid exceeds the sonic velocity, the fluid flow becomes *supersonic* and small pressure waves cannot be propagated upstream. Many numerical investigations in the aerodynamic and aerospace fields, the genesis of most CFD methods and techniques, involve *supersonic* fluid flows. The ratio of the fluid velocity to the sonic velocity is generally known as the dimensionless Mach number (*Ma*). If Mach number > 1, flow is *supersonic*; if Mach number < 1, flow is *subsonic*.

With the aim of providing the same genre of guidelines as for the examples of *subsonic* flows above, two examples of *supersonic* flows over a simple flat plate geometry and complex NACA 0012 wing configuration are demonstrated in this section. Salient aspects particularly in obtaining practical solutions for high-speed flows are described below.

8.4.5.1 Supersonic Flow over a Flat Plate

Supersonic flow over a flat plate is a classical *boundary layer* fluid dynamics problem. Despite its simplicity, an exact analytical solution remains *absent* for the fluid flow over this simple geometry. Consider the supersonic flow over a thin sharp flat plate at zero incidence and of length L, as sketched in Fig. 8.39. As the free stream fluid approaches the flat plate, a boundary layer develops at the leading edge of the flat plate. It is noted that the boundary layer is taken as the region of fluid close to the surface immersed in the flowing fluid. Away from the leading edge, the free stream fluid no longer "views" the sharp flat plate, but rather a fictitious curvature is formed due to the presence of the viscous boundary layer. If the Reynolds numbers are based on the distance from the leading edge of the plate, it can be appreciated that, initially, the value is low so that fluid flow close to the surface may be categorized as *laminar*. However, as the distance from the leading edge increases, so does the Reynolds number until a point must be reached where the flow regime becomes *turbulent*. In practice, the transition does not occur at one well-defined point, but rather, a transition zone is established between the laminar and turbulent flow regimes, as shown in Fig. 8.39. As also illustrated in the same schematic illustration, a curved induced shock wave is generated at the leading edge and the *shock layer* encapsulates the inviscid and viscous flows. In addition, *viscous dissipation* (dissipation of kinetic energy within the viscous flow) can cause high flow-field temperatures; high heat transfer rates are thus attained within the boundary layer.

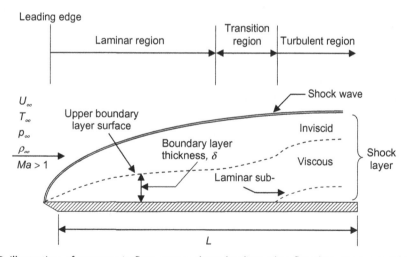

Fig. 8.39 Illustration of supersonic flow over a sharp leading-edge flat plate at zero incidence and development of boundary layer along the plate.

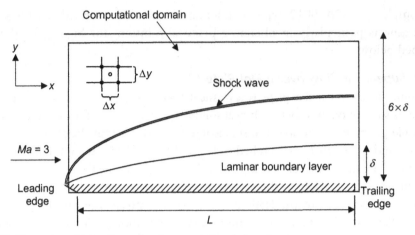

Fig. 8.40 Illustration of two-dimensional computational domain.

Problem considered: In this particular example the reader will gain an understanding of the flow physics by solving the complete Navier-Stokes equations. The flow over a flat plate is arguably the easiest application and it is packed with interesting fluid phenomena. This example consists of two parts.

Firstly, we will start from the very beginning by simply solving a laminar flow in two dimensions. This, of course, means that the length of the plate needs to be extremely small for the Reynolds number to be low. More importantly, the main aim of this problem is to demonstrate the practicability usage of a commercial computer code to sufficiently capture the desired flow physics. Fig. 8.40 depicts the size of the computational domain for the numerical calculations of $Ma = 3$ over the flat plate. The length of the plate has been chosen to be $L = 2.85 \times 10^{-5}$ m, which yields a Reynolds number of around 2000 based on the gas properties at reference sea level values. A uniform structure mesh of rectangular elements totaling 80×80 cells overlays the computational domain. The step size in the x direction (Δx) is 3.5625×10^{-7} m (0.000285/80). In the y direction it is imperative that the height of the domain must encapsulate the shock wave in order to obtain an accurate description of the fluid flow. As predicted by a Blasius calculation at the trailing edge, it is reasonable to assume that the computational domain is at least six times the height of the boundary layer δ, where δ is given by

$$\delta = \frac{5L}{\sqrt{Re_L}} = 3.186 \times 10^{-6} \text{ m} \tag{8.12}$$

The step size in the y direction (Δy) is thus 2.3898×10^{-7} m ($6 \times 3.186 \times 10^{-6}/80$). As a consequence of the stronger gradients in the direction normal to the plate, it makes rather perfect sense that the step size in the y direction is smaller than the x direction in order to better capture the flow field especially near the flat plate surface.

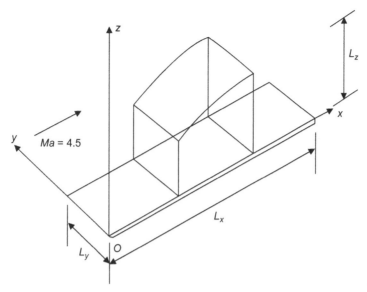

Fig. 8.41 Illustration of three-dimensional computational domain.

Secondly, with the current availability of computational speeds and resources such as the advent of computer clusters, the feasibility of employing direct numerical simulation techniques (see Chapter 7) is within reach to predict the onset of transition to turbulence for the fluid flow over this simple geometry. Here, the supersonic flow at $Ma = 4.5$ is solved. A solution to the three-dimensional complete Navier-Stokes equations is solved instead to fully describe the laminar-turbulent transition of the supersonic boundary layer, as shown by the schematic drawing in Fig. 8.41. Based on the local boundary layer length scale $l_s = 3.0486 \times 10^{-4}$ m, the non-dimensional dimensions of the computational domain are, respectively, $L_x/l_s = 3259$, $L_y/l_s = 65.6$, and $L_z/l_s = 438-587$. Owing to the increasing local resolution requirements for the streamwise evolution of the non-linear instability wave, the direct numerical simulation is performed in multiple stages involving overlapping sub-domains or boxes (see details in Table 8.6) to better manage the highly demanding computational resources. (Note the requirement for the increasing number of grid nodal points to resolve the wave propagation characteristics in the last column at high Reynolds numbers). The last box contains approximately a total of 65 million grid nodal points, a resolution that is deemed to be sufficient in capturing the structure of a fully developed turbulent flow. More details on the simulations carried out for the flow of this particular geometry can be found in Jiang et al. (2006).

CFD simulation: Not only does the advantage of employing the complete Navier-Stokes equations extend to investigations that can be carried out on a wide range of flight conditions and geometries, but also, in the process, the location of the shock wave as well as the physical characteristics of the shock layer can be precisely determined.

Table 8.6 Parameters Employed for Each Computational Box.

Box No.	Re	$N_x \times N_y \times N_z$	Points/Wavelength
1	400–1800	$864 \times 16 \times 132$	60
2	1600–2146	$864 \times 32 \times 132$	90
	1400–2099	$1024 \times 32 \times 160$	96
3 (Transition)	1968–2540	$1080 \times 48 \times 160$	96
	1968–2649	$1590 \times 102 \times 160$	112
	1968–2649	$1590 \times 256 \times 160$	112

Note: The ranges of Reynolds numbers evaluated above based on the local boundary layer length scale l_s. N_x, N_y, and N_z represent the total number of grid nodal points along the x, y, and z directions.

The Navier-Stokes equations for *compressible flows* have been described in Chapter 3. Note that for the energy conservation of *supersonic* flows, the specific energy E is solved instead of the usual thermal energy H applied in *subsonic* flow problems. In other words

$$E = \underbrace{e}_{\text{int ernal energy}} + \underbrace{\frac{1}{2}\left(u^2 + v^2 + w^2\right)}_{\text{kinetic enertgy}} \tag{8.13}$$

It is evident from above that the kinetic energy term contributes greatly to the conservation of energy because of the high velocities that can be attained for flows where $Ma > 1$.

The solution to supersonic flows requires nonetheless additional equations to close the fluid flow system. Firstly, the equation of state on the assumption of a perfect gas is employed, that is,

$$p = \rho R T$$

where R is the gas constant. Secondly, assuming that the air is calorically perfect, the following relation holds for the internal energy:

$$e = C_v T$$

where C_v is the specific heat of constant volume. Thirdly, if the Prandtl number is assumed a constant value of approximately 0.71 for calorically perfect air, the thermal conductivity can be evaluated by the following:

$$k = \frac{\mu C_p}{Pr}$$

The Sutherland's law is typically used to evaluate the viscosity μ, which is provided by

$$\mu = \mu_0 \left(\frac{T}{T_0}\right)^{1.5} \frac{T_0 + 120}{T + 120} \tag{8.14}$$

where μ_0 and T_0 are reference values at standard sea level conditions.

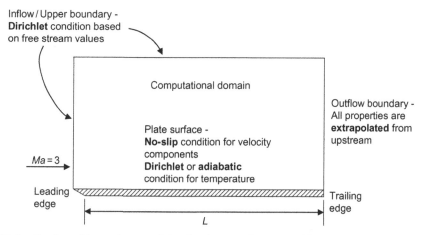

Fig. 8.42 Application of boundary condition for the two-dimensional laminar supersonic flow.

Fig. 8.42 illustrates the boundary conditions for the high-speed laminar fluid flow in two dimensions. At the left-hand side and upper boundaries of the computational domain, *Dirichlet* boundary conditions on the velocity, pressure, and temperature are imposed at their respective free stream values. On the right-hand side, it is imperative to note that the outflow condition in *supersonic* flows is not influenced by downstream conditions and thus differs from the usual *Neumann* boundary condition in *subsonic* flows, whereby all properties are now calculated based on an extrapolation from upstream in contrast to imposing the *zero* normal gradient constraint. Finally, the *no-slip* condition is prescribed for all the velocity components ($u = v = 0$). One of the most significant advantages of CFD is the ability to conduct numerical experiments in order to gain an understanding of the implications of changing a flow parameter. The impact between a constant wall temperature and an adiabatic wall will be assessed in the numerical calculations. For the constant-temperature wall boundary condition, it is assumed that the temperature is prescribed at the free stream value.

One of the many challenging aspects of direct numerical simulation is the requirement to impose appropriate flow conditions at the inflow boundary. To represent the actual characteristic of the flow through this boundary, *different single pair of oblique first-mode disturbances* has been introduced to parametrically study the sensitivity of the chaotic inflow conditions entering the computational domain influencing the internal flow. For the outflow boundary, all properties are extrapolated, as considered for the two-dimensional case. For the side boundaries encasing the flow along the spanwise *y* direction (see Fig. 8.41), *periodic* boundary conditions such as those exemplified for the compact heat exchanger problem above are employed. At the upper boundary, the *free-slip* condition is imposed, while the usual no-slip condition is prescribed with an adiabatic condition for the temperature at the surface of the plate.

Table 8.7 Data Used for the Laminar Flow Case.

Quantity	Dimension
Plate length L, m	0.0000285
Free stream air properties at sea level:	
Speed of sound U_∞, m/s	340.28
Pressure p_∞, N/m^2	101,325.0
Temperature T_∞, K	288.16
Density ρ_∞, kg/m^3	1.225
Dynamic viscosity μ, kg/m · s	1.7894×10^{-5}
Specific gas constant R, J/kg · K	287.0
Specific heat C_p, J/kg · K	1005.4
Prandtl number	0.71
Ratio of specific heats γ (C_p/C_v)	1.4

CFD results: The relevant data such as the free stream conditions and several thermodynamic constants for the laminar boundary layer simulation are detailed in Table 8.7. For the particular case considered, the commercial code ANSYS-CFX, Version 10, is used to obtain all the numerical results of the laminar flow condition.

Fig. 8.43 shows the steady-state temperature contours and velocity vectors for the supersonic flow over the flat plate. Note that the range of colors depicted by the velocity vectors has been chosen to correspond to the temperature distribution. For both the adiabatic and constant wall temperature cases, the distinctive regions occupied by the inviscid and viscous flows are clearly identified within the shock layer by the CFD methodology. The profile of the curved induced shock wave separating the silent zone (upstream free stream) from the shock layer is also well established. An adiabatic wall is seen to increase the boundary layer temperature by a substantial margin above the constant wall. This is primarily due to the result of a relatively lower density, which leads to a thicker boundary layer and consequently a broader shock layer, as illustrated in Fig. 8.43.

The normalized temperature and u component velocity profiles at the trailing edge are plotted in Fig. 8.44 and Fig. 8.45 to further confirm the aforementioned observation. In plotting the profiles the normalized y distance is adopted as suggested by Van Driest (1952), that is, $\bar{y} = y\sqrt{\text{Re}}/x$. The normalized temperature (T/T_∞) profiles adequately capture the leading edge shock wave as well as the classical boundary layer behavior near the plate surface. Since the temperature gradient is zero at the wall for the adiabatic case, temperatures within the thermal layer are expectedly to be higher than those of the constant wall temperature case. Essentially, a colder wall temperature suppresses the boundary layer flow. Similarly, a closer examination of the normalized velocity (U/U_∞) profiles

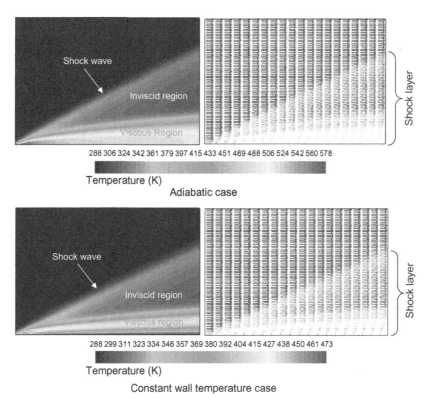

Fig. 8.43 Temperature contours and velocity vectors describing the distinct regions of the inviscid and viscous flows within the shock layer for the adiabatic and constant wall temperature conditions for $Ma = 3$.

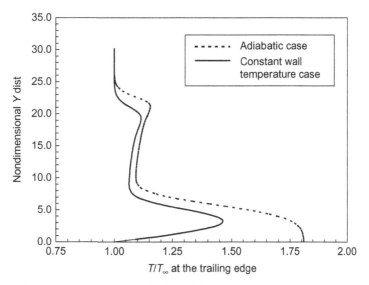

Fig. 8.44 Normalized temperature profiles for the adiabatic and constant wall temperature conditions for $Ma = 3$ at the trailing edge.

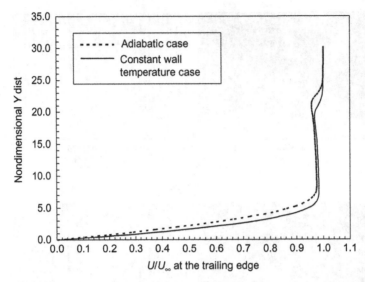

Fig. 8.45 Normalized *u* component velocity profiles for the adiabatic and constant wall temperature conditions for *Ma* = 3 at the trailing edge.

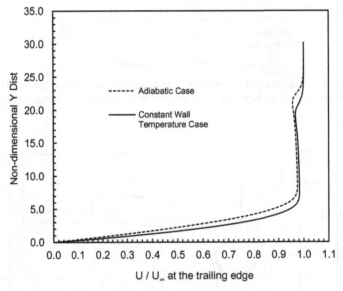

Fig. 8.46 Local Mach number profiles for the adiabatic and constant wall temperature conditions for *Ma* = 3 at the trailing edge.

near the surface, as exemplified in Fig. 8.45, also shows a thicker boundary layer for the adiabatic case. The local Mach number at the trailing edge is graphically presented in Fig. 8.46, of which the relative strength of the two leading edge shock waves clearly illustrates a stronger shock wave migrating across the flat plate when the adiabatic condition is imposed at the surface of the plate.

Nowadays, the increasing computational capabilities offer immense possibilities of actually revisiting a number of classical CFD problems such as the current boundary layer problem typified in this example through the use of DNS techniques. Particularly, the laminar investigation as described above can be further extended to study the onset of flow instability and better understand the transition from laminar flow to turbulent flow. For this flow problem, an in-house DNS computer code developed at the University of Texas at Arlington is employed to particularly predict the non-linear evolution of instability waves and onset of breakdown to turbulence in a Mach 4.5 flat-plate boundary layer flow after the occurrence of shock. To verify the DNS predictions, they are also subjected to intense scrutiny against the parabolic stability equation (PSE) calculations of NASA Langley, of which this work is based on the investigation performed by Jiang et al. (2006). For the remainder of the section, some results extracted from the work on the laminar-turbulent transition investigation are presented and discussed below.

One of the many challenges in DNS is to prescribe suitable inflow boundary conditions. Since DNS is a fully deterministic approach where all spatial and temporal scales are required to be computed, the specification of appropriate inflow conditions is of paramount importance. The comparison of wave propagation characteristics prior to the onset of transition based on various disturbance modes predicted by DNS is illustrated in Fig. 8.47. In Fig. 8.47, the DNS modal profiles of the fluctuating velocity components and temperature subject to different disturbance modes as a function of the local scaled wall normal coordinate are shown to be in excellent agreement with the results determined through the PSE calculations. For each of the disturbance modes affecting the numerical calculations, a successful cross-validation of the DNS solutions for the linear and early non-linear stage of disturbance growth is clearly demonstrated, which greatly

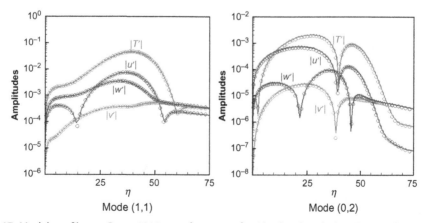

Fig. 8.47 Modal profiles at Re = 1500 as a function of η (the local scaled wall normal coordinate) subjected to two disturbance modes.

suggests the independent influence of the initial boundary conditions on the onset of laminar-turbulent transition.

It is noted that the prediction of the onset of breakdown to turbulence after the non-linear interaction is beyond the capability of PSE. However, since DNS solves the full unsteady Navier-Stokes equations without any *ad hoc* assumptions, the present methodology is able to provide solutions across a broadband of fluid dynamics ranging from laminar state to the early and later stages of transition as well as even to capture the manifestation of chaotic structures in a fully developed turbulent flow. To gain an insight of the complex flow structures occurring along the flat plate, the evolution of the contours for the density fluctuation and wall-normal vorticity along the increasing distance along the flat plate is illustrated in Figs. 8.48 and 8.49. As observed from these two figures, the flow remains quite laminar for $x < 4600$. Flow instability or transition

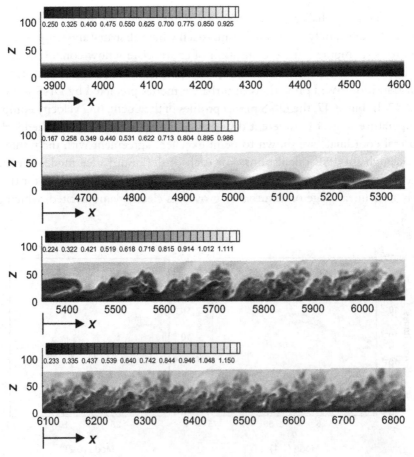

Fig. 8.48 Contours of density fluctuation in the x-z plane midway along the spanwise direction.

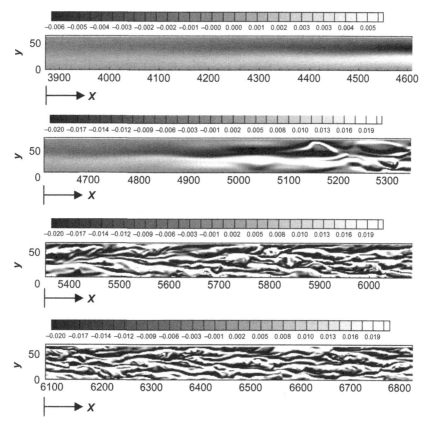

Fig. 8.49 Contours of wall-normal vorticity in the *x-y* plane at a vertical non-dimensional distance of $z^+ = 12.21$.

nonetheless begins to creep in at about $x = 4800$, with the increasing undulation persisting over the flow patterns after this point. Shortly after some distance downstream at $x = 5400$, the onset of breakdown of the flow to turbulence subsequently prevails and the fluid flow remains rather turbulent thereafter by the establishment of substantial density fluctuations within the fluid and vigorous coherent structures of wall–normal vorticities at $x = 6800$.

8.4.5.2 Subsonic and Supersonic Flows Over a Wing

An aerofoil can be defined as a streamlined body designed specifically to produce lift. It will also experience the counteracting influence of drag while placed in a fluid stream. The primary purpose of the construction of the aerofoil to be streamlined is to minimize the drag imposed on the body. As a measure of its usefulness, for example, as a wing section of an aircraft, the ratio of lift to drag must be sufficiently large in order for it to be capable of producing high lift at a small penalty of the drag. For an aircraft to remain in the airspace, the creation of lift on the wing surface is of paramount importance. Nevertheless, what is necessary to propel the craft's forward motion is the engine power that overcomes the drag.

Problem considered: In order to illustrate another CFD application example for high-speed flows, the fluid flowing past an NACA 0012 aerofoil is considered herein. This particular geometry has been specifically chosen because of the numerous aerodynamic investigative studies that have been carried out in research and design practices. There are a number of accepted terminologies concerning an aerofoil; it is best that familiarization with them is necessary in order to understand the discussion of the flow past such geometry.

Fig. 8.50 illustrates the schematic drawing of the aerofoil geometry and some of the pertinent terms relating to an aerofoil cross-section that will be constantly referred are:
- *Leading edge* – the front, or upstream edge, facing the direction of flow
- *Trailing edge* – the rear, or downstream edge
- *Chord line* – a straight line linking the centers of curvature of the leading and trailing edges
- *Chord, c* – the length of chord line between the leading and trailing edges
- *Span, b* – the length of the aerofoil in the direction perpendicular to the cross-section of the wing
- *Angle of attack (incidence), α* – the angle between the direction of the relative motion and the chord line

The CFD example in this section considers the subsonic and supersonic flows past an infinitely long aerofoil. The Reynolds Averaged Navier-Stokes (RANS) equations are solved alongside the shear stress transport (SST) turbulence model for the supersonic flow ($Ma = 2.5$) over the wing geometry. As only time-averaged results are of primary interest especially in RANS simulation and since the length b is infinite, such conditions or assumptions mean that the flow is truly two-dimensional; there is negligible spanwise variation of flow patterns and forces for a constant chord aerofoil. Numerical calculations

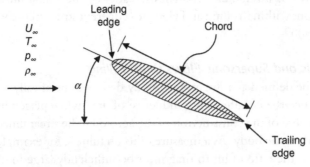

Fig. 8.50 Schematic drawing of an aerofoil.

are thus carried out in a two-dimensional fluid domain, which also include parametric investigations on the influence of different angles of attack affecting the expanding shock layers as the fluid travels past the aerofoil. This investigation forms the first part of this example. At flows that are below the speed of sound ($Ma = 0.2$ is considered for the present problem), currently available computational hardware permits the feasibility use of DNS techniques to fundamentally study the onset of flow instability and subsequent transition to turbulence subject to different inlet conditions. Here, the complete three-dimensional Navier-Stokes equations are solved to accommodate the spectrum of varying length scales that exist within the complicated fluid phenomenon spanning along the streamwise (x), spanwise (y), and vertical (z) directions. This study shapes the second part of this example.

CFD simulation: One challenging aspect of the present CFD example of the fluid flowing over the aerofoil involves the generation of appropriate meshes surrounding the geometry. Two different types of grid topologies that could be specifically used to solve the problem are illustrated in Figs. 8.51 and 8.52. One approach that can be suitably considered especially for RANS simulations is the block structured or multi-block mesh, as shown in Fig. 8.51. The entire fluid region is now subdivided into six contiguous blocks. This approach offers immense flexibility in generating different grids in each block particularly the increased mesh densities in blocks 2 and 5 capturing the developing boundary layer along the chord line between the leading and trailing edges and expanding shock layers along the aerofoil. From the viewpoint of practicality and general application, the attractiveness of multi-block meshing allows the ease of porting the mesh information into any commercial computer package. Alternatively, it is also possible to employ a single-block mesh to fit the entire fluid domain. The three-dimensional

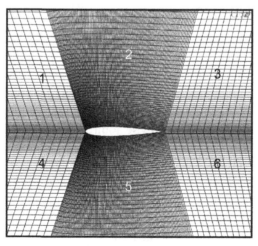

Fig. 8.51 A multi-block approach for the mesh surrounding the aerofoil for RANS calculations.

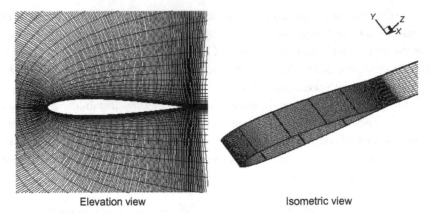

Elevation view   Isometric view

Fig. 8.52 A C-type mesh scheme adopted for DNS.

structured C-grid type encapsulating the aerofoil as seen in Fig. 8.52 typifies such grid topology and it has been purposefully employed for the DNS calculations. Numerical solutions are obtained in a mesh totaling 22,000 rectangular elements for the two-dimensional case and 1200 (streamwise) × 32 (spanwise) × 180 (vertical) hexahedral elements for the three-dimensional case, respectively. For the latter, the large overlay of elemental volumes ensures that the mesh distribution based on the non-dimensional spacing along the streamwise, spanwise, and vertical directions are below: $\Delta x^+ < 13$, $\Delta y^+ < 15$, and $\Delta z^+ < 1$.

The *full compressible* Navier–Stokes equations, namely Eqs (8.18)–(8.23), are applicable for the consideration of both the subsonic and supersonic flows in this CFD example. Note again that the two-dimensional forms are obtained by simplifying the three-dimensional governing equations, which involve the omission of the component variables in one of the coordinate directions. Boundary conditions as described by the boundary layer flow over the sharp-edged flat plate in the previous example can also be similarly employed in the present CFD example, in other words, *no-slip* condition for the velocity components on the wing surface, *free-slip* condition on the upper and lower boundaries, *Dirichlet* condition based on the free stream values at the inlet boundary, and an *extrapolation* of the upstream values at the outlet boundary. With regard to DNS, *periodic* boundary conditions are employed for the side boundaries encasing the three-dimensional flow along the spanwise y direction. For all the numerical results attained, *adiabaticity* is assumed for the wing surface.

CFD results: For the RANS simulations, the commercial CFD code ANSYS-CFX, Version 10, is utilized to obtain all the numerical results of the supersonic flow condition. The relevant input data such as the free stream conditions and several thermodynamic constants are the same as those tabulated in Table 8.3.

Fig. 8.53 illustrates the contours of the local Mach number and pressure for the fluid flow past the wing geometry based on a free stream $Ma = 2$ subject to 0° and 5° angles of

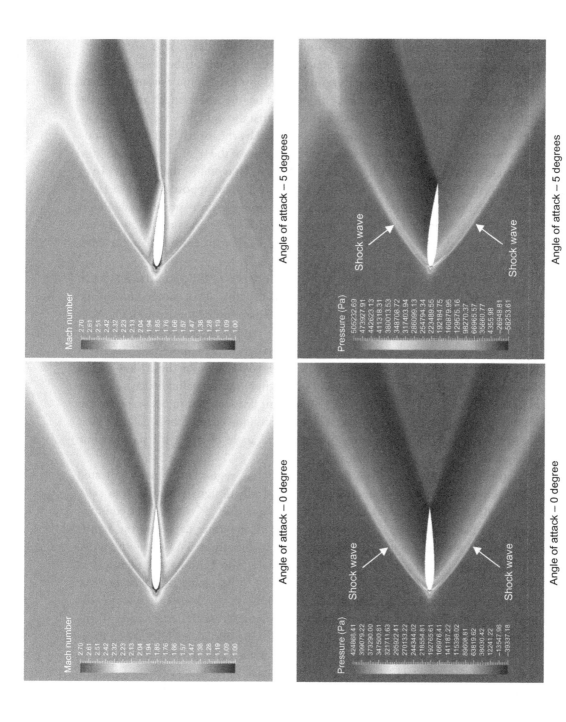

Fig. 8.53 Local Mach number and pressure contours for the supersonic flow based on the free stream $Ma = 2.5$ subject to 0° and 5° angles of attack past the wing geometry.

attack. From these results, the formation of shock waves above and below the wing surface is clearly identifiable as the fluid travels downstream from the leading edge to the trailing edge of the aerofoil. Because of complete flow symmetry, it is not entirely surprising for the case of the 0° angle of attack that the growth of the boundary layer is identical at the top and bottom surfaces; one would not expect any pressure gradient to be developed and hence the flow vorticities should be of equal strength and opposite rotation. However, at the 5° angle of attack, it is seen that the fluid flowing over the bottom surface of the aerofoil is slowed down, as evidenced by the local Mach number contours. The pressure is significantly increased from the case of the 0° angle of attack at the same flow region, which means that the pressure gradient is favorable, the boundary layer thickness is small, and hence the vorticity in it is also small. The opposite, however, prevails over the top surface whereby the boundary layer is thicker, the pressure gradient is adverse, and the vorticity in it is larger. This pressure difference gives rise to an upward resultant force, namely lift, which is of primary importance for an aerofoil when placed in a fluid stream.

The foregoing discussion indicates a strong dependence of lift upon the incidence angle. Well-documented observations have shown that there is also a compelling linkage between the creation of lift and the flow characteristic surrounding the aerofoil. Fig. 8.54 shows the velocity vectors for the two cases of the 0° and 5° angles of attack with the spectrum of colors associated with the pressure distribution. For these *small* angles of attack, there appears to be no separation of the boundary layer near the trailing edges. Nevertheless, a threshold limit exists whereby any further increase of incidence no longer produces an increase of lift. At this instance, significant flow separation occurs at the top surface and subsequently widens the wake behind the trailing edge of the aerofoil. This is known as the *stall* position and it constitutes a critical angle of attack for aerodynamic design since the lift drops rapidly thereafter.

At upstream, the flow away from the leading edge or stagnation point as indicated by "P" on the temperature contours in Fig. 8.55 is usually designated as the *silent zone*. Because the flow is supersonic, the disturbance generated at "P" is thus not communicated to any part of the zone. Further downstream, the zone depicting the significant changes of the fluid within the shock layer is sometimes known as the *action zone*. Here again, the upper and lower temperature distributions are identical at the top and bottom of the wing surface for the case of 0° angle of attack because of complete symmetry. For the case where the angle of attack increases to 5°, the symmetry is broken, registering a temperature gradient across the *action zone*. Temperatures are significantly higher over the bottom wing surface, while lower-than-expected temperatures are recorded over the top wing surface when compared to the case of 0° angle of attack.

For the remainder of this section, the non-linear evolution of instability waves and onset of breakdown to turbulence for the subsonic flow over the aerofoil based on the investigation performed by Shan et al. (2005) and Deng et al. (2007) through

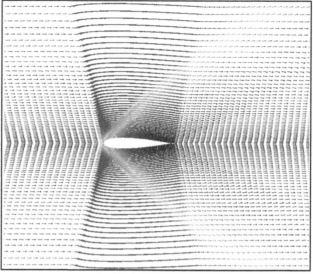

Angle of attack – 0 degree

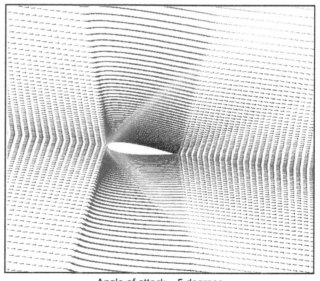

Angle of attack – 5 degrees

Fig. 8.54 Velocity vectors of the supersonic flow based on the free stream $Ma = 2.5$ subject to $0°$ and $5°$ angles of attack past the wing geometry.

DNS techniques are presented below. The boundary conditions for the inflow and out-flow are non-reflecting conditions based on the analysis of one-dimensional characteristic equations on the time-dependent Euler equations. These boundary conditions are different from the boundary conditions specified for the DNS of supersonic boundary layer flow over the flat plate in the previous section.

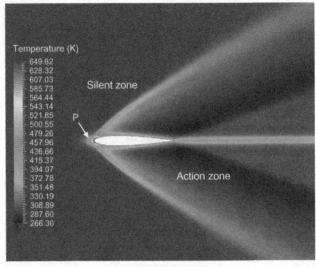

Angle of attack – 0 degree

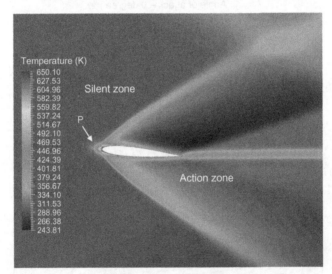

Angle of attack – 5 degrees

Fig. 8.55 Temperature contours of the supersonic flow based on the free stream $Ma = 2.5$ subject to 0° and 5° angles of attack past the wing geometry.

Fig. 8.56 shows a snapshot of segments representing instantaneous spanwise vorticity developing in the middle of the x-z plane at different times for the fluid flow over the aerofoil at $Ma = 0.2$ with an angle of attack of 4° investigated by Shan et al. (2005). As shown by the instantaneous spanwise vorticity at $t = 0.8794$, the presence of instability waves propagating downstream above the wing surface clearly establishes the onset of

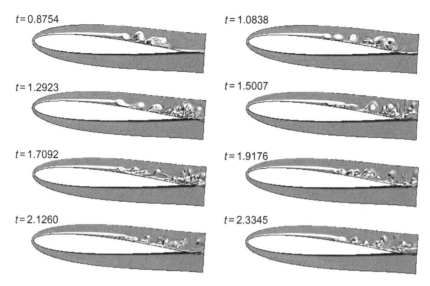

Fig. 8.56 Instantaneous spanwise vorticity at different times (t is non-dimensionalized by c/U_∞).

flow transition. As time progresses, these waves that are three-dimensional in nature continue to grow in the shear layer and they subsequently cause the vortex to break down near the trailing edge as indicated by the chaotic three-dimensional fluid flow at the later stages of the DNS calculations. In its fully turbulent state the presence of vortex shedding behind the trailing edge is associated with the rapid growth of the strong non-linear interactions of velocity fluctuations within the wake region.

Iso-surfaces of instantaneous vorticity along the x, y, and z directions are further illustrated in Fig. 8.57 to gain a better understanding of the flow transition. The appearance of undulated vorticity generated along the spanwise direction suggests that the fluid flow is three-dimensional and also depicts the propagation instability waves along the top surface of the aerofoil. In general, the results clearly show the simultaneous breakdown of the rolled-up shear layer along the chord line and the vortices are shed from the separated shear layer and become distorted while traveling downstream. More specifically, this shedding vortex that is quickly deformed and stretched actually conforms to a negative vortex being induced by the prime vortex that is situated alongside this negative vortex and eventually the prime vortex breaks down into smaller fragments corresponding to the flow transitioning to turbulence. The interaction between the streamwise and spanwise vortices leads to the development of a λ -shaped vortex, which then rolls up and breaks down. The boundary layer becomes fully turbulent after reattachment of the flow behind the trailing edge.

Recent results on the DNS performed by Deng et al. (2007) of flow separation control subject to pulsed blowing jets over the aerofoil at the same angle of attack are

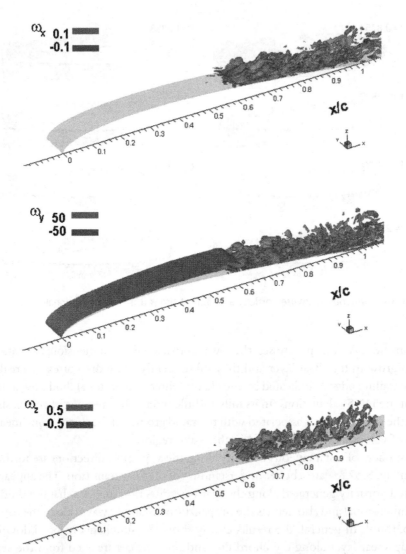

Fig. 8.57 Iso-surfaces of instantaneous vorticity components (from top to bottom: streamwise, spanwise, and vertical components).

presented below as a comparison case to the above. Here again, a snapshot of the instantaneous spanwise vorticity being developed in the middle of the x-z plane at different times is similarly illustrated in Fig. 8.58. The unsteady blowing enforced before the separation point with the creation of a large vortex shedding prevailing close to the leading edge of the aerofoil triggers the early transition of the boundary layer. Iso-surfaces of vorticity along the respective Cartesian directions in Fig. 8.59 further exemplify the breakdown of the separated shear layer and the development of the vortex structure. The fluid

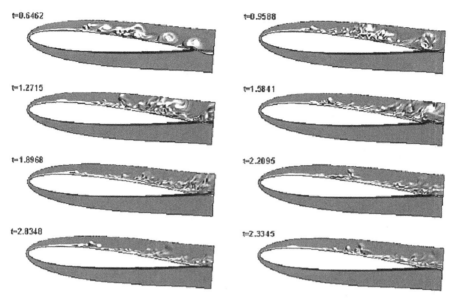

Fig. 8.58 Instantaneous spanwise vorticity at different times with pulsed blowing jets (t is non-dimensionalized by c/U_∞).

phenomenon reveals a number of interesting features. From the results obtained, it is clear that the pulsed blowing jets imposed at the inlet significantly affect the reattachment of the boundary layer shortly after separation. A shorter separation zone is developed, resulting in also a much reduced separation bubble, as evidenced by the vortical structures plotted in Fig. 8.59.

8.5 SUMMARY

The examples in this chapter have been purposefully selected to demonstrate the feasible application of the CFD methodology across a wide range of engineering disciplines and to provide some useful guidelines for handling some of these challenging flow problems in practice.

In the first example a detailed investigation of the appropriate turbulence models for ventilation design systems exemplifies how these models can be put to good use as a *design tool* in building design such as areas of architecture science and civil or construction engineering.

The ongoing relevance of CFD as a *research tool* is considered in the second example. For gas-particle flows, particles can either be treated as another continuum medium through additional conservation equations by the Eulerian approach or an ensemble of discrete solid particulates tracked by the Lagrangian approach within the parent phase fluid conservation equations. These approaches represent the currently adopted

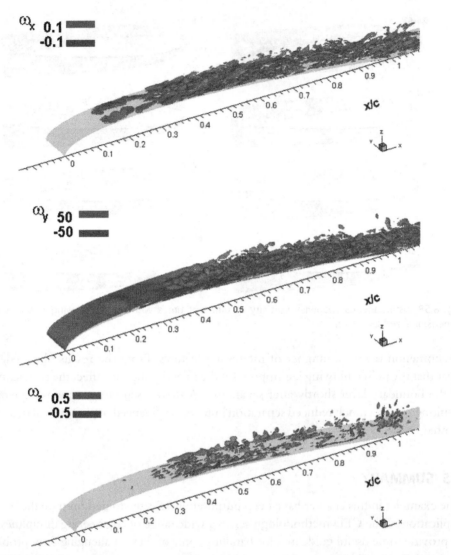

Fig. 8.59 Iso-surfaces of instantaneous vorticity components with pulsed blowing jets (from top to bottom: streamwise, spanwise, and vertical components).

state-of-the-art methods for handling such flows. The reader should take note that these so-called multi-phase models with appropriate constitutive closure relationships are not only restricted to gas-particle or gas-solid flows. They can also be readily applied to resolve the transport of gas-liquid, liquid-solid, or even gas-liquid-solid mixtures.

By the consideration of additional equations for the conservation of energy coupled with appropriate source terms in the momentum and turbulence equations, fluid flows coupled with heat transfer are illustrated through two practical examples in this chapter.

The wealth of information obtained from the fluid flow within the in-line or staggered arrangement of compact heat exchangers through CFD allows mechanical engineers to determine the required improvements in better augmenting the performance of the mechanical system. More importantly, CFD plays a key role in assisting in building energy-saving designs. It helps to identify the flow characteristics in a transparent glazing channel where radiation and natural convection heat transfer take place and enables the evaluation of natural ventilation rates that are enhanced by solar thermal energy.

The fifth example presents the practicality of adopting LES over the usual RANS in resolving the combustion and radiation processes associated with a free-standing fire. Capturing the fluctuating characteristics of a buoyant fire through LES signifies the advances of knowledge benefiting fire engineers as well as researchers from a combustion and science background particularly on the environmental consequences that could result from accidental large-scale fires in oil fields.

The wake flow over a moving mannequin, the sixth example, further establishes CFD as another effective research tool in understanding indoor aerodynamics with occupants and their consequent impacts on contaminant control. The use of dynamic mesh in achieving the challenging field containing moving objects also presents the potential to be integrated into more practical areas such as preventing cross-contamination between hospital wards or infection risk assessment in train and airline cabins etc. With dynamic mesh helping achieve more realistic scenarios, those simulation results can increase the accuracy of relevant assessments for public health.

Mature development of CFD meshing and numerical models as demonstrated in the seventh example represents the current state-of-the-art to probe the many associated complex flows inside a real human nasal cavity. CFD is increasingly being considered a viable tool in the field of biomedical engineering and the application of this methodology is steadily growing, especially the enormous interest in studying the deformation of vessel walls affecting the arterial blood flow inside a human vascular system.

Despite the long-standing application of the traditional theoretical boundary-layer-solution technique in aircraft designs in the aerodynamic industries, *supersonic* flows over varying aircraft geometries and flight conditions can now be handled through efficient CFD methods and models. For the last two fluid dynamics examples shown for this engineering discipline, complete two- and three-dimensional solutions through the full Navier-Stokes equations under *supersonic* conditions can be viably obtained nowadays through the commercial CFD computer packages. From a research-oriented investigation, the application of DNS to predict the onset of flow instability and transition to turbulence depicts the future direction of CFD in enhancing understanding of basic fluid dynamics and in all classical physics especially on the prediction of turbulence.

The windows of opportunity for CFD are indeed plentiful. CFD is aptly being considered the critical technology for fluid flow investigation in the 21st century. Relishing the many successful applications of CFD achieved thus far at this present time, the reader should also be well informed of some emerging new and innovative techniques that will continue to revolutionize the current CFD methodology. These will be further discussed in the next chapter.

REVIEW QUESTIONS

8.1 Why do engineers prefer to initially use the standard k-ε turbulence model as a starting point in solving their design problems, instead of using more complex models such as the LES approach? When would an LES approach be used?

8.2 For the internal pipe flow geometry shown below, formulate appropriate answers to the following questions:

   (a) Define and state all the boundary conditions for this problem.

   (b) What would be an approximately suitable length for L_1 and L_2 to achieve a fully developed flow at the outlets ($d = 0.2$ m) for this case?

   (c) Show where a fine mesh should be needed.

   (d) From your understanding of fluid mechanics, sketch roughly what the flow streamlines will look like for this flow problem.

   (e) Discuss what design ideas may be used to reduce the head loss in this pipe flow and how this can be implemented using CFD.

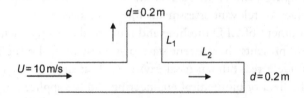

8.3 Explain the *Lagrangian* description of a fluid motion.

8.4 What is the *Eulerian* description of a fluid motion? How does it differ from the *Lagrangian* description?

8.5 A streak of dye is released into an internal flow and its motion is tracked and recorded. Is this a *Lagrangian* or *Eulerian* measurement?

8.6 A stationary pitot tube is placed in a fluid flow to measure pressure. Is this a *Lagrangian* or *Eulerian* measurement?

8.7 The flow in a 90° bend is shown below. Formulate appropriate answers to the following questions:

   (a) Would a structured or unstructured mesh be better for this geometry? Describe how you would create the mesh.

   (b) From your knowledge of the conservation equations of motion, sketch the expected flow streamlines for a flow that has a Reynolds number of 100 and a flow that has a Reynolds number of 10,000.

(c) If a particle with a Stokes number of 0.1 is released in a passive manner from the inlet under laminar conditions, sketch what would be the expected particle trajectory throughout the pipe. How would the trajectory be different if the flow is turbulent?

(d) What would the expected particle trajectory be if the particle Stokes number was instead equivalent to 20 for a laminar flow and a turbulent flow?

(e) If you solved this problem in the *Eulerian* reference, what kind of measurements can be made?

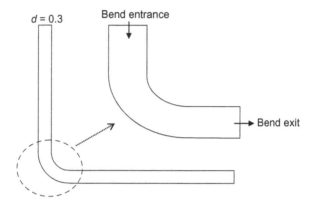

8.8 For heat transfer coupled with fluid flow, formulate appropriate answers to the following questions:

(a) Why is compact heat exchanger different from conventional heat exchanger?

(b) Symmetric and cyclic boundary conditions can usually be employed to simplify the flow geometry of a compact heat exchanger. Indicate the boundary conditions for the in-line and staggered arrangements shown below.

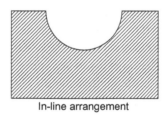

In-line arrangement

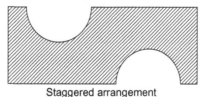

Staggered arrangement

(c) The Nusselt number is a ratio of two fluid properties. What are they?

(d) Explain why higher heat transfer rates are experienced in staggered arrangement over in-line arrangement.

(e) What is *conjugate heat transfer*?

(f) Why is it important to incorporate the buoyancy effect in natural convection? What additional modeling effort is required for the transport equations?

(g) What is *radiation heat transfer*? How does it affect the surface solid temperature?

8.9 For dynamic mesh modeling of human motion, formulate appropriate answers to the following questions:

(a) When is dynamic mesh required in simulations?

(b) Explain why there is a fine mesh surrounding the near regions of the human body.

(c) Indicate the regions of high and low pressure encountered around the human body.

(d) Compared to a scenario with a stationary mannequin facing incoming air at the same velocities of walking speeds, what is the necessity of adopting dynamic mesh to capture mannequin motion?

8.10 Two particle trajectory profiles are shown below for mono-sized particles released from the bottom inlet. The trajectories are colored by the particle velocity magnitude, where *red* represents the highest value and *blue* the lowest value. The initial particle velocity is a factor of ten of the air flow velocity, that is, $U^* = 10 = U_{particle}/U_{air\ flow}$.

(a) What would be the approximate particle Stokes number for the flow path in *a* and for the flow path in *b* shown in the figures below?

(b) What are the velocities of the particles in flow path *a* when they impact the upper walls? Compare this with the velocities of particles in flow path *b* at the same location.

(c) Explain the difference between particle trajectories obtained for flow path *a* (straight) and for flow path *b* (curve)? (Hint: compare the velocity profiles)

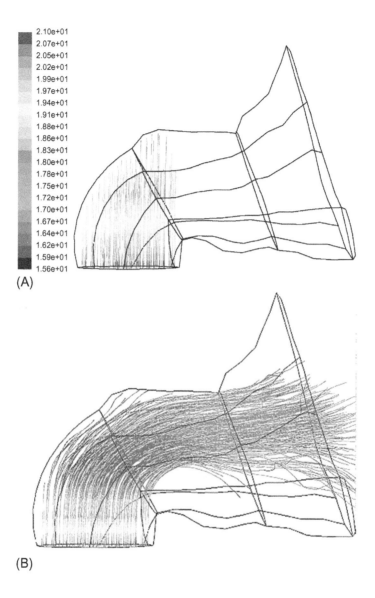

(A)

(B)

8.11 Formulate appropriate answers to the following questions for high-speed flows:
  (a) Why is the direct numerical solutions (DNS) technique very useful for studying *supersonic* flows?
  (b) How does the outflow condition in *supersonic* flows differ from the usual *Neumann* boundary condition in *subsonic* flows?
  (c) For the following supersonic flow over a flat plate, discuss what types of boundary conditions are suitable to capture the flow features.

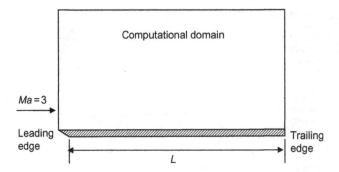

(d) Discuss the effects of using a constant wall temperature instead of an adiabatic wall in terms of the boundary layer development and its effect on the temperature profile in *supersonic* flows.

(e) Discuss the advantages and disadvantages of using DNS methods over theoretical approaches such as the parabolic stability equation (PSE) calculations.

(f) For the following flow over an aerofoil, discuss why the following grid is necessary.

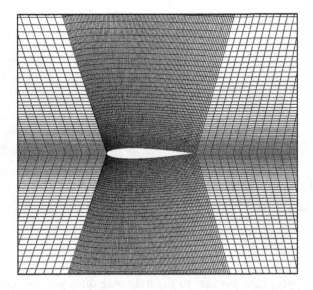

(g) When the angle of attack of the NACA 0012 aerofoil is at 5°, what happens to the pressure over the bottom surface and how does this affect the temperature (see Section 8.4.5.2)?

CHAPTER 9

Some Advanced Topics in CFD

9.1 INTRODUCTION

Over the last four decades, considerable progress has been made in the development of CFD. In many areas of applications the field of CFD is attaining a stage of maturity where most of the basic methodologies are rather well established and many have been implemented into a number of commercially available computer packages. Despite many significant achievements, there is still an impending need to develop and advance CFD to meet the demands bolstered by various engineering industries in resolving complex flow problems and within research communities especially in the area of biomedical research.

The materials that have been presented in this book thus far have served to present an introductory level to the use and application of CFD. With the advent of modern digital computers and growing computational power, the development of CFD can be considered by: (1) the improvement of algorithms to solve the Navier-Stokes equations and (2) the development of numerical models to solve sophisticated practical problems.

In the early stages heightened demand for numerical algorithms was central to the development of CFD as they serve to contribute to accuracy and calculation efficiency. Over the past 40 years, numerous algorithms have been developed to address consistency and stability issues in numerical analysis, beginning from basic flow problems in incompressible and compressible flows, and from explicit and implicit schemes. With the refinement of these algorithms, some specific techniques were proposed and developed, such as mesh-related techniques, to achieve more accurate and effective solutions.

With these well-developed algorithms laying the foundation of CFD, especially in commercially available computer packages, the focus in recent decades has shifted to the development of numerical models that can better resolve complex turbulent flows, multi-physics problems, coupling of fluid flow with chemical reactions and micro- or nano-scale fluid systems. Nowadays, advanced computational models and techniques are progressively being improved due to the rise in computational hardware and power. These have allowed more real-world fluid flow and heat transfer problems to be tackled, which were not possible in the past.

For the abovementioned reasons, some of these advanced topics will be briefly reviewed rather than further explored in this chapter. The rest of this chapter will now be devoted to advanced topics that are currently trending in the CFD discipline. For students who are keen on pursuing a research career or for those who are currently involved in research and development activities, this chapter is written specifically to

Computational Fluid Dynamics
https://doi.org/10.1016/B978-0-323-93938-6.00016-6

provide the latest development and address some of the important issues and challenges that are currently faced by many CFD researchers.

9.2 EARLY DEVELOPMENT OF CFD ALGORITHMS

9.2.1 Incompressible Flow

Incompressible flow, by definition, is an *approximation* of flow where the flow speed is insignificant compared to the speed of sound of the fluid medium. The majority of fluid and associated flow we encounter in our daily lives belongs to the incompressible category. Mathematically, the incompressible flow formulation poses unique and challenging issues not present in compressible flow equations because of the incompressibility requirement. Physically, incompressible flow is characterized by an elliptic behavior of the pressure waves, whereby the speed in a truly incompressible flow is *infinite*, which imposes stringent requirements on computational algorithms for satisfying incompressibility. Inherently, the major difference between an incompressible and compressible Navier-Stokes formulation is in the continuity equation. The incompressible formulation can usually be viewed as a singular limit of the compressible one where the pressure field is just considered part of the solution. The primary issue in solving the set of governing equations is to appropriately satisfy the mass conservation equation.

Based on the approach employing primitive variables, the two most commonly adopted methodologies are: (1) methods based on pressure iteration and (2) artificial compressibility method. For the various methodologies based on pressure iteration, the marker-and-cell (MAC) method developed by Harlow and Welch (1965) represents the first primitive variable method employing a derived Poisson equation for pressure. Through this method, the pressure is used as a mapping parameter to satisfy the continuity equation.

Ever since its enormous introduction, numerous variations of the MAC methods have surfaced, and successful computations have been made. Another pressure-based method that is also commonly used is the fractional-step procedure (Chorin, 1968; Yanenko, 1971). One important aspect of the fractional-step method that the reader should heed is that care needs to be exercised in evaluating the boundary conditions for the intermediate variables (Orszag et al., 1986). A logical way to overcome such difficulty is to employ the physical boundary conditions for the intermediate steps. This aspect has been thoroughly discussed by Rosenfeld et al. (1991).

The artificial compressibility approach involves modifying the continuity equation by adding a time derivative of the pressure term with an *artificial compressibility parameter*. In this formulation the continuity equation is modified by adding a time derivative of the pressure term with an *artificial compressibility parameter* β. Together with the unsteady momentum equations, this forms a hyperbolic-parabolic type of time-dependent system of equations. A series of pre-conditioning methods (Choy and Merkle, 1993; Turkel, 1999) and numerical schemes, based on the alternating direction implicit (ADI) procedure by Briley and McDonald (1977) and upwind differencing schemes implemented by Rogers and Kwak (1991), have been developed in conjunction with the compressible Navier-Stokes equations to solve the incompressible flow. Artificial compressibility relaxes the strict requirement to satisfy mass conservation in each step. Nonetheless, to utilize this convenient feature effectively, it is essential that the nature of artificial compressibility both physically and mathematically is properly understood. Useful guidelines for choosing the artificial compressibility parameter can be found in Chang and Kwak (1984).

In general, it is the authors' opinion that the above methods are well established and that they have been employed rather successfully to solve a wide range of fluid flow problems. Many review articles and books on CFD have been written on the state-of-the-art computational methods for viscous incompressible flow. Interested readers are strongly encouraged to refer to Gunzburger and Nicolades (1993), Hafez (2002), Loner et al. (2002), Gustafsson et al. (2002), and Kiris et al. (2002) on the latest trends and advances in these methods.

Despite the many significant advances achieved for the incompressible CFD methodologies, one aspect the authors still believe deserves some concerted attention is the *computational efficiency*, which is directed primarily toward the solution of the Poisson equation for pressure. The choice of the pressure solver can be a rather crucial consideration particularly to solve a large physical domain with high mesh density and/or march the solution through small time steps for a transient flow problem. In light of the computations in rectangular-type regions, one possible consideration is to employ a direct (non–iterative) fast Fourier transform (FFT)–based solver (Sweet, 1973) for the pressure equation. However, the application of this direct Poisson solver for general three-dimensional coordinates is not as straightforward and normally considered impractical; iterative solvers still remain the only feasible option to resolve such complex geometries. Traditional iterative solvers are well known to be inefficient due to their slow convergence, which therefore requires the development of more advanced solution methods. Currently, parallel computing presents one possible way of accelerating computations. To maximize the computing speed, iterative matrix solvers such as the parallelized version of generalized residual minimal equation solver (GRMES) by Saad and Schultz (1985), which essentially belongs to the family of Krylov methods, is a viable choice to solve the pressure equation. Another avenue that is also worth exploring is the application of the multigrid acceleration technique, as suggested by Kwak et al. (2005).

9.2.2 Compressible Flows

From a historical perspective, numerical methods for solving equations of compressible flows were developed because of their importance and relevance in aerodynamics and aerospace. For the airflow around an aircraft, enormously high speeds can be achieved as the aircraft travels through the earth's atmosphere, thus resulting in a very high Reynolds number; the turbulence effect is mainly concentrated within the thin boundary layers. Ignoring the frictional drag, the flow may be attributed to the presence of the *wave* drag due to shocks and *pressure* drag that is essentially inviscid in nature. The latter property has indeed necessitated much of the revolutionary-effort and evolution-effort focusing on special methodologies that have been designed specifically to obtain the solution of the inviscid Euler equations. As reviewed by Fujii (2005), CFD was first applied to simulate transonic flows in the 1970s. Embedded shock waves were adequately captured and the design process of commercial aircraft has drastically changed ever since then. In the middle of the 1980s CFD was employed to simulate hypersonic flows associated with the space transportation system development including reentry vehicles. Most of these can be readily solved through the inviscid Euler equations. Practical flow simulations in aerospace employing the full compressible Navier-Stokes equations first appeared in Fuiji and Obayashi (1987a, 1987b). For high-speed flows, the use of these equations to resolve the thin viscous boundary layers around airfoils still presents enormous challenges. The fine mesh requirement is unavoidable in order to adequately predict the near-wall turbulence for flow analysis and eventual control of flows. Finely distributed mesh resolution near walls greatly limits the time step size for computation; this in turn dramatically increases the computational burden to achieve a solution.

With regards to some specific methods for compressible flows, the earliest scheme was the method of MacCormack (1969), which is based on an *explicit* method and central differencing. It is still being used quite extensively even to this present day. To avoid the problem of oscillations especially due to the discontinuity at the shock front, the concept of artificial dissipation is introduced into the equations. A fourth-order dissipative term is the most common addition but higher-order terms have also been successfully applied. Another effective numerical method to solve the equations for compressible flows is the *implicit* method developed by Beam and Warming (1978), which is based on the approximate factorization of the Crank-Nicolson method. Like the MacCormack method, the addition of the explicit fourth-order dissipative term into the equations is imperative due to the central differencing consideration.

Recently, Roe (2005) in his review paper of "CFD – retrospective and prospective" identified some key issues to the seminal developments in CFD algorithms. One of them is to overcome the deficiency that deals with the basic matter of shock capturing when the shocks are very strong. Enormous interest has been placed to develop a class of upwind schemes of greater sophistication to produce a well-defined discontinuity without introducing an undue error into the smooth part of the solution elsewhere. In the next section the authors aim to provide the reader with a survey of the latest developments in high-resolution schemes. Instead of dwelling on the fine detail construction of these schemes herein, interested readers are strongly encouraged to refer to the literature for a more in-depth understanding. Another approach that is of significant interest in capturing the unsteady moving shock front is through the use of adaptive meshing, which will also be subsequently discussed below.

9.2.3 High-Resolution Schemes

Where a high degree of precision is required to resolve the presence of shocks or discontinuities, high-resolution techniques are utilized to attain the numerical solution of partial differential equations. High-resolution schemes are typically used to solve problems involving first-order partial differential equations with wave propagation, particularly non-linear wave propagation (Roe, 1997). Consider an inviscid Euler equation based on the one-dimensional wave equation; if the gradient $\partial u/\partial x$ is approximated using central differencing such as the traditional approach described above for the MacCormack (1969), the velocity u profile results in an oscillatory behavior near the discontinuous wave front. In some circumstances the numerical procedure can lead to an unstable and chaotic solution. The common remedy is to introduce an artificial dissipation term. Despite the numerical result exhibiting a *monotone* variation (no oscillations), the diffusive property remains undesirable.

To eliminate this undesirable property, some rather mathematically elegant algorithms have been developed during the past decades. These modern algorithms have included flux limiters by Sweby (1984) and Anderson et al. (1986) in the MINMOD and SUPERBEE schemes or slope limiters due to the monotone upwind scheme for conservation laws (MUSCL) family of methods that can be found in Van Leer (1974, 1977a, 1977b, 1979) and Godunov (1959) and approximate Riemann solvers (Toro, 1997). Among the many shock-capturing techniques in the literature, the total variable diminishing (TVD) algorithm is considered to be well suited for capturing shock waves. The first-order upwind scheme (Section 5.3), which does not result in oscillations in the vicinity of discontinuities, can be readily shown to obey the TVD condition. Nevertheless, it still suffers the same faith as the introduction of the artificial dissipation term as being very diffusive in the vicinity of discontinuities.

The greatest challenge in the ongoing research in this area is therefore to develop schemes that are *highly accurate in smooth regions of the flow* and have *sharp non-oscillatory transitions at discontinuities*. Colella and Woodward (1984) first attempted to overcome the latter aspect by introducing a piecewise parabolic method (PPM), which is a four-point centered stencil to define the interface value; the formulation of this value is then limited to control the oscillations. Leonard (1991) later combined this limiting approach with a higher-order (up to ninth-order) interface value. It is noted that the PPM can be considered an extension of Van Leer's MUSCL scheme, and MUSCL in turn is an extension of Godunov's approach. These limiting procedures nonetheless still cause the solution accuracy to degenerate because of the first-order approximation near extrema.

Along a different line of thought, ever since the development of the third-order essentially non-oscillatory (ENO) schemes by Shu and Osher (1988, 1989) and Harten (1989), improved schemes such as the fifth-order weighted-ENO (WENO) by Liu et al. (1994) and Jiang and Shu (1996) have surfaced to better define the interface value as a weighted average of the interface values from all stencils. The weights have been designed such that very high accuracy is achieved in smooth regions. The WENO schemes, however, still show a diffusive behavior in the vicinity of discontinuities; they smear the interface front as much as the ENO schemes. Later, the approach by Suresh and Huynh (1997) was developed to enlarge the TVD constraint for a better representation near extrema. This scheme, referred to as MP, where MP stands for Monotonicity Preserving, is fifth order and was developed to address the chronic problem of upwind methods, namely the narrow stencils that cannot distinguish between shocks and extrema. For such a scheme, Suresh and Huynh's (1997) result demonstrated that the unfiltered approach generally exhibits unwanted oscillations – a violation of the TVD condition – near the discontinuity locations. The ENO scheme though preserving the monotone variation of the velocity u fairs, however, no better than the unfiltered scheme in exceeding the limiting bounds: the realizability condition. Higher-order schemes such as the fifth-order methods of WENO and MP schemes clearly satisfy not only the monotonic variation of velocity u but also the realizability condition near the extrema. The superiority of MP over WENO is succinctly demonstrated especially near the discontinuities of the velocity u profile, where the latter scheme still shows some diffusion of the velocity u at these locations.

Recently, Daru and Tenaud (2004) extended the MP schemes of Suresh and Huynh (1997) to the seventh order for the solution of Euler and Navier-Stokes equations using local linearization and dimensional splitting in the multi-dimensional case. They commented that the schemes despite the lack of preservation of the formal high accuracy still gave very accurate results that compared well to the high-order WENO schemes at a lower cost. They also pointed out that the classical drawback associated with dimensional splitting, similar to the fractional method for incompressible flows, is the treatment of boundary conditions for the intermediate step boundary for bounded viscous flow calculations. Interested readers may wish to further investigate this pertinent aspect to improve these MP schemes.

From the brief survey conducted, it is apparent that the development of high-resolution TVD schemes for the numerical simulation of unsteady compressible flows remains an area of intense research. The current trend of research appears to be the development of higher-order schemes to better capture the strong shock profiles. Hill and Pullin (2004) constructed a hybrid method that combines a tuned center-difference scheme with the WENO method for LES of strongly compressible, shock-driven flows, which has been found to work very well for unstable gas-dynamic flows. Yu (2006) presented an efficient high-resolution shock-capturing hydrodynamics scheme to tackle challenges in astrophysical hydrodynamical flows, which requires the need to resolve complex flows including strong shock interaction. Subramaniam et al. (2019) presented a sixth-order weighted compact high-resolution (WCHR) scheme for resolving fine-scale aspects of compressible turbulent flows with shocks using a spatially implicit form.

9.3 ADVANCES IN NUMERICAL METHODS AND TECHNIQUES

9.3.1 Adaptive Meshing

For a strong migrating shock front, dynamic grid adaptation may be required to resolve the sharp discontinuities that exist within the flow domain as the solution changes continuously through time. There are a number of adaptation categories that have recently been discussed especially in Thompson et al. (1999) and Kallinderis (2000). Interested readers are encouraged to probe deeper into the literature for a better understanding of the various state-of-the-art techniques that are being applied.

One method that has been shown to perform well is the *r-refinement* technique. The basic idea is to devise an algorithm to physically relocate the original non-adapted mesh to the time-varying changes of the solution in the domain, based on some criteria for adaptation while maintaining the identity and data structure of the domain being solved. Many successful means are available for relocating nodes or re-meshing in order to perform adaptation. However, the success of any procedure depends on the criteria used to properly guide the node location. Benson and McRae (1991) employed a gradient-based algorithm to obtain a weight function. One such measure is the local gradient (curvature) of the dependent variable for resolution evaluation. Note that there are other approaches that could also be adopted for resolution evaluation. Nevertheless, let us consider the two-dimensional multi-block dynamic adaptive algorithm developed by Ingram et al. (1993) for the purpose of illustrating shock transition. By tracking the evolution of the outer shock from right to left and observing the grid orientation, topology, cell volume, etc., the mesh clearly reflects the continuous changes of the solution process, as depicted in Fig. 9.1. The sharp discontinuities are precisely captured through the concentrated mesh surrounding the shock front.

Another refinement method that is also very prevalent in the literature but not often applied to unsteady problems is the enrichment or *h-refinement* technique. The development of an adaptive re-meshing scheme for application to two-dimensional triangular and three-dimensional tetrahedral meshes has been reported in Hassan and Probert (1999). Here, the cell volume and shape are changed by the *insertion* or *deletion* of mesh nodes from the data structure resulting in an overall *increase* or *decrease* of the number of cells. It is certainly a more natural technique since the algorithms used for reconnecting the nodes are the same as those used during grid generation to improve distribution or to achieve some pre-determined conditions for the grid. Generally speaking, *h-refinement* is not strictly confined to only one single type of mesh structure. This rather flexible

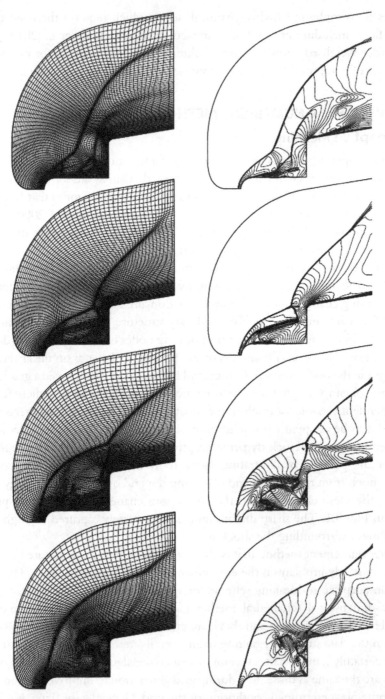

Fig. 9.1 Results extracted from Ingram et al. (1993) for the grid and density contours produced by *r-refinement* dynamic adaptation.

technique permits a combination use of both structured and unstructured meshes. Dynamic adaptation can also take place through a series of mesh layers fitted to the body.

Adaptive meshing is still in its infancy application in CFD. Many research issues remain outstanding and unresolved (McRae, 2000). Firstly, there is a further requirement to investigate the relationship between cell shape, size, orientation, etc., in particular for the *r-refinement* on a structured mesh concerning the grid skewness and aspect ratio in the vicinity of the shock front as illustrated by the problem in Fig. 9.1. Highly skewed cells can pose enormous problems to numerical methods; the sharp response of the flow behavior may lead to divergence of the solution procedure. Secondly, it must be recognized that even the best adaptive techniques with the best criteria (weight function) may not provide the same degree of resolution for all portions of the solution. In a recent article by Soni et al. (2001) it is stated in relation to the adaptive weight function that "determination of this function is one of the challenging areas of adaptive grid generation." Thirdly, the defining characteristic of the unstructured triangle/tetrahedral meshes for the *h-refinement* is reflected in their ability to resolve solutions based on the orientation of the cells' surfaces to the solution, which can be quite random locally. This does not pose any problems in the smooth regions but can create serious resolution difficulties in regions when the solution varies rapidly.

9.3.2 Moving Grids

Moving grids exist in many engineering applications. Take, for example, the simulation illustration of the flow in a diesel internal combustion engine in Chapter 1, where deforming grids have permitted the means of simulating the piston and valve motion to gain better insights into features that are present within in-cylinder flows. Another important example in Chapter 1 is the simulation of a rotating impeller that is common in gas-sparred stirred tanks. To resolve such a problem, one part of the grid can be taken to be attached to the impeller and allowed to freely move in time while the other encompasses the impeller and remains stationary. It is worth noting that such an approach can also be applied to the rotor-stator interaction in turbomachinery. As suggested by Lilek et al. (1997) and Demirdzic et al. (1997), the moving grid is allowed to *slide* along the interface without deformation between the part of the grid that is attached to the static stator and the other part that is attached to the rotating rotor. The grids do not have to match at the interface; this allows flexibility in employing different kinds of meshes and/or achieving the desired fineness in the respective domains.

Except for difficulties in ensuring exact conservation, there are essentially no limitations on the applicability of this approach. Farhat (2005) reviewed the application of CFD to the prediction of flows past flexible and/or moving deforming bodies in the aeronautical area. The driving application has centered on the non-linear computational aeroelasticity to address problems associated with local transonic effects, limit-cycle oscillation, high angle of attack flight conditions, and buffeting. Despite the achievements in the formulation and discretization of turbulence models and wall laws on dynamic meshes (Koobus et al., 2000) as well as successful simulations of vortex-dominated flows around maneuvering wings over a wide range of Mach numbers (Kandil and Chang, 1988), most successes moving grid applications according to Farhat (2005) have concentrated so far on either *complex geometries with inviscid flows* or *viscous flows with simple geometries*. Hence, to fill this gap, it is also the authors' opinion, as with Farhat's, that further research, applied to CFD on moving grids, needs to be undertaken to perfect the robustness of mesh motion algorithms and to combine them with the possibility of automatic partial re-gridding.

In all the problems discussed above the movement of the domain is pre-determined by external effects. For another class of problems such as free surface flows, the movement must be calculated as part of the solution process where the grid has to move with the boundary. In the majority of these cases the free surface interface can be a boundary between liquid and gas such as an ocean separated from the surrounding air atmosphere or liquid and water, for example, the freezing of water into ice. For such problems, the critical issues of this approach deal with the efficiency and stability of the numerical algorithm for the movement of the interface. As each grid point on the interface is moved through time, the surface grid may be rather irregular due to an uneven distribution and unconstrained movement. Some grid generation algorithms that retain the curvature information of the interface are generally required to redistribute the grid points to ensure numerical integrity and stability of the solution. For illustration, some results of a numerical study of three-dimensional natural convection and freezing of water in a cubical cavity by Yeoh et al. (1990) are discussed below. The distorted ice and water domains at a particular instant of time are shown in Fig. 9.2. The ice-water interface dramatically deforms under the influence of natural convection effects that are present in the water. The appearance of two counterclockwise flow regions shown in Fig. 9.3 is distinct for freezing of water because of its density extremum at 4°C. As expected, the flow inversions around this temperature significantly influence the development of the ice-water interface. Simulating such flow problems usually requires very small time steps to advance the interface as well as generate a new mesh using body-fitted grids in the respective domains at each time step. Both of these steps negatively impact the simplicity, robustness, and computational cost of the solution procedure. In an article Tezduyar (2001) indicates that the computational efficiency degenerates dramatically for cases involving large motion. The development of a more robust method is thus still required to fulfill the need to better treat this class of flow problems with moving interfaces or boundaries.

9.3.3 Parallel Computing and CUDA Environment

9.3.3.1 Parallel Computing

Central processing units (CPUs) in most single- or dual-processor computers (PCs and Workstations) are getting faster and more compact. To gain any significant increase in

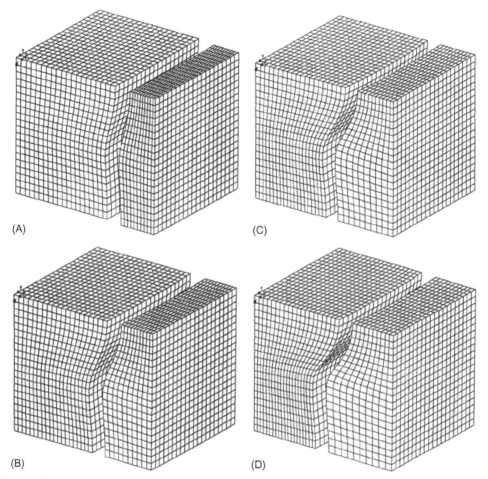

Fig. 9.2 Results extracted from Yeoh et al. (1990) for the distorted water (left) and ice (right) domains at: (A) 12.7 min, (B) 25.5 min, (C) 38.2 min, and (D) 54.1 min.

speed, consideration for future trends in high-performance CFD computing ultimately beckons in the wake of parallel computers.

The basic idea of parallel computing stems from the simple desire to perform simultaneous operations of multiple computational tasks on a computer system. Commercially available parallel computers generally comprise thousands of processors and gigabytes of memory and computing power measured in megaflops or even gigaflops. The advantage of parallel computers over vector supercomputers is *scalability*. These systems usually allow the accommodation of standard computer chips and therefore are cheaper to produce. Parallel computing, particularly in CFD, is a broad field of research and development. The authors will not venture into the details regarding the taxonomy of parallel computing architectures and programming paradigms, which are beyond the scope of this introductory book. Interested readers can refer to numerous articles in journal

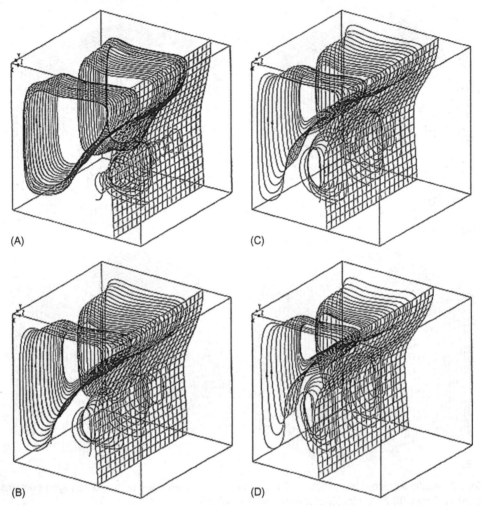

(A)

(B)

(C)

(D)

Fig. 9.3 Results extracted from Yeoh et al. (1990) for the isometric views of interface locations and particle tracks at: (A) 12.7 min, (B) 25.5 min, (C) 38.2 min, and (D) 54.1 min.

publications such as the *International Journal of High Speed Computing*, *The Journal of Super-computing*, or the *Journal of Parallel Programming* and a book by Simon (1992) for a more in-depth study particularly on the aforementioned aspects and the principal issues behind the utilization of parallel computing in CFD.

Rather, the authors would like to highlight outstanding issues that still need to be adequately addressed. For the effective utilization of parallel computing, we can first concentrate on the issues of *domain decomposition* and *load balancing*. The idea of partitioning data and computational tasks among multiple processors is commonly denoted as *domain decomposition*. The principal objective of *domain decomposition* is to maintain uniform computational activities on all processors and is known as *load balancing*. It appears straightforward enough but there are a number of factors that can complicate the aspect of load balancing. For example, combustion flows that involve chemical reaction rate source terms are computed only where the static temperature exceeds some threshold value. Another example is particle tracking, where particles can accumulate in a particular sub-region. All of these pose serious challenges in parallel computing that are in the midst of active research.

Another key concern is the impact of scaling on a given parallel computer architecture with increasingly larger numbers of processors. An important question arises: How does the *efficiency* of the computation depend on the number of processors? Generally speaking, the performance of parallel computing may be influenced by the cost of scheduling processors, communication between processors, and synchronization of time (i.e. the time required to allow processors to reach a common point following the execution of the parallel section of the code). Lastly, regarding the issue of *portability*, significant effort has been devoted to the development of standardized environments for the development of parallel computer codes. In recent years rigorous developments of a standard computer language FORTRAN to High-Performance FORTRAN (HPF) (Forum, 1993; Koelbel et al., 1994), standard heterogeneous, network-based parallel computing environments such as parallel virtual machine (PVM) (Sunderam et al., 1990; Mattson, 1995), and standard message-passing interface (MPI) (Forum, 1994) have been undertaken to expand the appeal and broaden the flexibility of parallel computing.

Parallel computing experienced some reservations as:

(1) lacking a decisive advantage performance over conventional serial (and vector) computers in many instances,

(2) difficult to program efficiently, and

(3) grossly lacking in portability.

With the rapid improvement of processor and memory technology, as well as current research in the HPC area, parallel computing will almost certainly become more efficient and cost effective (Dongarra et al., 2020). Álvarez-Farré et al. (2021) offered a way of reducing the complexity of both data structures and processing kernels as a natural strategy for increasing portability and keeping programming paradigms simple to implement. They used an algebraic method to increase code portability by replacing traditional stencil data structures and sweeps with algebraic data structures and kernels, and the discrete operators and mesh functions are then stored as sparse matrices and vectors, respectively.

It is also highly probable that modern parallel computers can come up to speed to equal or exceed, for example, the performance of even the largest multi-processor Cray supercomputers. Currently, the aerospace industry has taken the lead in applying parallel computing to practical analysis and design. Other engineering areas will eventually follow in similar footsteps as advancements in parallel computers are made more attractive for practical use and to better resolve more realistically complex fluid flow problems.

9.3.3.2 Graphics Processing Units and CUDA Environment

The use of graphics processing unit (GPU) platforms has been realized in a number of engineering applications especially in the field of fluid mechanics (Elsen et al., 2008; Kampolis et al., 2010), molecular dynamics (Anderson et al, 2008, Sunarso et al., 2010), wave propagation (Komatitsch et al., 2010), and building performance simulation (Tian et al., 2016).

CUDA (Compute Unified Device Architecture), OpenCL, and OpenACC are three commonly used programming models for the GPU. An illustration of the CUDA processing flow paradigm can be seen in the schematic representation in Fig. 9.4. GPU devices can carry out large numbers of simultaneous operations. Data level parallelism is thus exploited within the parallel devices. As data is transferred from the Main Memory of the Motherboard to the memory of the GPU, the CPU being the Host instructs the GPU to carry out the floating point arithmetic operations that are performed in each streaming processor (SP). Since this device has a large amount of SPs being grouped together in the streaming multi-processors (SMPs), they can collectively have significantly more gigaflops of operations than the current high-end CPUs. Once the calculations are completed, the result is transferred back to the Main Memory.

As each GPU applies the same functions on a large number of data, these data-parallel functions also known as kernels generate a large number of threads to exploit data parallelism – single instruction multiple thread (SIMT) paradigm. A thread is considered to be the smallest unit of processing that can be scheduled by the operating system.

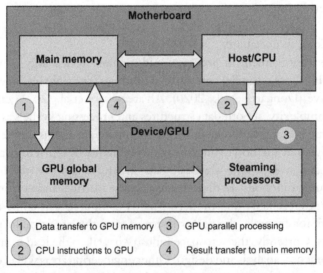

Fig. 9.4 CUDA processing flow paradigm. *(After Papadrakakis et al., 2011.)*

Threads in GPUs generally take only very few clock cycles to generate and schedule due to the underlying hardware support, whereas unlike CPUs thousands of clock cycles are required. These number of threads generated are generally organized in blocks and a grid which is defined by all threads generated by a kernel consists of a number of blocks being equal in size. Another form of thread grouping is by using warps, which are units of thread scheduling in SMPs. The purpose of warps is to ensure high hardware utilization where all processes are likely to have a workload at all times in order to avoid idle processors while waiting for the completion of operation.

One significant feature of the CUDA device is the ability to have a variety of different memories that can be utilized to achieve high performance. The global memory which resides in the device is responsible for the interaction with the Host/CPU. It is large in size and off-chip and the constant memory allows interaction with the Host to provide faster and more parallel data access paths for CUDA kernel execution. Also, other types of memories such as registers and shared memories could be assessed in a highly parallel manner in the device, which cannot be assessed by the Host. Registers are thread bound, which can be employed for holding variables that require frequent access but do not need to be shared with other threads. Shared memories are allocated to thread blocks which allow all threads in the block to access variables in the shared memory allocated locations for that block.

Through the consideration of domain decomposition methods, a new era of computing is emerging especially through the use of hybrid CPU-GPU systems with the purpose of exploiting all available processing power and CPU memory resources. Papadrakakis et al. (2011) have shown that the processing capabilities of hybrid CPU-GPU systems in conjunction with efficient domain decomposition methods could ensure high utilization time and minimize idle time especially by exploiting the intrinsic software and hardware features of the GPUs and the numerical properties of the solution methods. Recent developments have demonstrated the application of the CPU-GPU method in simulating insect flight (Yao and Yeo, 2017), complex particle-fluid flows (He et al., 2020), and free-surface flows with the mesh-free particle method (Crespo et al., 2011), etc. Furthermore, unlike previous studies that apply mostly to low-order methods, the CPU-GPU method is capable of handling complex configurations of large-scale models and high-order schemes. Xu et al. (2014) presented some novel techniques to achieve balances and scalable high-order CPU-GPU simulation.

Nevertheless, the difference in performance between the CPU and GPU is normally not the same when carrying out mathematical operations such as dot products, matrix-vector multiplications, or solving linear systems. Because of the difference in performance between the CPU and GPU, the arithmetic operations being executed and other parameters need to be treated carefully in order to achieve the required peak performance to solve large simulation-based applied science and engineering problems.

9.3.4 Immersed Boundary Methods

Immersed boundary methods were first introduced by Peskin (1972) to simulate cardiac mechanics and associated blood flow. One distinguishing feature of this approach is the ability to perform the entire simulation on a fixed Cartesian grid. A novel procedure was developed by directly imposing the effect of the immersed boundary on the flow; the requirement for the grids to conform to the complex geometrical structure of the heart was thus avoided. Since its inception, numerous modifications and refinements have been proposed and a number of variants of this approach now currently exist.

Consider the schematic drawing of the simulation of flow past a solid body in Fig. 9.5A. The conventional approach would be to employ structured or unstructured grids conforming to the particular body shape. This is achieved by first defining the surface grid covering the boundary Γ_b. The internal grids encompassing the regions occupied by the fluid Ω_f and the solid Ω_s are then generated after. For the finite-difference method, the partial differential governing equations are transformed into a curvilinear coordinate system that aligns with the grid lines. They are then discretized and solved in the computational domain with relative ease. For the finite-volume method employing a structured grid, the integral form of the governing equations is discretized and the geometrical information regarding the grid is incorporated directly into the discretization. If an unstructured grid is used instead, either the finite volume or finite element methodology can be adopted by incorporating all the relevant local cell geometry into the discretization. Both of these approaches do not have to resort to grid transformations.

On the other hand, let us consider a non-conformal body on a Cartesian grid in Fig. 9.5B. Here, the immersed boundary would still be represented through some means

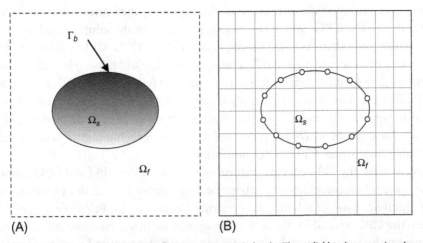

Fig. 9.5 (A) Schematic drawing showing flow past a generic body. The solid body occupies the volume Ω_s with boundary Γ_b. The volume of the fluid is denoted by Ω_f. (B) Schematic of body immersed in a Cartesian grid on which the governing equations are discretized.

of the surface grid covering the boundary Γ_b but the present Cartesian grid would be generated with no regard to this surface grid. Therefore the solid body cuts through this Cartesian volume grid. Because the grid does not conform to the solid boundary, the immediate task is to incorporate appropriate boundary conditions by modifying the equations in the vicinity of the boundaries. The governing equations can therefore be discretized using any technique, whether they may be finite difference, finite volume, or finite element, without the need to resort to coordinate transformation or complex discretization operators. This is the essence of the immersed boundary method.

Clearly, the imposition of suitable boundary conditions is not straightforward in immersed boundary methods. There are two approaches as indicated by Mittal and Iaccarino (2005) that are amenable to precisely accommodate the effects of the boundary condition on the immersed boundary. The modification can take place in the form of source terms (or forcing functions) in the governing equations in order to reproduce the effect of a boundary. In the first approach, which we can term *the continuous forcing approach*, the forcing function is incorporated into the continuous equations applied to the entire domain before discretization, whereas for the second approach, which can be termed *the discrete forcing approach*, the "forcing" is introduced after the equations are discretized. The continuous forcing approach is very attractive for flows with immersed elastic boundaries. For such flows, the method has a sound physical basis and is very simple to implement. Many successful applications in biology such as sperm motility (Fauci and McDonald, 1994) and multi-phase flows (Unverdi and Tryggvason, 1992) testify to the viability of such an approach where elastic boundaries abound. For the discrete forcing approach, it is, however, not as practical as the continuous forcing approach but it enables a sharp representation of the immersed boundary and this is especially desirable for high-Reynolds-number flows. It also allows direct control over the numerical accuracy, stability, and discrete conservations of the solver. This method has been used rather successfully for simulating compressible flow past a circular cylinder and an airfoil (Ghias et al., 2004), flow through a rib-roughened serpentine passage (Iaccarino et al., 2003), flapping foils (Mittal et al., 2002), and objects in free fall through a fluid (Mittal et al., 2004). It is apparent that the choice of either employing the continuous or discrete forcing approaches is very problem dependent on the particular fluid flow process. More details on the mathematical basis and numerical considerations for both of these approaches can be found in the article by Mittal and Iaccarino (2005).

As a demonstration of the application of the immersed boundary method, Fig. 9.6 illustrates the flow structures in the near wake behind a sphere. Numerical simulations performed by Yun et al. (2006) are compared against the experimental flow visualization. The large eddy simulation (LES) results are shown to be in good agreement with the experiment, thereby validating the fidelity of the immersed boundary method.

The popularity of immersed boundary methods is increasing at a tremendous rate. The few applications highlighted above do not even begin to scratch the surface of

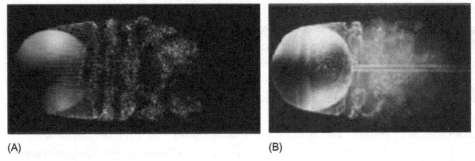

(A)   (B)

Fig. 9.6 Results extracted from Yun et al. (2006) for the flow structures in the near wake behind a sphere: (A) Reynolds number $Re = 10^4$ (immersed boundary simulation) and (B) Reynolds number $Re = 1.5 \times 10^4$ (experiment).

the many applications that have been simulated using these methods. In due time concerted ongoing improvements on the issues of accuracy and efficiency and the development of innovative and creative approaches such as adaptive grid meshing with immersed boundary method (Roma et al., 1999) will see increased applications in new areas of complex turbulent flows, fluid-structure interaction (FSI), multi-material and multiphysics simulations, and also well-established areas such as biological and multi-phase flow applications. For example, with a focus on solution strategies for simulating FSI issues, Sotiropoulos and Yang (2014) compared immersed boundary approaches for imposing boundary conditions, efficient iterative algorithms for solving the N-S equations in the presence of dynamic immersed borders, and strong and loose coupling FSI strategies. They also provided current findings from the use of such tools to investigate a variety of issues, including vortex-induced vibrations, aquatic swimming, insect flight, human walking, and renewable energy. Griffith and Patankar (2020) discussed immersed approaches for both elastic and kinematically specified structures. Benchmark issues revealed the efficiency of these methods, and selected applications at Reynolds numbers up to roughly 20,000 demonstrated their importance in biological and biomedical modeling and simulation.

9.3.5 High-Order Methods

After decades of research and development, RANS solvers with first- and second-order accuracy have gained wide acceptance as a rapid, efficient, and reliable tool. Despite their extensive use, they are still lacking in resolving flow problems associated with vortex-dominated flows, significant flow separations, and aero-acoustic problems (Wang et al., 2013). High-order methods have thus received significant interest because of their ability to provide more accurate solutions than low-order approaches, though they are generally more difficult and less robust to implement.

The term *high order* can be unanimously agreed as third order or higher as solutions from most commercial computer packages are only accurate to first- or second-order accuracy (Wang et al., 2013). Although high-order methods require more CPU time to compute, they produce significantly more accurate results for the same computational cost; hence high-order methods are usually considered more efficient; for example, based on the cost to achieve the same error, a high-order method can achieve the error threshold on a coarser mesh. Popular high-order methods for CFD include Discontinuous Galerkin (DG), Spectral Difference (SD), and Spectral Volume (SV) methods. The DG technique has risen to prominence as one of the most promising approaches due to its ability to attain highly accurate numerical solutions from increasingly complex physical models. Flux Reconstruction (FR) or Correction Procedure utilizing Reconstruction (CPR) technique based on a differential formulation provides a unifying framework for various high-order systems (Huynh et al., 2014).

Although second-order approaches have long been the option for many CFD applications, high-order approaches have their own niche applications such as determining the aerodynamic performance of aircraft vehicles being influenced by unstable vortices in vortex-dominated flows. Such solutions are of crucial importance to the aerospace industry especially when precise aerodynamic designs are paramount (Wang, 2011). Apart from fluid dynamics, high-order schemes have also been receiving an increasing amount of attention in magnetohydrodynamics (MHD), which plays an important role in fields such as astrophysics, plasma physics, and space physics (Wu and Shu, 2019), where the need for highly accurate solutions with reduced computational effort is preferred (Mignone et al., 2010).

High-order methods are also applied in conjunction with other advanced CFD techniques. For example, the use of overset grid that can handle complex geometry and moving systems through high-order interpolation methods. Lee et al. (2012) developed a high-order interpolation method for overset grid that can deal with problems with co-existing non-linear and linear characteristics, such as shock-vortex interaction. Freret et al. (2017) integrated a high-order discretization approach with an adaptive mesh refinement technique to accurately capture divergent spatial and temporal scales as in space-physics problems.

Although high-order methods are well known to be less robust and slower to converge, high-order approaches could eventually become the workhorse for future CFD applications especially the use of methods such as h- and p-adaptations (mesh and polynomial order adaptations) in the context of adaptive discretization. Dong and Karniadakis (2004) discovered that parallel computing using threads can effectively facilitate p-refinement to conduct adaptive discretizations for high-order algorithms, as the number of threads per process can be varied proportionately to maintain a constant wall clock time every step. This represented a significant breakthrough that had the potential to transform how parallel computers implement high-order techniques. Subsequently, more studies on parallel computing's efficiency with high-order methods have been explored, such as for under-resolved turbulence simulation (Wang et al., 2020), to capture fine structures in complex flows (Chalmers et al., 2019), and the application of a spinning golf ball (Crabill et al., 2018).

9.3.6 Reduced-Order Modeling Approaches

The increase of sophistication in real-life problems has given rise to a number of coupling methods with CFD, arising from problems associated with multi-physics (e.g. FSI) and multi-scale. Although coupled systems generally use far fewer computational resources and time than stand-alone CFD computer codes, the gains by the CFD portion remain limited. To maintain the accuracy of full non-linear methods at a lower and manageable cost, reduced-order modeling (ROM) approaches have been developed to address the prohibitive costs associated with real-time analyses for parametric design and optimizations.

Reduced-order models (ROMs) are simplified, high-fidelity, sophisticated models, based on projecting the equations characterizing the fluid problem onto a low-dimensional basis while retaining the core physics and dynamics of high-fidelity CFD models (Hesthaven et al., 2016). Compared to high-fidelity models, ROM has fewer degrees of freedom; thus it is computationally more efficient. Herein, engineers can swiftly investigate a system's dominant effects and its behavior using minimal computational resources. Because of the need for quicker design cycles, ROMs have grown rather popular in the product development business.

Applications of ROM approaches have been widely found in control or sensitivity assessments requiring results of a large number of simulations for various parameter values (Star et al., 2021), such as to reduce the cost of shape optimization around a gas turbine blade (Ferrero et al., 2020) and aeroelastic structural optimization (Li et al., 2019). ROM also shows excellent performance when coupling with other systems. In the coupling between CFD and building energy simulation (BES), smaller time steps are required by CFD compared to the larger time steps usually adopted for BES. To tackle that issue, Fast Fluid Dynamics (FFD) was proposed to couple with BES to provide highly efficient computations. Even though FFD has been reported to be about 50 times faster than CFD, it is still not fast enough for some control applications (Kim et al., 2015). Hiyama et al. (2010) have reported that a CFD analysis for 100 min transient calculation took about 23 h, while the same analysis with reduced-order indoor models took only milliseconds. Kim et al. (2015) then developed a ROM-CFD coupled model for simulating the dynamics of indoor environments and building envelopes, which allowed for the analysis of spatial differences in zone air temperature at a modest computational cost of roughly 1 s every day of simulation. For fluid-structure interaction (FSI) problems, Adaptive Reduced-Order Models (AROMs) have shown significant improvement in computational performance. While the traditional approach normally takes roughly 77% of the entire simulation time to solve the structural problem, the ROM has resulted in a four-fold speedup (Thari et al., 2021).

Among the approaches of ROMs, the proper orthogonal decomposition (POD) is one of the most effective ways to derive the reduced basis for high-dimensional non-linear flow systems or heat conduction problems. Cai et al. (2019) proposed a POD-ROM with a fast calculation algorithm to predict the dynamics for flow-heating coupling problems, which greatly assisted in engineering applications for optimization and flow control. Hijazi et al. (2020) proposed POD-Galerkin mixed-ROM to simulate turbulent flows by introducing a reduced version of the eddy viscosity concept. Because of the proven accuracy of these approaches, they represent enormous potentials to be utilized in conjunction with artificial neural networks to improve the evaluation of the eddy viscosity coefficients.

9.3.7 Liutex and Third Generation of Vortex Identification Methods

Vortex is intuitively recognized as the rotational/swirling motion of fluids. Vortex identification algorithms that detect and visualize the vortical flow behavior are an essential part of understanding the underlying physical mechanism of the flow field. This section introduces a new physical quantity to represent fluid rotation or vortex: Liutex, discovered by Liu et al. at the University of Texas at Arlington (UTA) in 2018. Historically, there are three generations of vortex identification methods.

In 1858 Helmholtz defined vortex as vorticity tubes, which can be classified as the first generation of vortex identification. Such correlation between vortex and vorticity has nonetheless been found to be rather weak, especially in the near-wall boundary region. Alternate vortex identification criteria including the Q, Δ, λ_2, and λ_{ci} methods have been developed, which can be classified as the second generation of vortex identification. They are all based on the eigenvalues of the velocity gradient tensor, which are scalars and threshold dependent. Nevertheless, this second-generation vortex identification remains inadequate due to the absence of accounting for the stretching and shearing of vortices that are not only time dependent but also space dependent. Liutex, the third generation of vortex identification, is defined as a vector rather than a scalar that uses the real eigenvector of velocity gradient tensor as its direction and twice the local angular speed of the rigid rotation as its magnitude. The main idea of Liutex is to extract the rigid rotation part from fluid motion to represent the vortex. Several vortex identification methods have been developed by Liu and the UTA Team including Liutex vector, Liutex vector lines, Liutex tubes, Liutex iso-surfaces, Liutex-Omega methods, objective Liutex, and, more recently, Liutex core-line methods.

The vector for Liutex can be defined as

$$\vec{R} = R\,\vec{r} \quad and \quad \vec{\omega}\cdot\vec{r} > 0$$
$$R = \vec{\omega}\cdot\vec{r} - \sqrt{\left(\vec{\omega}\cdot\vec{r}\right)^2 - 4\lambda_{ci}^2} \tag{1}$$

where $\vec{\omega}$ is vorticity, $\vec{r}$ is the real eigenvector of the velocity gradient tensor $\nabla\vec{v}$, and λ_{ci} is the imaginary part of $\nabla\vec{v}$ complex eigenvalue (see Gao et al., 2018; Liu et al., 2018, 2019; Wang et al., 2020).

1. Modified Liutex–Omega method (MLOM) (Fig. 9.7): $\widetilde{\omega}_R = \dfrac{\left(\vec{\omega}\cdot\vec{r}\right)^2}{2\left[\left(\vec{\omega}\cdot\vec{r}\right)^2 - 2\lambda_{ci}^2 + 2\lambda_\sigma^2 + \lambda_r^2\right] + \varepsilon}$

   where λ_r is the real eigenvalue of $\nabla\vec{v}$. MLOM is an iso-surface method but is threshold insensitive and is capable of capturing both strong and weak vortices. Normally, $\widetilde{\omega}_R = 0.52 - 0.6$, and ϵ is a small positive number (see Liu et al., 2019).
2. Liutex core-line method (LCLM) (Fig. 9.8): the main idea herein is to locate the local maxima of Liutex, where $\nabla R \times \vec{r} = 0\,(\nabla R//\vec{r} - \text{parallel})$. LCLM yields a unique solution, threshold-free, and has both rotation axis and rotation strength (see Gao et al., 2019).

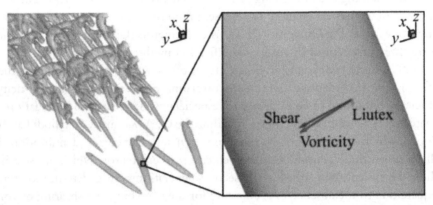

Fig. 9.7 Vorticity and Liutex vectors.

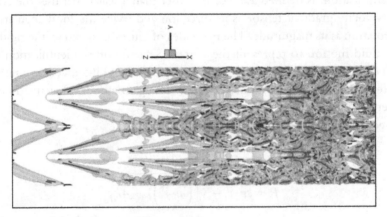

Fig. 9.8 Liutex core-lines for flow transition.

9.4 ADVANCES IN COMPUTATIONAL MODELS

9.4.1 Direct Numerical Simulation

Turbulence in fluids is a non-linear phenomenon and it comprises a wide spectrum of spatial and temporal scales. Turbulent flow poses significant challenges to modelers, as already evidenced by the range of turbulence models that have been developed and applied. Decades of active research have certainly translated some of these models as commonplace in many practical and design analyses. One significant area of activity in the modern simulation strategies for turbulent flow is the return to direct simulation of turbulence to embrace increasing realism (complexity and Reynolds numbers) in the flow as computers increase in performance.

As described in Chapter 6, direct numerical simulation (DNS) refers to computations where all relevant spatial and temporal scales are adequately resolved *for the given application*. A valid simulation must accommodate all the range of length scales including the smallest scales of which the viscosity is active. This means that it is important to capture all of the kinetic energy dissipation within the turbulent flow. Estimates for the smallest scales are available from the so-called Kolmogorov micro-scales. From dimensional analysis, assuming dependence only upon viscosity ν and dissipation rate of kinetic energy ε, the length micro-scale η can be expressed as:

$$\eta = \left(\frac{\nu^3}{\varepsilon}\right)^{1/4} \tag{9.11}$$

If the integral length and velocity macroscales of the problem are respectively L and u and we further assume the dissipation scales in the same way as production, that is, u^3 / L, it can be shown according to Tennekes and Lumley (1976) that

$$\frac{L}{\eta} = \mathrm{Re}_L^{3/4} \tag{9.12}$$

Here, $\mathrm{Re}_L = uL/\nu$ is a Reynolds number based on the magnitude of the velocity fluctuations and the integral length scale. The difference between the largest and smallest length scales in turbulence thus increases as the Reynolds number increases; the number of grid points required to resolve turbulence increases as $\mathrm{Re}_L^{9/4}$. If the time step restriction is factored in, the computational cost scales as the cube of the Reynolds number. Clearly, flows with very high Reynolds numbers are impossible to simulate. Nevertheless, some phenomena in turbulence appear to asymptote in the range of high Reynolds numbers asymptote such as free shear layer rates and near-wall behavior and so numerical simulations that can be feasibly performed at "high enough" Reynolds numbers are still required to capture these phenomena in order to enhance understanding and contribute to model development.

Some numerical issues pertaining to DNS simulations are discussed herein. The key issue that typically dictates a DNS simulation is the mesh resolution within the flow domain. Some applications, such as those requiring statistics involving higher derivatives, will definitely need greater resolution than others. Detailed grid refinement studies have shown that there are some "rules of thumb" that can be adopted in DNS calculations. One rule of thumb that the user may adopt is that grid spacing away from the wall should be in the order of 5η, which has been found to be sufficient for most purposes in the prediction of the mean flow, second moments of turbulence, and all the terms in the kinetic energy transport equation. To suggest another rule of thumb, take, for example, calculations being performed for free shear layer flows where the ability to preserve 6 decades of roll-off in the energy spectrum could expect good second moment turbulence statistics to be attained during the DNS calculations.

Another consideration for DNS is the need to implement suitable numerical methods to obtain accurate realization of a flow that contains a broad spectrum of time and length scales. DNS generally requires an accurate time history. Since small time steps are naturally adopted, *explicit* methods that are usually stable because of the strict CFL requirements have been the preferred time-advance methods in most simulations instead of *implicit* methods. There are, of course, notable exceptions especially fine meshes being concentrated near solid walls to capture the essential small-scale flow structures that may cause instability due to the explicit determination of the viscous terms in these regions. This can be easily overcome by treating these terms in an implicit manner to promote stability during transient calculations. The most commonly used time-advance methods have been the second-order Adams-Bashforth and the third- or fourth-order Runge Kutta methods. The reader may wish to refer to Rai and Moin (1991) for a greater understanding of the use of these methods applied to the direct simulation of turbulent flows in a channel. In practice, Runge Kutta methods allow a larger time step for the same accuracy and thus compensate for the increased amount of computation. To treat the viscous terms, the Crank-Nicolson method (see Appendix C) is often applied in conjunction with these time-advance explicit methods.

DNS also requires appropriate *discretization* methods. Unquestionably, the prevalent method for DNS is the spectral method. However, this method places restrictions on the type of geometry and grids that can be efficiently handled. In consideration of the commonly adopted finite difference or finite volume approaches, it is imperative to reiterate the importance of employing an energy-conservative spatial differencing scheme. In many Reynolds-Averaged Navier-Stokes (RANS) calculations, upwind differencing schemes are commonly advocated since the dissipation that is introduced tends to stabilize the numerical methods and bound the scalar solutions. However, when these schemes are used in DNS, the dissipation produced is often much greater due to the physical viscosity resulting in a solution that has little connection to the flow physics of the problem being solved. To overcome the issue of false diffusion, "diffusion-free" central differencing schemes are used instead in many DNS simulations. Accuracy ranging from second to fourth order has been applied. Nonetheless, Rai and Moin (1991) have found that the fifth-order upwind-biased scheme has proven to yield good first- and second-order statistics that agree well with experimental data. It appears that the high-order accurate upwind-biased scheme is a good candidate for direct simulations of turbulent flows with complex geometries provided that the mesh is sufficiently fine. However, the problem of false diffusion may still resurface due to the inadequate resolution of the flow domain.

Generating initial and boundary conditions for the flow domain in DNS can also be challenging. In comparison to RANS computations, where only mean conditions are required, DNS must contain all the details of the three-dimensional velocity field including the complete velocity field on a plane (or surface) for the inflow conditions of a turbulent flow at each time step. Since the memory of the initial and boundary conditions may well be required through some considerable time, they can have a significant effect on the results at the later stages of the flow simulations. Certain boundary conditions that are applicable to RANS cannot be readily applied in DNS such as symmetry boundary conditions. For outflow boundaries, it is important to impose boundary conditions that do not allow the pressure waves to be reflected off these boundaries and back into the interior of the domain. Despite all attempts to prescribe the initial and boundary conditions to be as realistic as possible, a DNS simulation must take its course for some lengths of time so that the flow develops with the correct characteristics of the physical flow. The best way to ascertain the flow development is to actually monitor some flow quantity; the choice depends on the flow that is being simulated. Initially, the quantity may reveal some systematic decreasing or increasing trends but when the flow is fully developed, the value will show statistical fluctuations with time. At this stage, statistical averaging over time can thus be performed on the results (e.g. the mean velocity and fluctuations).

Lately, more sophisticated DNS applications such as a by-pass transition in which passing wakes trigger turbulence spots in a flat-plate boundary layer or vortex structure identification have been performed (Geurts, 2001). The reader should be aware that such simulations usually amount to hundreds of computational hours to arrive at a solution. It is not entirely surprising that the number of grid nodal points that can be used is limited by the processing speed and memory of computers. Nonetheless, as computers become quicker and larger memories are made available, solutions to more complex and higher Reynolds number flows are becoming more attainable. DNS allows the provision of acutely detailed information about the flow. This wealth of information covering a wide range of length and time scales can be used to attain a better understanding of the flow physics such as the mechanisms of turbulence production, energy transfer, and dissipation in turbulent flows or the effects of compressibility on turbulence. Interested readers can refer to the following journal articles (Hattori and Nagano, 2012; Ström and Sasic, 2013; Sun et al., 2015; Magolan et al., 2017; Armengol et al., 2019; Hurtado Rosas and Wang, 2022; Rosa et al., 2022). Significant insights into the physics of the flow through DNS can at times be better realized, which may not have been possible through experiments.

The rapid growth of computational capabilities presents significant opportunities for DNS's application in industrial problems. In recent years DNS has taken a more prominent role in combustion modeling (Chen, 2011; Bisetti et al., 2014; Trisjono and Pitsch, 2015). Progress in computational power allows for increased grid size, a larger number of time steps, ensemble averaging (Shalaby and Thévenin, 2010), and a more complete temporal development of the turbulent flame with high statistical reliability. In two-phase flows (Lebas et al., 2009) it is now possible to simulate the primary breakup of a diesel spray with Eulerian-Lagrangian Spray Atomization (ELSA) model. Special attention has been devoted to describing realistically the primary break-up via DNS of two-phase flows based on LS/VOF/GF (Ménard et al., 2007). Particle-resolved direct numerical simulation (PR-DNS) for understanding gas-solid flow physics and obtaining quantitative information for model development can be found in the review by Tenneti and Subramaniam (2014). Reckinger et al. (2016) used DNS to investigate compressible Rayleigh-Taylor instability (RTI), which necessitates efficient numerical methods, advanced boundary conditions, and consistent initialization to capture the system's wide range of scales and vortex dynamics while minimizing the computational impact of acoustic wave generation and subsequent interaction with the flow.

9.4.2 Large Eddy Simulation

Large eddy simulation (LES), also briefly described in Chapter 7, is essentially a simulation that directly solves the large-scale motion but approximates the small-scale motion. The establishment of the LES method has its roots in the prediction of atmospheric flows since the 1960s and, like DNS, has grown in importance as computers have increased in size and performance.

As aforementioned, LES requires a flow field where only the large-scale components are present. This can be achieved through a *filtering* process. By eliminating the small-scale eddies through localized filter functions, either a Gaussian or top-hat filter, only the large or resolved scale field remains to be solved. The instantaneous flow variables are then decomposed into its filtered or resolvable component (essentially a local average of the complete field) and a *subgrid-scale* (SGS) component that accounts for the scales not resolved by the filter width. Within the finite volume method, it is rather sensible and natural to define the filter width as an average of the grid volume. In a rough sense the flow eddies larger than the filter width are *large eddies*, while eddies smaller than the filter width are *small eddies*, which require modeling. When filtering is performed on the incompressible Navier-Stokes equations, a set of equations very similar to the RANS equations of momentum and energy in Chapter 3 are obtained. Similar to RANS, there are additional terms where a modeling approximation must be introduced. In the context of LES these terms are the *SGS* turbulent stresses and heat fluxes, which require *SGS* models to close the set of equations.

The most widely used *SGS* model is the one proposed by Smagorinsky (1963). Since it is prescribed through the eddy-viscosity assumption, it therefore shares many similarities to the formulation of the Reynolds stresses as obtained through the RANS approach. For the unresolved *SGS* turbulent stresses, these are modeled accordingly as:

$$\tau_{ij} - \frac{\delta_{ij}}{3}\tau_{kk} = -2\nu_T^{SGS}\overline{S}_{ij}, \qquad \overline{S}_{ij} = \frac{\partial \overline{u}_i}{\partial x_j} + \frac{\partial \overline{u}_j}{\partial x_i} \qquad (9.13)$$

where ν_T^{SGS} is the *SGS* kinematic viscosity and $\overline{S}_{ij}$ is the strain rate of the large-scale or resolved field. The form of the *SGS* eddy viscosity μ_T^{SGS} noting that $\nu_T^{SGS} = \mu_T^{SGS}/\rho$ can be derived by dimensional arguments and is given by:

$$\mu_T^{SGS} = C_s^2 \rho \Delta^2 |\overline{S}_{ij}|, \qquad |\overline{S}_{ij}| = \sqrt{2\overline{S}_{ij}\overline{S}_{ij}} \qquad (9.14)$$

where Δ is denoted by the grid filter width and the model constant C_s varies between 0.065 and 0.3 depending on the particular fluid flow problem. There is a difference in the way the turbulent viscosity is evaluated between the LES and RANS approaches. From Eq. (9.14), LES determines the turbulent viscosity directly from the filtered velocity field. However, by referring to Eq. (3.42) in Chapter 3, RANS requires the turbulent viscosity to be evaluated through the flow field containing two additionally derived variables, which are the turbulent kinetic energy k and its rate of dissipation ε values. Over the past decade, other basic subgrid-viscosity models such as the Structure Function model by Métais and Lesieur (1992) and the Mixed Scale model by Sagaut (2004) that exhibit a triple dependency on the vorticity of the resolved scales have also been proposed in addition to the basic Smagorinsky model. The reader should realize that all of these models have been designed assuming that the simulated flow is turbulent, fully developed, and isotropic and therefore do not incorporate any information related to an eventual departure of the simulated flow from these assumptions. To obtain an automatic adaptation of the models for inhomogeneous flows, simulations of engineering flows are more likely to be based on the dynamic formulations of the basic versions of these models, which are briefly discussed below.

One possibility for designing a self-adaptive SGS model is the dynamic procedure originally developed by Germano et al. (1991). All the basic developments rest upon the usage of the Germano et al. (1991) identity[1]:

$$\tilde{L}_{ij} = \overline{\overline{u}_i \overline{u}_j} - \tilde{\overline{u}}_i \tilde{\overline{u}}_j = \tau_{ij}^T - \tilde{\tau}_{ij} \qquad (9.15)$$

where $\tilde{(\,)}$ represents a second filtering operation, called a "test" filter, with a larger filter width than the grid filter, and $\overline{(\,)}^T$ indicates a quantity computed using the test-filtered

---

[1] This identity is also known as the Leibniz identity in classical mechanics.

velocity. The dynamic procedure can in principle be applied to any subgrid model. By equating L_{ij} to a term $C_s^2 M_{ij}$, where M_{ij} is given by

$$M_{ij} = -2\left(\tilde{\Delta}^2 \left|\tilde{\bar{S}}_{ij}\right|\tilde{\bar{S}}_{ij} - \bar{\Delta}^2 \overline{\left|\bar{S}_{ij}\right|\bar{S}_{ij}}\right) \tag{9.16}$$

with $\tilde{\Delta}$ as the test filter width and combined with the procedure of Lilly (1992), the model constant C_s can be obtained by minimizing the square of the error term $L_{ij} - C_s^2 M_{ij}$ to yield

$$C_s^2 = \frac{\langle L_{ij} M_{ij}\rangle}{\langle M_{ij} M_{ij}\rangle} \tag{9.17}$$

The above-described procedure is the basic formulation of the dynamic Smagorinsky model. Following similar arguments, other dynamic models such as the Mixed Smagorinsky Scale-Similarity models developed by Zang et al. (1993) and Shah and Ferziger (1997) and the Lagrangian Dynamic model of Meneveau et al. (1996) have been used with considerable success and certainly broadened the accessible applications for LES.

The numerical models and boundary conditions required for LES are essentially the same as those used in DNS. In regions close to the solid surfaces the dynamic Smagorinsky model is well suited for such flows since it automatically decreases the model constant automatically in the correct manner near the solid wall. Nonetheless, it is also still possible to employ wall functions of the kinds used in RANS modeling while adopting the original Smagorinsky model or other basic SGS models in some turbulent flow problems. The Van Driest (1956) wall damping function that reduces the near-wall eddy viscosity according to the normal distance from the wall has proven to be a successful recipe. A variation to the van Driest wall damping function such as that formulated by Piomelli et al. (1987) can also be used to suppress the near-wall eddy viscosity. Both of these models are available in the commercial code ANSYS-CFX. Another approach that is based on the renormalization group (RNG) theory by Yakhot and Orszag (1986), where the eddy viscosity reduces according to the SGS Reynolds number, has also been proposed. This model is adopted in the commercial code ANSYS-FLUENT. The use of wall functions in LES has been shown to work rather well for attached flow problems (Piomelli et al., 1989). Further details can be found in several recent review articles. Piomelli and Balaras (2002) examined the available approaches for performing high-Reynolds-number LES at a reasonable cost without bypassing the wall layer. Piomelli (2008) evaluated three models: equilibrium laws, zonal models, and hybrid approaches, to assess the current state-of-the-art of wall-layer modeling.

Like DNS, most LES schemes that model real flows correctly are still very computationally expensive. Owing to computational limitations, a possibly damaging aspect prevalent in most LES schemes is that the simulations are still not being performed on a fine enough grid and consequently do not capture some important dynamics at high wave numbers, which have not been filtered or modeled. Numerical truncation errors begin to overcome the SGS model and the original physics may be lost. Lately, a way to address some of the drawbacks has been to utilize the numerical truncation errors directly to act as an SGS model implicitly instead of the explicit models proposed above. This model is known as the monotone integrated large eddy simulation, also better recognized by the acronym MILES. The key success of this approach is how to construct a numerical method that will depict the proper SGS model and recent investigations have shown it is feasible to employ high-resolution shock-capturing methods (see Section 9.2.3). At this particular juncture, it is not the authors' intention to present the formal procedure behind the MILES-type algorithm, which is beyond the knowledge at this introductory level. Interested readers are strongly encouraged to refer to Garnier et al. (1999), Sagaut (2004), and Hahn and Drikakis (2005) for a greater understanding of the background theory behind MILES and their respective applications in turbulent flow simulations.

9.4.3 RANS-LES Coupling for Turbulent Flows

The idea behind this methodology is to address the urgent need to solve high-Reynolds-number flows especially for wall-bounded flows where the requirement of a very fine mesh near walls still precludes a full LES simulation. It is important that the near-wall turbulent structure in the viscous and buffer sub-layers consisting of high-speed in-rushes and low-speed ejections (often called the streak process) are properly resolved within the near-wall region. At low to medium Reynolds numbers, this streak process generates the major part of the turbulence production; these structures must be fully accommodated in LES to obtain an accurate representation of the phenomena. For wall-bounded flows at high Reynolds numbers of engineering interest such as bluff bodies, the computational resource requirement of accurate LES can be prohibitively large. Because of the enormous amount of work that has been devoted to the development of RANS turbulence models for near-wall predictions, it is therefore possible to take advantage of these models, which are less expensive to capture the near-wall turbulent structures instead of the full resolution through LES. Nevertheless, it is well known that the RANS approach is unable to account for the spectrum of unsteadiness that is associated with the turbulent fluctuations away from the wall. LES, because of its explicit cutoff frequency, can choose the degree of accuracy of the description of the turbulent fluctuations; accurate results can be obtained at an affordable cost without imposing a very fine mesh resolution. Hence the idea of combining both methods where the inner near-wall (the unsteady RANS) region is handled by the RANS approach while the outer region of the bulk flow is solved through LES makes rather perfect sense. In the LES region the mesh resolution is dictated mainly by the requirement of resolving the largest turbulent scales rather than the near-wall turbulent processes.

The RANS-LES coupling of turbulent flows can be handled through a number of different approaches. For attached flow problems using LES, as described in the previous section, wall law functions employed to reduce the computational cost are in some way representative of the zonal use of RANS and LES. Nonetheless, the application of these functions to simulate separated flows presents enormous problems and they also tend to be lacking in universality when complex geometries are handled. The use of a RANS model in the near-wall zone by adopting a reduced RANS model (e.g. of mixing-length type), solving the turbulent boundary layer equations in an embedded near-wall mesh (Balaras et al., 1996; Wang and Moin, 2002), or even applying a full RANS model designated as hybrid RANS-LES, are the possible alternatives to circumvent the problems associated with wall functions. The Non-Linear Disturbance Equations (NLDEs) approach initially proposed by Morris and his co-workers (1997, 1998) and recently reviewed by Labourasse and Sagaut (2004) represents another RANS-LES technique that can be applied to handle a range of complex flows. Like the hybrid RANS-LES, it solves the near-wall flow field through the RANS model and the energy and dynamics of the main unsteadiness flow in the far field through LES but with an additional subgrid model to handle the unresolved fluctuations (in the zones of interest). For the remainder of this section, the authors would like to focus on the basic concepts and advancements made to the hybrid RANS-LES models.

Hybrid RANS-LES models can be divided into two categories. The first category is the zonal decomposition into the physical space, where the computational domain is divided into sub-domains. Some of these sub-domains are computed using a RANS approach and the other through LES. In the unsteady RANS region two-equation models, whether they may be the *standard k-ε model, k-ω model,* or other developed turbulent models, can be employed. Davidson and Peng (2003) applied the *k-ω model* in the near-wall flow field. The transition between the domains is prescribed through a predefined switching plane, which is explicitly provided. The location of this plane or interface can be determined in many ways. It can be chosen using some blending function (sharp or smooth) (Davidson and Peng, 2003; Temmerman et al., 2005), comparing the unsteady RANS and LES turbulent scales, computed from turbulence/physics requirements (Tucker and Davidson, 2004) or solving different partial differential equations to automatically locate the interface (Tucker, 2003). Although encouraging results have been obtained for selective comparative cases with the application of this zonal hybrid RANS-LES, proper treatment of the interface bridging the RANS and LES regions is crucial to the success of this model, which remains an area of active research. For the second category, hybrid computations are performed without predefining a switching plane. This approach appears to be preferable to the first category models. However, the possibility that a subgrid model with a RANS is required limits this method. Similar to the NLDE, where the same numerical scheme is used throughout the whole domain, instability modes such as wiggles can appear in the RANS zone, which are some of the main disadvantages of this approach; thus no clear advantages can be found in adopting this second category over the first category models. Examples of these models are Speziale's very large eddy simulation (VLES) (1998), detached eddy simulation (DES) (Spalart et al., 1997), and the renormalization group approach (de Langhe et al., 2005a, 2005b). Some other hybrid methods such as Scale Adaptive Simulation (SAS) for flows in which strong instabilities of the flow exist, and stress blended eddy simulation (SBES), which is ideal for situations involving a mix of boundary layers and free shear flows, can be found in the literature (Menter and Egorov, 2010; Menter, 2018; Menter et al., 2021).

The hybrid RANS-LES coupling is an excellent candidate for massively separated flow problems. Typical successful applications that have been found in Strelets (2001) and Squires and Constantinescu (2003) are flow around a sphere or cylinder, NACA 0012 airfoil, backward-facing step, and a landing gear truck. For flow over periodic hills, overall energy and mean velocity profiles are well predicted, with 20 times fewer grid points than those needed for resolving using LES. Generally speaking, the mesh density in hybrid RANS-LES is comparable to those used in RANS and it provides better results than LES on the same coarse grids. It is also superior to a full RANS simulation in predicting the energy of the main unsteadiness in the flow. The current main research issue in this methodology is turbulent synthetization. Interested readers can refer to Battan et al. (2004) for a better understanding of the required turbulent closures of this issue.

9.4.4 Multi-Phase Flows

In Chapter 8 we observed how CFD could enhance the understanding of physical processes associated with gas-particle flows in simple or complex geometries such as in a 90° duct bend or physiological respiratory human airway system. There are nonetheless many other engineering applications that often involve multi-phase flows; examples are gas bubbles in liquid or liquid droplets in gas, fluidized bed combustion, liquid fuel injected as a spray in combustion machines, etc.

Multi-phase flow investigations are certainly at the crossroads of intense research in the climate of significant advancements being achieved in computing power and performance. Two commonly adopted approaches that have been briefly described in Chapter 8 to compute two-phase flows are the Eulerian-Eulerian and Eulerian-Lagrangian methods. The Eulerian approach is usually adopted for the carrying or continuous phase fluid. The distinction results in the dispersed phase, where it may be handled either through the discrete approach in the Lagrangian framework or the continuous Eulerian approach similar to the carrying phase fluid. When the mass loading is small, this phase can be represented by a finite number of particles whose motion can subsequently be computed in a Lagrangian manner. These particles are injected into the flow domain and their trajectories are determined from the pre-computed velocity field of the background fluid. This depicts the *one-way* coupling approach. When the mass loading is substantial, the influence of particles on the fluid motion needs to be properly accounted. A *two-way* coupling requires the computation of particle trajectories and fluid flow to be simultaneously carried out; each particle contributes to the momentum, mass, and energy of the parent fluid in each control volume cell it passes through. On the range of Eulerian-Eulerian and Eulerian-Lagrangian investigations of gas-particle flow applications, the reader may wish to refer to the current works by Tian et al. (2005, 2006).

Interaction between particles such as collision or agglomeration and between particles and walls can exist where appropriate models to account for all these effects are areas of ongoing research. These subject matters are certainly complicated, and the reader can refer to the book by Crowe et al. (1998) for a broader understanding. Recently, attempts to address particle impaction particularly on curved wall boundaries have been performed and some encouraging results are found in the articles by the present authors: Morsi et al. (2004) and Tu et al. (2004b). Ongoing research is still being carried out to address many of the complex issues associated with particle-wall behavior. Robust models for practical applications remain elusive at this present stage.

For flows with large mass loadings when the phase change occurs, the two-fluid model (Eulerian-Eulerian approach) is applied to both phases. Here, both phases are treated as continua with separate velocity and temperature fields. To account for interfacial effects between phases, appropriate constitutive relationships for the inter-phase exchange terms are required to close the system. The principles of two-fluid models have been described in detail in recent texts by Kolev (2005) and Ishii and Hibiki (2006) for the application to gas-liquid flows. In the context of a phase change multi-phase flow sub-cooled boiling flow presents enormous challenges in modeling and simulation. The two-fluid model that is applied to handle both continua phases occupied by the gas bubbles and liquid requires additional wall models to predict the nucleation of bubbles generated from heated walls. Based on a number of experimental observations, bubbles departing from the heated walls have been found to form bigger or smaller bubbles either through merging, shearing off, or disappearing due to condensation in the bulk sub-cooled liquid flow. The bubble mechanistic behaviors of coalescence and breakage handled through the *population balance approach* form an integral part in addition to the two-fluid model in determining the bubble size in the dispersed phase. The accurate determination of this bubble diameter is required because it strongly governs the drag as well as the non-drag characteristics of the fluid flow. Much research has been performed on this particular type of multi-phase flow and the reader can refer to the latest development of models in Yeoh and Tu (2005, 2006a, 2006b, 2008). The methods required to compute these flows are similar to those described in this book except for the addition of models such as the wall and bubble mechanistic models and boundary conditions. For multi-phase flow using population balance modeling (PBM), Yeoh et al. (2014) introduced the scale-up of CFD and PBM to a variety of engineering and industry applications in the chemical, pharmaceutical, energy, and petrochemical sectors.

It is also worth mentioning another class of multi-phase flows, which is flows with free surfaces like channel flows, flows around ships, mold filling including solidification or melting, etc. The free surface is generally treated in addition to the *interface-tracking* methods (moving grids) by another category of methods known as *interface-capturing*. These methods share some similarities like the immersed boundary methods, where the computation is also performed on a fixed grid; the shape of the free surface is determined by solving an additional transport equation for the void fraction of the liquid phase such as the volume-of-fluid (VOF), which was first introduced by Hirt and Nichols (1981) to indicate whether the control volume cell is filled by liquid and/or gas. The critical issue in this type of method is the discretization procedure for the advective term in the equation. Low-order schemes have a tendency to smear the interface and introduce artificial mixing of the two fluids. Like shocks in compressible flows, the smearing effect degrades the profile and high-resolution schemes are generally preferred. For a sharper interface, the interface-capturing methods based on the *level-set formulation* present another alternative. Here, a step-wise variation of the fluid properties is enforced. Nevertheless, this causes problems when computing viscous flows and the need to arbitrarily introduce some finite thickness across the interface region is necessary to promote smooth but rapid change of the properties. More details on the level-set methods can be found in Sethian (1996).

Multi-phase CFD modeling is maturing with stronger predictive capabilities that the industry may employ for an improved design and dependable process control, due to the increasing power of modern computers. This field has rapidly evolved in recent years with more advanced physical models, more accurate numerical schemes, and faster algorithms. Recent implementations incorporate additional physical effects into the numerical models, for example, friction and cohesive forces (Makkawi et al., 2006), heat and mass transfer (Deen et al., 2014), and multi-physics field coupling (Wang et al., 2018), (Xiao et al., 2020). Readers can refer to Yeoh and Tu (2019) for the latest research and theories covering the latest development in modeling multi-phase flows.

9.4.5 Combustion

The interaction between turbulence and combustion is an area of great research interest and its complexity defies an analytic solution or even a basic understanding in most cases. This is especially true in the case of turbulent non-premixed or diffusion flames in which turbulence and chemical reactions are strongly coupled within the thin reaction zone of the flame.

There are many difficulties associated with handling turbulent combustion flows. Firstly, there are various degrees of mixing on combustion since turbulence has a wide range of time and length scales. Secondly, the chemical reactions of combusting species are highly non-linear; they strongly affect the flow variables such as temperature and density, which can change rapidly by a factor of eight through the flame region. Thirdly, a combustion process usually comprises a large number of species partaking in many reaction steps, which may result in having very different complicated transport properties. Fourthly, the combustion time and length scales are typically smaller than the smallest turbulent scales, making them difficult to resolve by numerical or experimental means.

Nevertheless, all types of combustion are governed by the same set of conservation laws of mass, momentum, energy, and chemical species at the fundamental level. If these laws of governing equations can be solved numerically without any approximations, then a full set of information about the flow and combustion is known. This is the basic idea behind the direct simulation of combusting flows. In most cases the chemical scales are much smaller than the smallest scales of turbulence, which are the Kolmogorov scales. This is why turbulent combustion presents an even bigger challenge than non-reactive turbulent flow. Conducting DNS studies on turbulent flames can be extremely difficult depending on the degree of complexity that needs to be imposed on the simulations. If detailed chemistry is to be incorporated, then the computational cost can increase dramatically, in some circumstances to epic proportions. Imagine that even for the simple combustion process of methane, more than 200 reaction steps are required to adequately represent the chemistry process. It is therefore not surprising that DNS of turbulent combustion is still restricted to small-scale laboratory flames.

To realize the feasibility of resolving complex reacting flows in practical combusting systems, simplifications to the combustion process have resulted in a number of widely applied models. A short description of some of these models for non-premixed and premixed combustion is briefly discussed below.

By representing the combustion of fuel as a global one-step, infinitely fast chemical reaction, the simple chemical reacting system (SCRS) assumes the oxidant reacts with the fuel to form products at stoichiometric proportions. The intermediate reactions are ignored since we are only concerned with the global nature of the combustion process. With this model, the mass fractions of the reactants and products accompanied by the inert species can be expressed as fixed algebraic state relationships in terms of a passive scalar called the mixture fraction f. As a consequence, it is only necessary to solve one extra transport equation for f rather than individual transport equations for each mass fraction. To account for the fluctuations of mass fractions due to turbulence, the average scalars of these variables can be obtained by weighting the instantaneous value with a probability density function for the mixture fraction f. Clipped Gaussian and Beta functions are typical probability density functions that have been applied to provide the best results. Interested readers are encouraged to refer to Lockwood and Naguib (1975) and Pope (1976) for further details.

The eddy break-up concepts introduced by Spalding (1971) and Magnussen and Hjertager (1976) present an alternative approach to the SCRS where the rate of consumption of fuel is solved as a function of local flow properties. Here, the mixing-controlled rate of reaction is expressed in terms of the turbulence time scale. The model considers the slowest rate as the reaction rate of fuel depending on the minimum dissipation rates of fuel, oxygen, and products. Within this model, it is also possible to accommodate kinetically controlled reaction terms expressed by the Arrhenius kinetic rate expression to govern the reaction rate of fuel in addition to the mixing-controlled rate of reaction. The implementation is straightforward and it has been shown to yield reasonably good predictions, but the quality of the predictions depends greatly on the turbulence models used.

In addition to the development of the SCRS and eddy break-up models, another popular combustion model is the consideration of laminar flamelets. This approach is based on the assumption that these flamelets are reaction-diffusion layers in quasi-steady-state that are continuously displaced and stretched within the turbulent medium. These layers are assumed to be thinner than all the turbulent scales, so their internal structures have the compositional structure of laminar flames. Like the SCRS, a transport equation for the mixture fraction is solved. However, the instantaneous species mass fractions are now deduced from the laminar state relationships, which can be taken from experimental measurements. The species fluctuations can also be accounted for through the probability density function described above to obtain the average variables. The provision of this model allows the inclusion of detailed chemistry. Application of the eddy-break up and laminar flamelet models for the predictions of velocity and temperature fields in a compartment fire has been reported in Yeoh et al. (2002); the reader may wish to refer to this article for a better understanding of simulating combusting flames in the context of fire dynamics.

Many of the traditional combustion models developed above have been derived on the basis that the flames are under near-equilibrium conditions. To predict highly non-equilibrium flame events such as ignition, lift-off, or extinction, it would be possible to modify the state relationships to include the scalar dissipation rate dependence and distinguish between the burning and extinguishing flamelets. Peters (2000) proposed a strained laminar flamelet model to accommodate such effects. Another possible approach is through the compositional probability density function transport model (Dopazo, 1993) that particularly simulates finite-rate chemical kinetic effects in turbulent reacting flows. This transport equation can be solved either through a Lagrangian approach using stochastic methods as suggested by Pope (1994) or in an Eulerian framework using stochastic fields developed by Valino (1998). It is clear that combustion modeling is still very much an area of active research.

With the advancements in computer speed and parallel architectures, time-accurate LES of combusting flames is becoming ever more feasible and prevalent. The flow unsteadiness within the flame has been found to be better captured using the LES approach and this has been succinctly demonstrated by the recent study of a free-standing buoyant flame as described by the CFD application example in Chapter 8 and the investigation of a strongly radiating non-premixed turbulent jet flame by Desjardin and Frankel (1999). More applications of LES for gas and spray combustion can be found in recent literature (Kurose and Makino, 2003), (Gharebaghi et al., 2011). Hu et al. (2013) conducted an LES of swirling combusting gas-particle flows with results validated by detailed experiments. Zhou (2016) developed LES for single-phase and two-phase combustion to predict the instantaneous turbulence and flame structures with better statistical results.

9.4.6 Fluid-Structure Interaction

Interaction between a flexible structure and the surrounding fluid promotes a variety of phenomena with applications such as stability analysis of airplane wings, design of bridges, and flow of blood through arteries (to be discussed in more detail later). In this section the authors would like to discuss the various techniques behind the emerging area of computational aeroelasticity (CAE) for performing analyses on a range of aeroelastic applications, where the basic approaches are equally applicable to other applications related to FSI problems.

Despite the many advancements in CFD methods and structural dynamic tools that have been irrespectively developed, many approaches to fluid-structure studies still seek to synthesize independent computational approaches to the fluid dynamics and structural dynamic sub-systems. Such approaches are known to be fraught with complications associated with the interaction between the two simulation modules. As reviewed by Kamakoti and Shyy (2004), the task is therefore to choose appropriate models for fluid and structure based on its application and to develop an efficient interface to couple the two models. CAE can be broadly categorized into three approaches: *fully coupled*, *closely coupled*, and *loosely coupled*.

In the *fully coupled* model the governing equations are reformulated by combining the fluid and structural equations of motion that are solved and integrated in time simultaneously. This kind of method poses severe limitations because of the need to solve the equations in two different reference systems: fluid equations as Eulerian and structural equations as Lagrangian. The stiff set of equations for the structural system makes it virtually impossible to solve the equations using monolithic computational schemes for large-scale problems. For the class of methodologies that belong to the *loosely coupled* model, unlike the *fully coupled* analysis, the structural and fluid

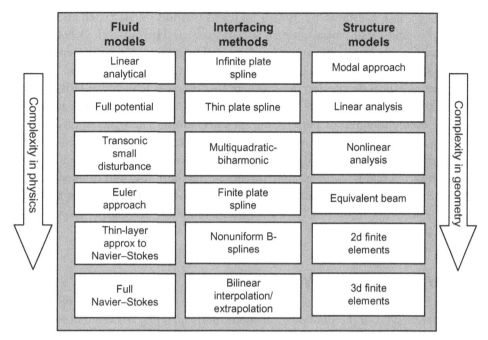

Fig. 9.9 Sample fluid and structure solvers along with select interfacing methodologies for aeroelastic simulation taken from Kamakoti and Shyy (2004).

equations are solved separately. This results in two different computational grids that may not be coincident at the interface or boundary. To establish a link between the respective regions, an interfacing technique is developed to exchange information back and forth between these modules. Fig. 9.9 illustrates a sample for the fluid and structure solvers along with selective interfacing methodologies for aeroelastic simulation. This method provides the flexibility of choosing different solvers but the coupling procedure leads to a loss in accuracy as the modules are only updated after sufficient convergence is achieved in each of the respective regions. It is usually limited to small perturbations and problems with moderate non-linearity. In the *closely coupled* model the approach not only paves the way to use different solvers for the fluid and structure modules but also couples the solvers in a tight fashion through *one single module,* with the exchange of information taking place at the boundary. The information exchanged here are the surface loads that are mapped from the CFD surface grid onto the structure dynamics grid; and the displacement field that is mapped from the structure dynamics grid onto the CFD surface grid. This implies that a moving boundary technique is required for the CFD surface mesh so that the moving mesh algorithms such as the spring analogy and transfinite interpolation–based method by Hartwich and Agrawal (1997) or moving mesh partial differential equation by Huang et al. (1994), Huang (2001), and Cao et al. (1999) are applied to enable re-meshing of the entire CFD domain as the solution marches in time.

Based on the short description in relation to the advances achieved in FSIs, what the authors have aimed to provide the reader is the culmination of some suitable solvers and interfacing methodologies for performing FSI calculations. The materials presented in this section are by no means exhaustive. Numerous research issues remain open and unresolved. Particularly, future challenges lie in the development of higher-order methods to better capture the highly complex flow physics in the fluid flow and detailed structure modeling including non-linear effects. Ultimately, the robust application of this methodology for accurately modeling large systems still lingers in the wake of more efficient procedures.

9.4.7 Physiological Fluid Dynamics

There has been an enormous interest in biomedical or bio-fluid engineering toward a better understanding of the physiological applications spanning the respiratory and cardiovascular systems in the human body. Owing to the significant progress that has been made in the development and application of computational methods for modeling airflow and blood flow in the respective respiratory and cardiovascular systems, the discoveries achieved are transforming our fundamental understanding of the fluid mechanical behaviors within. When linked with medical imaging techniques such as magnetic resonance imaging (MRI), the ability to construct anatomically accurate geometric models are providing new insights into the airflow and blood flow velocities and pressure fields in humans in health and disease. Some examples in relation to biomedical CFD have been demonstrated in Chapters 1 and 7. Despite the wealth of data that has been attained through three-dimensional computational models, significant work still remains (Grotberg and Jensen, 2004; Taylor and Draney, 2004).

For the cardiovascular system, one prognosis for the possible cause of heart attack or stroke could be the dangerous manifestation of self-excited oscillation that arises in arterial stenoses according to Ku (1997). The resulting static and dynamic loading on the diseased arterial wall may well be sufficiently vigorous or sustained to fracture the stenosis' plaque cap, thereby causing fragments to be swept downstream in the arteries, a possible precursor of heart attack or stroke. In an attempt to quantify phenomena difficult to describe using in vitro techniques, computational models based on CFD methodologies have been used. Perktold et al. (1991) have adopted the finite element method to simulate the pulsatile flow of a Newtonian fluid in a model representing the carotid bifurcation using rigid wall approximations. This assumption, though justifiable for arteries where the wall motion is small (as may be expected in steady flow), may not be appropriate for arteries experiencing large deformation (e.g. pulsatile flow in younger subjects). Rappitsch and Perktold (1996) have described the transport of albumin in a stenosis model, while Lei et al. (1997) focused on the application of computational models to the design of end-to-end anastomoses.

Although a detailed description of blood flow regions such as the complex three-dimensional flow in the event of an aneurysm can be handled by current CFD models, the future challenge remains in the development of closed-loop circulatory models that not only combine these *local* models (as described above) but also with *global* models that can accurately describe the pressure-flow relations in a vascular network comprised of millions of blood vessels or pressure-volume models of the heart. Because of the availability of MRI to construct subject-specific geometric models, cardiovascular surgery planning using computational methods in predicting changes in blood flow resulting from possible therapeutic intervention for individual patients is becoming a reality (Taylor et al., 1999). The recent development of this cardiovascular surgery planning as demonstrated by Wilson (2002) has led to the creation of a plan being specifically constructed for a patient with atherosclerosis in the lower extremities. Here, three-dimensional blood flow simulations were carried out to compute the flow velocities in simulating the scenario for the clearance of a dye in the blood stream. Such simulations have been deemed necessary as they provided useful information in identifying sites of relative flow stasis that may be prone to clot formation or thickening of the inner layer of blood vessels, where such diagnoses have provided treatment targeting specifically for patients with cardiovascular disease.

Another challenge of the cardiovascular fluid investigations is the development of models to combine molecular- and cellular-level phenomena with macroscale fluid mechanics. These models can allow the interaction between atherogenic agents and the artery wall and the response of the artery wall to hemodynamic forces from cellular to tissue scales to be studied. For such flows, the explicit inclusion of the particle nature of matter is required through specially developed spatial scale-multi-scale models, which are discussed in the next section.

For the respiratory system, the airways can experience deformation to some degree and FSIs underlie a number of important pulmonary conditions. Of particular significance are the expiration and inspiration flow limitations. For the latter aspect, it can lead to flow-induced instabilities that can generate snoring and upper-airway obstruction, contributing to sleep apnea. The airways of the lung also provide another source of pertinent problems where multi-phase fluid mechanics has important biological applications, involving flexible tubes with a liquid lining or liquid occlusion. The consideration of how surface tension, elastic forces, and airflow act together in controlling the configuration of a deformable airway and its internal liquid lining is critical in determining the conditions leading to airway closure; the liquid lining forms a plug occluding (and collapsing) the airway and inhibiting gas exchange, a prognosis for the possible occurrence of an asthma attack.

Many three-dimensional models have been developed to describe the three-dimensional FSI of an internal three-dimensional Navier-Stokes flow. It has been shown that non-axisymmetric buckling of a tube contributed to non-linear pressure-flow relations can exhibit flow limitations through pure viscous mechanisms (Heil and Pedley, 1996; Heil, 1997). This work was extended to computations that described three-dimensional flows in non-uniformly buckled tubes at Reynolds numbers of a few hundred (Hazel and Heil, 2003). The development of models for bio-fluid mechanics in flexible structures is at the crossroads of intense research and the authors have just barely scratched the surface in describing some applications to respiratory systems. Understanding the physical origin and nature of these phenomena to the vessel's biological function or dysfunction remains computationally and experimentally as well as analytically challenging.

Most physiological fluid investigations have thus far concentrated on the assumption of rigid wall approximations to the geometrical model. Realistically speaking, pulsatile flows in the cardiovascular system and airflow through the airways in the respiratory system require flexible wall structures to appropriately represent the actual flow characteristics. Future fluid-structure analyses will become a much required feature in carrying out investigations on the respiratory system and/or cardiovascular system.

For readers wishing to engage in the emerging field of computational hemodynamics modeling, Tu et al. (2015) discussed suitable geometric and mathematical models that can be used to study fluid and structural mechanics in the cardiovascular system. The article focused on the basic knowledge and techniques for reconstructing geometric models from medical imaging, mathematics that describes the fluid and structural mechanics, and corresponding numerical/computational methods to solve the transport equations. Tu et al. (2012) introduced recent advances in medical imaging and computational fluid dynamics (CFD), with details of recent advances in the modeling of the respiratory system for researchers, engineers, scientists, and health practitioners. Wong et al. (2013) provided a road map for applying the stages in conceptualization, evaluation, and testing of biomedical devices in a systematic order of approach, leading to solutions for medical problems within safety limits.

9.5 OTHER NUMERICAL APPROACHES FOR COMPUTATION OF FLUID DYNAMICS

Unlike conventional numerical schemes based on the discretization of macroscopic continuum equations commonly used in many engineering applications, there are other alternative numerical approaches that have been developed lately for simulating fluid flows and modeling physics in fluids. The authors would like to review three promising methods: Lattice Boltzmann, Monte Carlo, Particle, and Discrete Element methods. For the particle method, an overview of the vortex methods and smooth particle hydrodynamics for the simulation of the continuum phenomena is discussed and described.

9.5.1 Lattice Boltzmann Method

Lattice Boltzmann method (LBM) is a methodology based on the microscopic particle models and mesoscopic kinetic equations. According to Kadanoff (1986), it has been found that the macroscopic behavior of a fluid system is generally not very sensitive to the underlying microscopic particle behavior if only collective macroscopic flow behavior is of interest. The fundamental idea behind the LBM is to construct simplified kinetic models that incorporate only the essential physics of microscopic or mesoscopic processes so that the macroscopic averaged properties obey the desired macroscopic equations. This subsequently avoids the use of the full Boltzmann equation (LBE) and one also avoids following each particle as in molecular dynamics simulations.

It is worth noting that even though LBM is based on a particle representation, the principal focus remains on the averaged macroscopic behavior. The kinetic nature of the LBM introduces three important features that distinguish this methodology from other numerical methods. Firstly, the convection operator of the LBM in the velocity phase is linear. The inherent simple convection when combined with the collision operator allows the recovery of the non-linear macroscopic advection through multi-scale expansions. Secondly, the incompressible Navier-Stokes equations can be obtained in the nearly incompressible limit of the LBM. The pressure is calculated directly from the equation of state in contrast to satisfying a Poisson equation with velocity strains acting as sources. Thirdly, the LBM utilizes the minimum set of velocities in the phase space. Because only one or two speeds and a few moving directions are required, the transformation relating the microscopic distribution function and macroscopic quantities is greatly simplified and consists of simple arithmetic calculations.

For the LBM, the LBE based on the lattice gas automata can be constructed via simplified, fictitious molecular dynamics in which space, time, and particle velocities are all discrete. Using the commonly adopted approach of the single relaxation time collision model of Bhatnagar-Gross-Krook (1954), the LBE can be expressed as:

$$\frac{\partial f}{\partial t} + \boldsymbol{e} \cdot \nabla f = \frac{f - f^{eq}}{\lambda} \tag{9.18}$$

where f is the particle distribution function, $\boldsymbol{e}$ is the particle velocity, λ is the relaxation time due to particle collision, and f^{eq} is the equilibrium Boltzmann-Maxwell distribution function. It can be demonstrated that the particle velocity space *e* can be discretized into a small set of discrete vectors e_i such that the macroscopic conservation laws are satisfied (He and Luo, 1997). Eq. (9.18) can thus be discretized along each velocity direction e_i at each lattice by the following:

$$f_i(x + e_i \delta t, t + \delta t) - f_i(x, t) = -\frac{1}{\tau}[f_i(x, t) - f_i^{eq}(x, t)] \qquad (9.19)$$

Here, $f_i(x, t)$ denotes the density function along velocity e_i at lattice position x and time t, $\tau = \lambda / \delta t$ represents the relaxation parameter, and δt is the time step. With the LBM, the macroscopic density ρ and flow velocity u can be defined in terms of the particle distribution function as:

$$\rho = \sum_i f_i, \quad \rho u = \sum_i f_i e_i \qquad (9.20)$$

It is shown above that the numerical algorithm of the LBM is relatively simple when compared with conventional Navier-Stokes methods. In terms of computational effort the LBM consists of mainly two operations: collision on the right-hand side of Eq. (9.19) and streaming on the left-hand side. The collision operation is completely local and since streaming can be easily achieved by a simple *shift* operation that is offered as an intrinsic function by most compilers, the LBM is well suited for parallelism. Because of the availability of very fast and massively parallel machines, LBM fulfills the requirement in a straightforward manner. Eq. (9.19) can, however, be also interpreted as a special finite difference form of the continuous form of Eq. (9.18), which means that the time step is limited by the lattice size due to the explicit nature of the lattice. Hence the computational efficiency of LBM remains an important issue to be evaluated particularly for steady-state flows.

Lattice Boltzmann simulations of fluid flows have been performed for flows with simple geometries including driven-cavity flows, flow over a backward-facing step, flow around a circular cylinder, and flows with complex geometries. For the simulation of fluid turbulence, the development of a SGS LBE turbulent model has provided the means of possibly employing LBM for the investigation of turbulent flows in industrial applications of practical interest. LBM also provides an alternative for simulating complicated multi-phase and multi-component fluid flows. The method has proven to overcome difficulties associated with the conventional macroscopic approaches in modeling interface dynamics and important related engineering applications that include flow through porous media, boiling dynamics, and dendrite formation. Interested readers are encouraged to refer to the list of articles for the range of applicability to the LBM: Hou et al. (1995, 1996), Benzi et al. (1996), Luo (1997), He and Doolen (1997), Shan (1997), Spaid and Phelan (1997), Chen and Doolen (1998), and Yang et al. (2000), among others. In recent applications LBM has been used in many numerical fluid problems like nanofluids (Kefayati et al., 2011; Kamyar et al., 2012; Bahiraei, 2014); turbulent flows (Choi and Lin, 2010); compressible flow (Li et al., 2007); particle fluid flow (Xiong et al., 2014); etc.

9.5.2 Monte Carlo Method

Lately, the direct simulation Monte Carlo (DSMC) has been proposed in addition to the LBM for the computation of fluid dynamics because of the practical scientific and engineering importance of solving high Knudsen number (Kn) flows. This dimensionless parameter Kn, by definition, generally characterizes the ratio of the molecular mean free path to the characteristic length. As the mean free path becomes comparable with or even larger than the characteristic length, especially for high Kn, the particle nature of matter must be explicitly addressed since the continuum fluid approximation breaks down particularly in the area of micro-scale or nano-scale fluid systems. Table 9.1, as illustrated below, shows the hierarchy of mathematical models to describe the interactions of atoms, ions, and molecules. These models can range from very fundamental solutions of sets of elementary interactions of particles (such as molecular dynamics models) to approximation of systems in which the individual particles are replaced by continuum fluid elements (such as the Navier-Stokes equations). Molecular dynamics is the most fundamental level of this hierarchy; the LBM and DSMC lie somewhere in between the molecular dynamics and continuum fluid consideration.

DSMC is a direct particle simulation method based on kinetic theory. The fundamental idea behind this method is to track a large number of statistically representative particles. The particles' motion is later used to modify their positions, velocities, or even chemical reactions in reacting flows. The primary approximation of the DSMC method is to uncouple the molecular motions and the intermolecular collisions over small time intervals. Here, particle motions are modeled deterministically while the collisions are treated statistically.

The core of the DSMC procedure consists of four primary processes. Firstly, the simulated particles are moved within a time step. Boundary conditions are enforced through

Table 9.1 Levels of Models of Many-Body Interactions

Equation	Solution Method
Newton's Law $f = ma$	Molecular dynamics (deterministic, particle-based, prescribed inter-particle forces)
Liouville equation $F(x_i, v_i, t), i = 1, N_p$	Monte Carlo methods (statistical, particle-based methods) – DMSC
Boltzmann equation $F(x, v, t)$ binary collision (low density)	Direct solution – LBM
Navier-Stokes equation $\rho(x, t), u(x, t)$ short mean free path	Direct solution: finite difference, finite volume, spectral methods, and so on (continuum flow methods)

Taken from Oran et al. (1998).

modeling molecule-surface interactions, which may include physical effects such as chemical reactions, three-body collisions, and ionized flows. Secondly, indexing and cross-referencing of the particles are performed in this step. This is a prerequisite for the next two steps: simulating collisions and sampling the flow field. The key to the practicality of DSMC for large-scale processing is the accurate and fast indexing and tracking of the particles. Thirdly, the step of simulating collisions sets DSMC apart from other deterministic simulation methods such as molecular dynamics. The currently preferred model is the *no-time counter technique* by Bird (1994), used in conjunction with the sub-cell technique of Bird (1986). This sub-cell method ensures that collisions only occur between near neighboring particles by calculating local collision rates based on individual cells but restricts possible collision pairs to sub-cells. Fourthly, the sampling of the particles provides the macroscopic flow properties. The spatial coordinates and velocity components of molecules in a particular cell are used to calculate macroscopic quantities at the geometric center of the cell. Interested readers may wish to refer to Muntz (1989), Cheng (1993), Cheng and Emmanuel (1995), Oran et al. (1998), Wu and Lian (2003), Sun et al. (2009), Gallis et al. (2014), and Aufiero and Fratoni (2017) for more details pertaining to the recent advances and developments of DSMC.

DSMC is a time-accurate explicit procedure. The method has shown to be a good candidate for unsteady flow applications in computing non-equilibrium structure of shocks and boundary layers as well as hypersonic viscous flows and high-temperature rarefied gas dynamics. The latter application has sparked strong interest especially in the production of high-speed aerospace planes and space transportation systems. In the field of material processing the use of DSMC in handling the growth of thin films for a variety of vapor-phase processing and plasma-etch is also steadily growing. To investigate the potential of DSMC as a predictive tool, results obtained by DSMC have also been compared against experiments with Navier-Stokes calculations in the low Kn regime. Computational studies have demonstrated that DSMC solutions approach the Navier-Stokes solutions at this limit. Limitations of current computer technology though gradually diminishing may still inhibit DSMC to be extensively applied to the computation of fluid dynamics. DSMC is, however, being seriously considered to resolve more specialized flows such as flows of low-speed and high-Knudsen-number; examples can be found in studies by Pan et al. (1999), Ewart et al. (2008), and Shariati et al. (2019), among others.

9.5.3 Particle Methods

The key characteristic of particle methods especially for the simulation of the continuum phenomena though deceptively simple can be obtained through the solution of ordinary differential equations (ODEs) that determine the trajectories and the evolution of properties carried by the particles. These methods amount to the solution of a system of ODEs in the general form given by:

$$\frac{d\boldsymbol{x_p}}{dt} = \boldsymbol{u_p}\left(\boldsymbol{x_p}, t\right) = \sum_{q=1}^{N} \boldsymbol{K}\left(\boldsymbol{x_p}, \boldsymbol{x_q}; \omega_p, \omega_q\right) \tag{9.21}$$

$$\frac{d\omega_p}{dt} = \sum_{q=1}^{N} \boldsymbol{F}\left(\boldsymbol{x_p}, \boldsymbol{x_q}; \omega_p, \omega_q\right) \tag{9.22}$$

where $\boldsymbol{x_p}$ and $\boldsymbol{u_p}$ denote the locations and velocities for the N particles; ω_p denotes the particle properties such as density, temperature, and vorticity; and $\boldsymbol{K}$ and $\boldsymbol{F}$ represent the dynamics governing the simulated physical system. Particles trajectories are implemented with a Lagrangian formulation and the key common characteristic of the two popular particle methods such as vortex methods (VMs) and smooth particle hydrodynamics (SPH) involves the approximation of the Lagrangian form of the Navier-Stokes equations by replacing the derivative operators through equivalent integral operators, which are in turn discretized on the particle locations.

Particle methods are often defined as *grid-free/meshless* methods, making them an attractive alternative to *mesh-based* methods such as conventional Navier Stokes methods that have been described rather extensively. When applied to the Lagrangian formulation of the convection-diffusion equations, particle methods enjoy an automatic adaptivity of the computational elements as dictated by the flow map. Nonetheless, truncation errors introduced in the methodology can result in the creation and evolution of spurious vortical structures through particle distortion; these methods have to be conjoined with a grid to provide consistent, efficient, and accurate simulations. The grid aims to restore regularity in the particle location via re-meshing but does not detract from the inherent adaptive character of these methods.

Vortex particle methods have been employed since it was first introduced by Rosenhead (1930) to describe the evolution of vortical structures in incompressible flows. Based on the vortex-blob approximation of Krasny (1986), the field is recovered at every location of the domain by considering the collective behavior of all computational elements. When particles do overlap, the scales of the physical quantities that are resolved are determined by the finite particle core size rather than the inter-particle distance. Viscous effects are simulated using the method of particle-strength exchange (PSE) (Koumoutsakos, 2005). The kinematic boundary conditions such as no-through flow at solid boundaries are enforced by boundary integral methods, whereas viscous boundary conditions such as non-slip are imposed by translating into vorticity fluid boundary conditions (Cottet and Poncet, 2004) complementing the viscous part of the equations.

VMs have certainly come of age through their extensive range of engineering applications. For the DNS of flow past an impulsively started cylinder for a range of Reynolds numbers, VMs have demonstrated the capability of automatically adapting computational elements in regions of the flow where increased resolution was found to be necessary to capture the unsteady separation phenomena (Koumoutsakos and Leonard, 1995). Whereby these parts are usually not known *a priori*, suitable criteria need to be devised to add computational elements in these critical regions for mesh-based methods. In VMs the computational elements are inherently linked to the flow physics they represent and no such additional criteria are necessary. The adaptivity and robustness of VMs have also enabled simulations of reacting flows to be carried out (Knio and Ghoniem, 1992). In comparison to mesh-based methods, the capture of highly anisotropic diffusion phenomena was better attained through the Lagrangian particle methods while accurately transporting the scalar fields. Recently, a novel particle-level set method for capturing interfaces has been proposed by Hieber and Koumoustsakos (2005). It has been well known since the pioneering work of Krasny in the 1980s (Krasny, 1986) that particle methods are well suited for interface capturing. Comparisons on a set of benchmark problems with existing level set formulations have demonstrated the promising prospect of this proposed method in achieving superior results with a reduced number of computational elements required for interface capturing.

The method of SPH was first introduced by Lucy in the late 1970s and further developed by Monaghan (1988) for grid-free astrophysics simulations. In SPH the inter-particle distance is taken instead to identify the core size. Many simulations using SPH have been conducted by extending its application range from gas dynamics in astrophysics to Newtonian and viscoelastic flows such as that found through the important works of Monaghan (1988) and Ellero et al. (2002). Like VMs, SPH also experiences the problem of particle distortion. Several techniques have been proposed to compensate for this problem (e.g. dynamic conditions by Ellero et al., 2002). Inspired by techniques in VMs, the introduction of regularization of particle distortion in SPH via re-meshing has managed to improve the method to second-order accuracy; this detracts, however, the characterization of the method being totally grid-free. The so-called meshless methods based on Galerkin-type have been explored to increase the capabilities of SPH for flow simulations. Some works by Duarte and Oden (1996) and Belytschko et al. (1996) on the unifying methods of Moving Least Squares, Reproducing Kernel Particle, and Element-free Galerkin in the context of SPH are noted. For an extended understanding of the background theory and formulation, numerical implementation and challenges, and extension of the particle methods to simulate molecular phenomena, readers are encouraged to refer to an excellent article written by Koumoutsakos (2005), where the subject of multi-scale flow simulations using particles has been thoroughly reviewed.

As SPH has developed considerably in the past decade, interested readers can refer to several recent review articles on the diversification of SPH. Violeau and Rogers (2016) assessed some recent trends in the SPH with a particular focus on its potential use in modeling free-surface flows. Wang et al. (2016) provided some of the recent advances and applications in SPH for simulating the multi-phase flow. Shadloo et al. (2016)

summarized the motivations for using the SPH method in an industrial context, presenting the current state-of-the-art of this method's application to industrial problems and drawing general conclusions about its pros and cons while emphasizing the remaining challenges in making it a hands-on computational tool. Ye et al. (2019) reviewed the recent developments of SPH in methodology and applications for different complex fluid flows, including biological flows, microfluidics and droplet dynamics, non-Newtonian fluid flows, free surface flows, multi-phase flows, and flows with FSI. Lind et al. (2020) presented the progress of SPH toward high-order converged simulations which are necessary to resolve multi-scale aspects of turbulent flows.

9.5.4 Discrete Element Method

The discrete element method (DEM) was originally developed to simulate granular flow problems such as those prevalent in rock mechanics (Cundall and Strack, 1979). Similar to particle methods, it solves the motion of each individual element by solving the linear momentum equation as well as the angular momentum subject to forces and torques arising both from particle interaction and those imposed on the particles by the surrounding fluid. In comparison to other particle simulation methods such as molecular dynamics, dissipative particle dynamics, or Brownian dynamics, DEM differs primarily by the different particle interaction laws as well as by the imposition of random forcing to mimic the collisions or interaction with molecules of the surrounding fluid.

DEM can be employed for a wide range of particle sizes. While other particle simulation methods are only applicable where the particle diameters are much less than the characteristics length scale of van der Waals forces ($\sim$10 nm), DEM has the propensity of handling collision of larger-size particles. Since it is now necessary to model the detailed mechanics by which particles interact during collision, particle contact models for DEM are therefore considerably more complex than any of the other particle simulation methods. Two different approaches are adopted to characterize the interaction of two particles during collision, which can be referred to as the hard-sphere and soft-sphere models.

The soft-sphere model considers the inter-particle forces between particles, namely, the normal, tangential, damping, and sliding forces, which can be modeled using springs, dashpots, and sliders. As reviewed in Kruggel-Emden et al. (2009), the treatment of the normal force acting between two particles can be considered based on the degree of overlap and displacement rate. Four main groups for the evaluation of the normal force are: continuous potential models, linear viscoelastic models, non-linear viscoelastic models, and hysteretic models. For molecular dynamics simulations on the atomic or molecular level, continuous potentials between particles are widely used such as those of the Lennard-Jones potential (Verlet, 1967). The most frequently employed models in DEM are nonetheless linear (Tsuji et al., 1993). In this model a mean coefficient of restitution and a mean collision time is defined; the related spring stiffness and the damping coefficient are then computed. Limitations of a constant coefficient of restitution and a constant duration of contact of the linear model can be overcome by applying non-linear spring damper models. Several force laws proposed by Kuwabara and Kono (1987), Tsuji et al. (1992), and Lee and Herrmann (1993) have been developed by extending the original approach by Hertz (1882). In order to include the effect of plasticity and to avoid the usage of the velocity-dependent damping, hysteretic models, which may be linear or non-linear, have been proposed. Notable contributions to the development of these models are: Walton and Braun (1986), Sadd et al. (1993), Thornton (1997), Vu-Quoc and Zhang (1999), and Tomas (2003). In all hysteretic models the materials in contact result in permanent deformation.

For the evaluation of the tangential force for the soft-sphere model, a wide variety of different complex models are also available to describe the force-displacement behavior. Two main groups can be defined; these are: linear and non-linear models. For linear tangential models, the earliest approach that has found wide application is the model proposed by Cundall and Strack (1979). It determines the tangential force according to a linear elastic spring unless the Coulomb force is exceeded. Ever since then, several versions of linear tangential displacement models have been developed. Besides the linear models, a number of non-linear tangential models have also been proposed especially for elastic materials following the Hertz (1882) theory (Maw et al., 1976; Vu-Quoc and Zhang, 1999; Di Renzo and Di Maio, 2005). In addition to the complex models for determining the normal and tangential displacements during collisions, one important feature of the time-driven, soft-sphere model is the ability to handle problems whereby particles can remain in contact for a prolonged time period to simulate particle agglomeration.

The hard-sphere model, which is based on the proposal by Hoomans et al. (1996), assumes that interactions between particles are modeled by binary collision dynamics. Main assumptions concerning the particle shape, deformation history during collision, and nature of collisions are: particles are quasi-rigid and particle shape is retained after impact, collisions are quasi-instantaneous, particle contact occurs at a point which is in contrast to the soft-sphere model where an overlap displacement is allowed, particles are in free flight in between collisions and interaction forces are impulsive, and all other finite forces are negligible during collisions. One characteristic feature of the hard-sphere model is that a sequence of collisions is processed one collision at a time. Another important feature is that the simulations are performed with realistic values of the key parameters: restitution and friction coefficients. Particle collisions are solved using particle impulse equations. Because of the limitation of collision dynamics that has been assumed, it is not surprising that the model is much more computationally efficient than the soft-sphere model since the particle collision time scale does not need to be resolved. One characteristic feature of the soft-sphere model is the very small time required (often $<10^{-6}$ s).

Fundamentally, DEM can be employed to model simple particle systems of practical interest such as those occurring in microfluidics, particle filtration, or blood flow problems in arteries and arterioles. DEM can also be useful in the development of constitutive models for the kinetic theory approach and population balance models, and to better understand the microstructure of large-scale systems. Owing to the exponential growth of computer power and speed as well as more refined collisions models, applications of DEM have increased dramatically, for example, in milling processes (Mishra, 2003a, 2003b), environmental particulate flows (Richards et al., 2004), granular mixing (Bertrand et al., 2005), and fluidized-bed processes (Deen et al., 2007). There have also been recent attempts to extend DEM for systems containing small particle sizes at the sub-micron or even nano-scale range. Fanelli et al. (2006) have developed a model for contact forces of nano-clusters by direct consideration of the van der Waals attractive and Born repulsive forces and applied DEM to investigate nanoparticle dispersion. Luan and Robbins (2005) have compared the predictions of contact models with molecular dynamics simulations for nano-scale particles and found that good agreement was achieved for the normal force predictions but noted that atomic scale roughness leads to significant variation in sliding resistance force between the particles. For more introduction to the principles, applications, advantages, and disadvantages of different fluid flow/DEM coupling methods, and speculation on future development trends, interested readers can refer to a recent review by Wang et al. (2022).

9.5.5 Machine Learning–Accelerated CFD

Many research and engineering problems require fluid modeling and, depending on the complexity of the problems to be solved, they can be computationally expensive. On the other hand, machine learning (ML) models may be able to provide a quick approximation of the physics though at the sacrifice of accuracy. Combining CFD and ML is becoming a new research field that has the potential to tackle previously unsolved challenges in a range of application sectors, changing the focus of CFD from time-consuming simulations to in-depth analysis of features or fast predictions (Jagode et al., 2020). Recent research has shown that incorporating ML into traditional fluid simulations can enhance accuracy and performance, even when working with samples that are drastically different from the training data. ML may make meshing easier, reduce human interaction, improve turbulence prediction accuracy, and enable quick data visualization and analysis, all while enhancing fluid simulation accuracy and generalization performance.

The application of ML in CFD encompasses a wide range of subjects, the majority of which necessitates a thorough understanding of computer techniques. A recent review by Brunton et al. (2020) has provided a historical overview of how the development of numerical methods for fluid dynamics since the 1940s was linked to ML algorithms, and it discussed the applications of key ML approaches for understanding, modeling, optimizing, and controlling fluid flows. Wang and Wang (2021) concentrated the recent

AI applications onto five CFD domains: aerodynamic models, turbulence models, some specialized flows, and mass and heat transfer.

ML has been used to augment turbulence models in CFD since it can drastically reduce the overall cost of a three-dimensional, time-dependent simulation. Using high-fidelity DNS data, Luo et al. (2020) utilized a physics-informed neural network (PINN) method to calibrate five parameters in the RANS turbulence model. The mean absolute inaccuracy of the velocity profile between RANS and DNS was reduced by 22% when the neural network inferred parameters were applied. Kochkov et al. (2021) adopted end-to-end deep learning to enhance approximations in CFD for 2D turbulent flows and obtained the same accuracy as DNS and LES with a 40–80 times speedup without sacrificing any loss of accuracy. Their method paves the way for ML to solve large-scale physical modeling such as plane design and climate prediction. Zhao et al. (2020) presented a novel CFD-driven ML framework to train Reynolds stress closures for RANS models and applied the method to turbine wake mixing cases, which obtained improved predictive accuracy.

ML applications in fluid dynamics are thriving in the wake of the advancement of computer hardware and ML methodologies. For areas where CFD analysis can be expensive, data-driven models are replacing, improving, or aiding CFD simulations. Such examples can be found in the area of the built environment (Ding and Lam, 2019; Calzolari and Liu, 2021), natural convection (Xu et al., 2022), nano-fluid heat transfer (Mohammadpour et al., 2022), and chemical—biomass fast pyrolysis (Lu et al., 2022). Ravindran and Kokjohn (2021) have also employed Gaussian Process Regression (GPR)–based ML in conjunction with CFD to improve the cold-start performance of a DISI engine. The computing time was dramatically decreased from 6 days to a few seconds for the GPR model, while accuracy remained within acceptable limits. While these ML models cannot fully displace CFD simulations, they can play a significant role in speeding up the resolution of CFD-based problems. More examples can be found in Yan et al. (2019), Ringstad et al. (2021), Sui et al. (2022), etc.

9.6 SUMMARY

New areas of applications in fluid mechanics have brought about the development of many novel and innovative techniques. The advanced topics presented in this chapter have thus aimed to explore these latest trends in CFD research and development. It has never been the intention of the authors from the onset to provide a critical and thorough review of each of the topics presented in this chapter but rather to provide a broad overview of the research methodologies adopted and key issues that still need to be addressed and resolved. Nevertheless, the significant advances achieved in numerical methods and computational models, and successful applications of some unconventional approaches, particularly the Lattice Boltzmann Method, Direct Simulation of Monte Carlo, Vortex Method and Smooth Particle Hydrodynamics in simulating fluid flows, and Discrete Element Method in simulating flows of a wide range of particles sizes, should leave us in a sense of awe at the sheer enormity of CFD in the many purposeful investigations of fluid-related problems that have been covered within this chapter. Flow systems involving migration of strong shocks in transonic and hypersonic flows, complex bubble mechanistic behavior in multi-phase flow structures, or even complicated flows that exist in our human body systems that comprise trillions of cells and billions of blood vessels capable of dramatic redistribution of blood and growth over billions of cardiac cycles, are just some of the challenging fluid dynamics problems that further accentuate the demands of CFD research. Like our predecessors, namely Newton, Euler, Bernoulli, Poiseuille, Young, Lighthill, and many others, the study of fluid mechanics has certainly generated decades of intense fascination. As the future unfolds, CFD methods and models will continue to remain at the forefront of intensifying research and development so long as the vast majority of fluid flows and processes remain unresolved.

REVIEW QUESTIONS

9.1 Simplify the general continuity equation below to a steady incompressible flow equation:

$$\frac{\partial \rho}{\partial t} + \frac{\partial(\rho u)}{\partial x} + \frac{\partial(\rho v)}{\partial y} + \frac{\partial(\rho w)}{\partial z} = 0$$

9.2 How is the pressure term used to satisfy the continuity equation in the marker-and-cell (MAC) method?

9.3 Discuss briefly the idea behind the fractional step procedure.

9.4 What types of applications and situations involve compressible flows?

9.5 What difficulties arise from modeling a transient supersonic flow around an airfoil?

9.6 What is the greatest difficulty that has to be overcome with compressible flows?

9.7 What techniques can be used to minimize oscillations that occur in compressible flow due to discontinuities at the shock front?

9.8 How does a higher-order scheme such as a fifth-order scheme deal with discontinuities?

9.9 Under what circumstances would adaptive meshing be used? What would happen if a fixed mesh is employed instead?

9.10 Discuss briefly the concept behind the *r-refinement* grid adaptive technique.

9.11 What kinds of applications commonly use moving grids?

9.12 Explain how a moving grid can be applied to simulate a screw supercharger shown below. What parts would need to remain stationary and what parts would be allowed to move?

9.13 What is the main advantage of the numerical solution of using multi-grid methods in terms of the handling of *high-* and *low-frequency* errors?

9.14 Explain what *domain decomposition* and *load balancing* are in parallel computing.

9.15 What is the immersed boundary method and how is this different from using a boundary-fitted grid?

9.16 What is a Direct Numerical Simulation (DNS)? How does it differ from a Reynolds-Averaged Navier-Stokes (RANS) approach in terms of its handling of turbulence?

9.17 What are the Kolmogorov micro-scales? How do these scales impact the mesh design?

9.18 Why can't DNS be used to solve high-Reynolds-number flows at the moment?

9.19 What is the main concept behind LES in turbulent modeling?

9.20 What are the subgrid-scale (SGS) models in LES? How are they used to define small scales of turbulence?

9.21 In the RANS-LES coupling approach, in which region of the mesh would you apply the RANS model and in which region would you apply the LES model?

9.22 Why would you use a RANS-LES coupling approach to model high-Reynolds-number turbulent flows?

9.23 What is the difference between *one-way coupling* and *two-way coupling* in multi-phase flows?

9.24 Would you use an *Eulerian-Eulerian* or *Eulerian-Lagrangian* for a multi-phase flow that had a high mass loading for the secondary phase (i.e. not the continuous phase)?

9.25 Combustion is a complex phenomenon to model. What types of considerations must be made in modeling combustion?

9.26 Explain fluid-structure interaction (FSI) modeling. In what applications can this be used?

9.27 What is the key requirement (and the most complex) that enables the interaction between the fluid and structure in FSI?

9.28 What advanced techniques would be required to simulate airflow through the respiratory system into the lungs? What about pulsating blood flow through veins and arteries?

9.29 Briefly discuss the concept of the lattice Boltzmann method.

9.30 Briefly discuss the concept of the Monte Carlo method.

9.31 Briefly discuss the concept of the particle method.

9.32 Briefly discuss the concept of the discrete element method.

APPENDIX A

Full Derivation of Conservation Equations

The full derivation of the conservations equations for momentum and energy is presented in this appendix.

The concept of *substantial derivative* is described herein. It is conveniently acceptable to collect all the density terms together by expanding Eq. (3.10) by the chain rule. This gives

$$\frac{\partial \rho}{\partial t} + u\frac{\partial \rho}{\partial x} + v\frac{\partial \rho}{\partial y} + w\frac{\partial \rho}{\partial z} + \rho\left(\frac{\partial u}{\partial x} + \frac{\partial v}{\partial y} + \frac{\partial w}{\partial z}\right) = 0 \qquad (A.1)$$

or

$$\frac{D\rho}{Dt} + \rho\left(\frac{\partial u}{\partial x} + \frac{\partial v}{\partial y} + \frac{\partial w}{\partial z}\right) = 0 \qquad (A.2)$$

where D/Dt is the substantial derivative in Cartesian coordinates. The time derivatives of $D\rho/Dt$ and $\partial \rho/\partial t$ are physically and numerically different quantities. The reader should note that the former is the time rate of change following a moving fluid element, while the latter is the time rate of change at a fixed location. Also, if we consider the general variable property per unit mass denoted as ϕ, the substantial derivative of ϕ with respect to time, written as $D\phi/Dt$, is

$$\frac{D\phi}{Dt} = \frac{\partial \phi}{\partial t} + u\frac{\partial \phi}{\partial x} + v\frac{\partial \phi}{\partial y} + w\frac{\partial \phi}{\partial z} \qquad (A.3)$$

The above equation defines the rate of change of the variable property ϕ per unit mass. As in the case of mass conservation, we are interested in developing equations for rates of change per unit volume. The rate of change of the variable property ϕ per unit volume can be obtained by multiplying the density ρ with the substantial derivative of ϕ that is given by

$$\rho\frac{D\phi}{Dt} = \rho\frac{\partial \phi}{\partial t} + \rho u\frac{\partial \phi}{\partial x} + \rho v\frac{\partial \phi}{\partial y} + \rho w\frac{\partial \phi}{\partial z} \qquad (A.4)$$

It is recognized that Eq. (A.4) represents the *non-conservation form* of the rate of change of the variable property ϕ per unit volume.

The mass conservation equation derived in Section 3.2.1 defines the sum of the rate change of density and is called the advection term, which is

$$\frac{\partial \rho}{\partial t} + \frac{\partial(\rho u)}{\partial x} + \frac{\partial(\rho v)}{\partial y} + \frac{\partial(\rho w)}{\partial z} = 0$$

The generalization of these terms for the variable property ϕ in *conservation form* can be expressed as

$$\frac{\partial(\rho\phi)}{\partial t} + \frac{\partial(\rho u\phi)}{\partial x} + \frac{\partial(\rho v\phi)}{\partial y} + \frac{\partial(\rho w\phi)}{\partial z} \tag{A.5}$$

The above formula expresses the rate of change of ϕ per unit volume with the addition of the net flow of ϕ out of the fluid element per unit volume. It is now rewritten to illustrate the relationship between the conservative form of Eq. (3.16) and non-conservative form of Eq. (3.15):

$$\frac{\partial(\rho\phi)}{\partial t} + \frac{\partial(\rho u\phi)}{\partial x} + \frac{\partial(\rho v\phi)}{\partial y} + \frac{\partial(\rho w\phi)}{\partial z} =$$

Invoking the continuity equation

$$\rho\frac{\partial\phi}{\partial t} + \rho u\frac{\partial\phi}{\partial x} + \rho v\frac{\partial\phi}{\partial y} + \rho w\frac{\partial\phi}{\partial z} + \phi\underbrace{\left[\frac{\partial\rho}{\partial t} + \frac{\partial(\rho u)}{\partial x} + \frac{\partial(\rho v)}{\partial y} + \frac{\partial(\rho w)}{\partial z}\right]}_{=\,0} = \rho\frac{D\phi}{Dt} \tag{A.6}$$

Both of these forms can be used to express the conservation of a physical quantity. For brevity, the non-conservative form is adopted to derive the conservations equations for momentum and energy.

For the conservation of momentum, the net force in the x direction is the sum of the force components acting on the fluid element. Considering the velocity component u as seen in Fig. 3.4, the surface forces are due to the normal stress σ_{xx} and tangential stresses τ_{yx} and τ_{zx} acting on the surfaces of the fluid element. The net force in the normal x direction is:

$$\left[\sigma_{xx} + \frac{\partial\sigma_{xx}}{\partial x}\Delta x\right]\Delta y\Delta z - \sigma_{xx}\Delta y\Delta z \tag{A.7}$$

while the net tangential forces acting along the x direction are respectively given by

$$\left[\tau_{yx} + \frac{\partial\tau_{yx}}{\partial y}\Delta y\right]\Delta x\Delta z - \tau_{yx}\Delta x\Delta z \tag{A.8}$$

and

$$\left[\tau_{zx} + \frac{\partial\tau_{zx}}{\partial z}\Delta z\right]\Delta x\Delta y - \tau_{zx}\Delta x\Delta y \tag{A.9}$$

The total net force per unit volume on the fluid due to these surface stresses should be equal to the sum of Eqs (A.7), (A.8), and (A.9) divided by the control volume $\Delta x\,\Delta y\,\Delta z$:

$$\frac{\partial\sigma_{xx}}{\partial x} + \frac{\partial\tau_{yx}}{\partial y} + \frac{\partial\tau_{zx}}{\partial z} \tag{A.10}$$

It is not too difficult to verify that the total net forces per unit volume on the rest of the control volume surfaces in the y direction and z direction are given by:

$$\frac{\partial \tau_{xy}}{\partial x} + \frac{\partial \sigma_{yy}}{\partial y} + \frac{\partial \tau_{zy}}{\partial z} \tag{A.11}$$

and

$$\frac{\partial \tau_{xz}}{\partial x} + \frac{\partial \tau_{yz}}{\partial y} + \frac{\partial \sigma_{zz}}{\partial z} \tag{A.12}$$

Combining Eq. (A.10) with the substantial derivative of the horizontal velocity component u and body forces, the x momentum equation becomes

$$\rho \frac{Du}{Dt} = \frac{\partial \sigma_{xx}}{\partial x} + \frac{\partial \tau_{yx}}{\partial y} + \frac{\partial \tau_{zx}}{\partial z} + \sum F_x^{body\ forces} \tag{A.13}$$

In a similar fashion the y-momentum and z-momentum equations, using Eqs (A.11) and (A.12), can be obtained through

$$\rho \frac{Dv}{Dt} = \frac{\partial \tau_{xy}}{\partial x} + \frac{\partial \sigma_{yy}}{\partial y} + \frac{\partial \tau_{zy}}{\partial z} + \sum F_y^{body\ forces} \tag{A.14}$$

and

$$\rho \frac{Dw}{Dt} = \frac{\partial \tau_{xz}}{\partial x} + \frac{\partial \tau_{yz}}{\partial y} + \frac{\partial \sigma_{zz}}{\partial z} + \sum F_z^{body\ forces} \tag{A15}$$

If the fluid is taken to be Newtonian and isotropic – since all gases and majority of liquids are isotropic – the normal stresses σ_{xx}, σ_{yy}, and σ_{zz} appearing in Eqs (A.13)–(A.15) can be formulated in terms of pressure p and normal viscous stress components, τ_{xx}, τ_{yy}, and τ_{zz} acting perpendicular on the control volume. The remaining terms contain the tangential viscous stress components as also described from Eqs (A.13) to (A.15). In many fluid flows a suitable model for the viscous stresses is introduced, which can be expressed as a function of the local deformation rate (or strain rate). Assuming that the fluid is Newtonian and isotropic since all gases and majority of liquids are isotropic, the rate of linear deformation on the control volume $\Delta x\, \Delta y\, \Delta z$ caused by the motion of fluid can usually be expressed in terms of the velocity gradients. The normal stress relationships can be expressed as

$$\sigma_{xx} = -p + \tau_{xx} \qquad \sigma_{yy} = -p + \tau_{yy} \qquad \sigma_{zz} = -p + \tau_{zz} \tag{A.16}$$

According to *Newton's law of viscosity*, the normal and tangential viscous stress components are given by

$$\tau_{xx} = 2\mu\frac{\partial u}{\partial x} + \lambda\left[\frac{\partial u}{\partial x} + \frac{\partial v}{\partial y} + \frac{\partial w}{\partial z}\right] \quad \tau_{yy} = 2\mu\frac{\partial v}{\partial y} + \lambda\left[\frac{\partial u}{\partial x} + \frac{\partial v}{\partial y} + \frac{\partial w}{\partial z}\right]$$

$$\tau_{zz} = 2\mu\frac{\partial w}{\partial z} + \lambda\left[\frac{\partial u}{\partial x} + \frac{\partial v}{\partial y} + \frac{\partial w}{\partial z}\right]$$

$$\tau_{xy} = \tau_{yx} = \mu\left(\frac{\partial v}{\partial x} + \frac{\partial u}{\partial y}\right) \quad \tau_{xz} = \tau_{zx} = \mu\left(\frac{\partial w}{\partial x} + \frac{\partial u}{\partial z}\right) \tag{A.17}$$

$$\tau_{yz} = \tau_{zy} = \mu\left(\frac{\partial w}{\partial y} + \frac{\partial v}{\partial z}\right)$$

The proportionality constants of μ and λ are the (first) dynamic viscosity that relates stresses to linear deformation and the second viscosity that relates stresses to the volumetric deformation, respectively. To this present day, not much is known about the second viscosity. Nevertheless, Stokes hypothesis of $\lambda = -2/3\mu$ is frequently used and it has been found for gases to be a good working approximation.

When we combine Eqs (A.16) and (A.17) with Eqs (A.13)–(A.15), the equations for the velocity components $u, v,$ and w in three dimensions can be rewritten as

$$\rho\frac{Du}{Dt} = -\frac{\partial p}{\partial x} + \frac{\partial}{\partial x}\left[2\mu\frac{\partial u}{\partial x} + \lambda\left(\frac{\partial u}{\partial x} + \frac{\partial v}{\partial y} + \frac{\partial w}{\partial z}\right)\right]$$

$$+ \frac{\partial}{\partial y}\left[\mu\left(\frac{\partial u}{\partial y} + \frac{\partial v}{\partial x}\right)\right] + \frac{\partial}{\partial z}\left[\mu\left(\frac{\partial u}{\partial z} + \frac{\partial w}{\partial x}\right)\right] + \sum F_x^{body\ forces} \tag{A.18}$$

$$\rho\frac{Dv}{Dt} = -\frac{\partial p}{\partial y} + \frac{\partial}{\partial y}\left[2\mu\frac{\partial v}{\partial y} + \lambda\left(\frac{\partial u}{\partial x} + \frac{\partial v}{\partial y} + \frac{\partial w}{\partial z}\right)\right]$$

$$+ \frac{\partial}{\partial x}\left[\mu\left(\frac{\partial u}{\partial y} + \frac{\partial v}{\partial x}\right)\right] + \frac{\partial}{\partial z}\left[\mu\left(\frac{\partial v}{\partial z} + \frac{\partial w}{\partial y}\right)\right] + \sum F_y^{body\ forces} \tag{A.19}$$

$$\rho\frac{Dw}{Dt} = -\frac{\partial p}{\partial z} + \frac{\partial}{\partial z}\left[2\mu\frac{\partial w}{\partial z} + \lambda\left(\frac{\partial u}{\partial x} + \frac{\partial v}{\partial y} + \frac{\partial w}{\partial z}\right)\right]$$

$$+ \frac{\partial}{\partial x}\left[\mu\left(\frac{\partial u}{\partial z} + \frac{\partial w}{\partial x}\right)\right] + \frac{\partial}{\partial y}\left[\mu\left(\frac{\partial v}{\partial z} + \frac{\partial w}{\partial y}\right)\right] + \sum F_z^{body\ forces} \tag{A.20}$$

For the conservation of energy, the rate of work done on the control volume $\Delta x\ \Delta y\ \Delta z$ is equivalent to the product of the force and velocity component, which in the x direction is the velocity component u. From Fig. 3.5, the work done by the normal force in the x direction is

$$\left[u\sigma_{xx} + \frac{\partial(u\sigma_{xx})}{\partial x}\Delta x\right]\Delta y\Delta z - u\sigma_{xx}\Delta y\Delta z \tag{A.21}$$

while the work done by the tangential forces in the x direction are respectively given by

$$\left[u\tau_{yx} + \frac{\partial\left(u\tau_{yx}\right)}{\partial y}\Delta y\right]\Delta x\Delta z - u\tau_{yx}\Delta x\Delta z \tag{A.22}$$

and

$$\left[u\tau_{zx} + \frac{\partial\left(u\tau_{zx}\right)}{\partial z}\Delta z\right]\Delta x\Delta y - u\tau_{zx}\Delta x\Delta y \tag{A.23}$$

The net rate of work done by these surface forces acting in the x direction divided by the control volume $\Delta x\,\Delta y\,\Delta z$ is given by

$$\frac{\partial\left(u\sigma_{xx}\right)}{\partial x} + \frac{\partial\left(u\tau_{yx}\right)}{\partial y} + \frac{\partial\left(u\tau_{zx}\right)}{\partial z} \tag{A.24}$$

Work done due to surface stress components in the y direction and z direction can also be similarly derived and these additional rates of work done on the fluid are:

$$\frac{\partial\left(v\tau_{xy}\right)}{\partial x} + \frac{\partial\left(v\sigma_{yy}\right)}{\partial y} + \frac{\partial\left(v\tau_{zy}\right)}{\partial z} \tag{A.25}$$

and

$$\frac{\partial\left(w\tau_{xz}\right)}{\partial x} + \frac{\partial\left(w\tau_{yz}\right)}{\partial y} + \frac{\partial\left(w\sigma_{zz}\right)}{\partial z} \tag{A.26}$$

For heat added, the net rate of heat transfer to the fluid due to the heat flow in the x direction is given by the difference between the heat input at surface at x and heat loss at surface $x + \Delta x$ as depicted in Fig. 3.5:

$$\left[q_x + \frac{\partial q_x}{\partial x}\Delta x\right]\Delta y\Delta z - q_x\Delta y\Delta z \tag{A.27}$$

Similarly, the net rates of heat transfer in the y direction and z direction may also be expressed as

$$\left[q_y + \frac{\partial q_y}{\partial y}\Delta y\right]\Delta x\Delta z - q_y\Delta x\Delta z \tag{A.28}$$

and

$$\left[q_z + \frac{\partial q_z}{\partial z}\Delta z\right]\Delta x\Delta y - q_z\Delta x\Delta y \tag{A.29}$$

The total rate of heat added to the fluid divided by the control volume $\Delta x\, \Delta y\, \Delta z$ results in:

$$\frac{\partial q_x}{\partial x} + \frac{\partial q_y}{\partial y} + \frac{\partial q_z}{\partial z} \tag{A.30}$$

Combining Eqs (A.24), (A.25), (A.26), and (A.30) with the substantial derivative for a given specific energy E of a fluid, the equation for the conservation of energy becomes:

$$\rho \frac{DE}{Dt} = \frac{\partial(u\sigma_{xx})}{\partial x} + \frac{\partial(v\sigma_{yy})}{\partial y} + \frac{\partial(w\sigma_{zz})}{\partial z}$$
$$+ \frac{\partial(u\tau_{yx})}{\partial y} + \frac{\partial(u\tau_{zx})}{\partial z} + \frac{\partial(v\tau_{xy})}{\partial x} + \frac{\partial(v\tau_{zy})}{\partial z} + \frac{\partial(w\tau_{xz})}{\partial x} + \frac{\partial(w\tau_{yz})}{\partial y} \tag{A.31}$$
$$- \frac{\partial q_x}{\partial x} - \frac{\partial q_y}{\partial y} - \frac{\partial q_z}{\partial z}$$

Thus far, we have not defined the specific energy E of a fluid. Often the energy of a fluid is defined as the sum of the internal energy, kinetic energy, and gravitational potential energy. We shall regard the gravitational force as a body force and include the effects of potential energy changes as a source term. In three dimensions the specific energy E can be defined as

$$E = \underbrace{e}_{internal\ energy} + \underbrace{\frac{1}{2}\left(u^2 + v^2 + w^2\right)}_{kinetic\ energy} \tag{A.32}$$

For compressible flows, Eq. (A.32) is often re-arranged to give an equation for the *enthalpy*. The specific enthalpy h_{sp} and the specific (total) enthalpy h of a fluid are defined as

$$h_{sp} = e + \frac{p}{\rho} \quad \text{and} \quad h = h_{sp} + \frac{1}{2}\left(u^2 + v^2 + w^2\right)$$

Combining these two definitions with the specific energy E, we obtain

$$h = e + \frac{p}{\rho} + \frac{1}{2}\left(u^2 + v^2 + w^2\right) = E + \frac{p}{\rho} \tag{A.33}$$

APPENDIX B

Upwind Schemes

The first-order upwind scheme has been described in Chapter 5, Section 5.3. Here, we concentrate on the formulation of the second-order upwind and third-order QUICK schemes, as illustrated below. As an improvement to the first-order upwind scheme, the idea is to incorporate additional variables located at the neighboring grid nodal points indicated by the properties at points WW and EE, as shown in Fig. B.1 in order to evaluate the interface values at the cell faces of w and e.

For the second-order upwind scheme, assuming uniform distribution of the grid nodal points, additional information on the fluid flow is introduced into the approximation by the consideration of an extra upstream variable point, $viz.$,

$$\phi_w = \frac{3}{2}\phi_W - \frac{1}{2}\phi_{WW}$$
$$\phi_e = \frac{3}{2}\phi_P - \frac{1}{2}\phi_W \qquad \text{if } u_w > 0 \quad \text{and} \quad u_e > 0 \qquad \text{(B.1)}$$

$$\phi_w = \frac{3}{2}\phi_P - \frac{1}{2}\phi_E$$
$$\phi_e = \frac{3}{2}\phi_E - \frac{1}{2}\phi_{EE} \qquad \text{if } u_w < 0 \quad \text{and} \quad u_e < 0 \qquad \text{(B.2)}$$

For the third-order QUICK scheme, a quadratic approximation is introduced across two variable points at the upstream and one at the downstream depending on the flow direction. The unequal weighting influence of this particular scheme still hinges on the

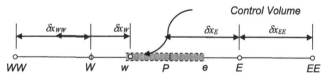

Fig. B.1 A schematic representation of a control volume around a node P in a one-dimensional domain with surrounding grid nodal points of WW, W, E, and EE.

knowledge biased toward the upstream flow information. The interface values ϕ_w and ϕ_e based on a uniform grid nodal point distribution can be evaluated as:

$$\phi_w = -\frac{1}{8}\phi_{WW} + \frac{6}{8}\phi_W + \frac{3}{8}\phi_P$$

$$\phi_e = -\frac{1}{8}\phi_W + \frac{6}{8}\phi_P + \frac{3}{8}\phi_E$$

$$\text{if } u_w > 0 \quad \text{and} \quad u_e > 0 \qquad \text{(B.3)}$$

$$\phi_w = -\frac{1}{8}\phi_E + \frac{6}{8}\phi_P + \frac{3}{8}\phi_W$$

$$\phi_e = -\frac{1}{8}\phi_{EE} + \frac{6}{8}\phi_E + \frac{3}{8}\phi_P$$

$$\text{if } u_w < 0 \quad \text{and} \quad u_e < 0 \qquad \text{(B.4)}$$

APPENDIX C

Explicit and Implicit Methods

The first-order explicit and implicit methods have been described in Chapter 5, Section 5.3. Here, we further concentrate on the formulation of the second-order *explicit* Adams-Bashford and *semi-implicit* Crank-Nicolson methods, as illustrated below.

As illustrated from the sketch in Fig. C.1, the extension of the first-order *explicit* method to the second-order *explicit* Adams-Bashford requires the values not only at time level n but also at time level $n-1$. The unsteady one-dimensional convection-diffusion of Eq. (5.58) can be recast in the form of:

$$
\begin{aligned}
\frac{\phi_P^{n+1} - \phi_P^n}{\Delta t} &= \frac{3}{2}\frac{\partial \phi}{\partial t}\bigg|^n - \frac{1}{2}\frac{\partial \phi}{\partial t}\bigg|^{n-1} \\
&= \frac{3}{2}\bigg[-u_e A_E \frac{1}{2}(\phi_W + \phi_P) + u_w A_W \frac{1}{2}(\phi_P + \phi_E) \\
&\quad + \Gamma_e A_E \frac{1}{\rho}\left(\frac{\phi_E - \phi_P}{\delta x_E}\right) - \Gamma_w A_W \frac{1}{\rho}\left(\frac{\phi_P - \phi_W}{\delta x_W}\right) + \frac{S_\phi}{\rho}\Delta V\bigg]^n \\
&\quad - \frac{1}{2}\bigg[-u_e A_E \frac{1}{2}(\phi_W + \phi_P) + u_w A_W \frac{1}{2}(\phi_P + \phi_E) \\
&\quad + \Gamma_e A_E \frac{1}{\rho}\left(\frac{\phi_E - \phi_P}{\delta x_E}\right) - \Gamma_w A_W \frac{1}{\rho}\left(\frac{\phi_P - \phi_W}{\delta x_W}\right) + \frac{S_\phi}{\rho}\Delta V\bigg]^{n-1}
\end{aligned} \tag{C.1}
$$

For the second-order Crank-Nicolson method, this special type of differencing in time requires the solution of ϕ_P^{n+1} to be obtained by averaging the properties between

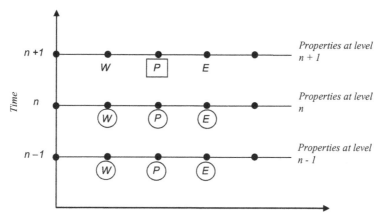

Fig. C.1 An illustration for the second order Adams-Bashford.

time levels n and $n + 1$. There are many versions of the Crank-Nicolson form in CFD. Some may only include the implicit evaluation of the diffusive term while others may even choose to consider the convective term in addition to the diffusive term to be implicitly determined but the source term remains in the previous time level n regardless. The authors present the latter form as depicted below:

$$
\begin{aligned}
\frac{\phi_P^{n+1} - \phi_P^n}{\Delta t} = & -\frac{u_e^n A_E}{2}\left[\frac{1}{2}\left(\phi_W^{n+1} + \phi_W^n\right) + \frac{1}{2}\left(\phi_P^{n+1} + \phi_P^n\right)\right] \\
& + \frac{u_w^n A_W}{2}\left[\frac{1}{2}\left(\phi_P^{n+1} + \phi_P^n\right) + \frac{1}{2}\left(\phi_E^{n+1} + \phi_E^n\right)\right] \\
& + \frac{\Gamma_e^n A_E}{\rho}\left[\frac{\frac{1}{2}\left(\phi_E^{n+1} + \phi_E^n\right) - \frac{1}{2}\left(\phi_P^{n+1} + \phi_P^n\right)}{\delta x_E}\right] \\
& - \frac{\Gamma_w^n A_W}{\rho}\left[\frac{\frac{1}{2}\left(\phi_P^{n+1} + \phi_P^n\right) - \frac{1}{2}\left(\phi_W^{n+1} + \phi_W^n\right)}{\delta x_W}\right] + \frac{S_\phi^n}{\rho}\Delta V
\end{aligned}
\tag{C.2}
$$

The above equation represents an example of a *semi-implicit* approach. Like the *fully implicit* approach, it also requires a simultaneous solution of the unknowns at *all* grid nodal points in the respective difference equations for a given time level of $n + 1$.

APPENDIX D

Learning Program

The materials presented in this book have been partially designed from teaching the course of *Introduction to Computational Fluid Dynamics* for senior undergraduate students in the School of Aerospace, Mechanical and Manufacturing Engineering at RMIT University, Australia. This learning program can be adopted by an instructor to conduct either a one-semester (6–8 h/week) or two-semester (3–4 h/week) CFD course in any engineering department. For example, the instructor may wish to assign 3 or 4 h per week to Lectures and 3 or 4 h per week to CFD Labs for a one-semester course. The appropriate allocation of hours for Lectures or CFD Labs is entirely up to the instructor. He/she may reduce the number of hours for Lectures and concentrate more on CFD Labs to allow students to attain more practical experiences in handling real fluid flow problems through CFD methods.

This program consists of the student's own reading of the relevant chapters described below, working out the assignments, and completing a final CFD project, as will be presented in Appendix E. The teaching approach has worked very well in facilitating students to better engage in various real problem-based assignments and projects. From the students' perspective, an air of excitement is exuberated from the beginning of the course through working on relatively easier problems in early assignments toward solving a real-life fluid dynamics problem chosen by the student at the final stage of this course. Details with regard to mathematical formulations are kept to a minimum and computer programming is avoided during the teaching of this course. Rather, the basic and practical knowledge of the ability of analyzing CFD solutions is emphasized within this program. The lectures are thus directed primarily to present the major theories and methodologies used in CFD and guide the students as to where their learning efforts should be concentrated. The CFD Labs, as indicated below, are to facilitate the learning of the basic theories, development of analysis capability, and ability to resolve practical engineering problems through the use of CFD software. During the CFD Labs, tutorials will also be used to introduce and discuss the problems in assignments, and to provide the hands-on assistance and feedback to the students. The final CFD project allows the student to attain experience in the application of CFD methods and analysis to real-world engineering problems.

Learning Program for a One-Semester CFD Course

Week 1	Lecture:	Introduction to CFD and CFD Procedure
	Reading:	Chapters 1 and 2
	CFD Lab:	Introduction to CFD Software
		Assignment 1 (Introduction)
Week 2	Lecture:	Basic CFD Equations
	Reading:	Chapter 3
	CFD Lab:	Assignment 1 (Discussion)
Week 3	Lecture:	Mesh Generation and Boundary Conditions
	Reading:	Chapters 4 and 7
	CFD Lab:	Assignment 1 (Finalization)
Week 4	Lecture:	Basic Numerical Methods
	Reading:	Chapter 5
	CFD Lab:	Assignment 2 (Introduction)
Week 5	Lecture:	Basic Numerical Techniques
	Reading:	Chapter 5
	CFD Lab:	Assignment 2 (Discussion)
Week 6	Lecture:	CFD Solution Analysis
	Reading:	Chapter 6
	CFD Lab:	Assignment 2 (Finalization)
Week 7	Lecture:	Turbulence Modeling
	Reading:	Chapter 7
	CFD Lab:	Assignment 3 (Introduction)
Week 8	Lecture:	Practical Guidelines and Case Study
	Reading:	Chapter 7
	CFD Lab:	Assignment 3 (Discussion)
Week 9	Lecture:	CFD Applications
	Reading:	Chapter 8
	CFD Lab:	Assignment 3 (Finalization)
Week 10	Lecture:	Invited Seminar: Engineering Design and Optimization Using CFD
	Reading:	Chapter 8
	CFD Lab:	Introduction to CFD Project
Week 11	Lecture:	Advanced CFD Topics
	Reading:	Chapter 9
	CFD Lab:	CFD Project (Discussion)
Week 12		CFD Project (Feedback)
Week 13		CFD Project (Finalization)
Week 14		Revision and Final Examination

APPENDIX E

CFD Assignments and Guideline for CFD Project

A sample of three assignments and a guideline for a CFD project clarifying the aim and objectives are described in this appendix. For students who do not possess their own project topics, they are most welcome to select one from the attached project topics (CFD Projects A–C) exemplified herein.

E.1 ASSIGNMENT 1

E.1.1 Background and Aim

The backward-facing step is commonly used as a benchmark for validating numerous flow characteristics, including flow recirculation and reattachment, and testing of numerical models and methods. This problem has numerous applications in industry, such as for HVAC, combustion chamber, etc.

The aims of this problem are:

1. To learn the process of creating and exporting a mesh by using any available mesh generation software packages. For this assignment, the mesh generator in ANSYS Workbench is employed.
2. To learn how to set suitable boundary conditions and numerical models using any available CFD software packages. ANSYS-FLUENT is used to solve the flow problem in this assignment.
3. To explore the post-processing facilities of the CFD code to analyze the numerical results.
4. To formulate concise, professional reports.

E.1.2 Problem Description

The student is required to compute laminar flows through a backward-facing step, as detailed below (Fig. E.1). The coordinates given for the geometry are normalized against the characteristic length scale. For the case of a backward-facing step, the characteristic length scale is the *step height* (in this case a length of unity is assumed). The normalized fluid properties at the *velocity inlet* are given as:

Inlet velocity: $u_x = 1$ and $u_y = 0$

Fluid properties: Density, $\rho = 1$, Dynamic viscosity, $\mu = 1/Re$, where Re is the Reynolds number

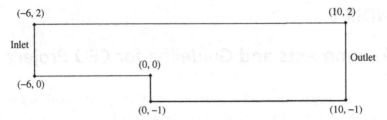

Fig. E.1 A schematic illustration of two-dimensional geometry of a backward-facing step.

The outlet boundary is defined as an outflow condition, while the no-slip condition is invoked for the rest of the computational walls. Turbulence is ignored and no heat transfer exists within the system. It is noted that the dimensions of the computational domain may need to be altered to ensure the flow is sufficiently developed at the outlet boundary.

E.1.3 Instructions

1. Initiate ANSYS Workbench to create a mesh for the backward-facing step. Assign appropriate boundary conditions to the computational domain. Structured mesh is preferred but the user may alternatively generate the geometry with an unstructured mesh. Ensure that proper mesh quality is achieved. Provide explanations for areas that require further mesh refinement. Export the two-dimensional mesh to ANSYS-FLUENT, Version 17.
2. Using ANSYS-FLUENT, solve the simulation to obtain the velocity and pressure contours and the velocity vectors for $Re = 100$. Ensure that the flow is fully developed (ensuring no flow reversal) or close-to-developed flow at the outlet boundary. Discuss any observed flow characteristics using the physical parameters.
3. Repeat steps 2 and 3 for other meshes of varying densities. Use the same flow settings for $Re = 100$ to determine the sensitivity of the mesh to the reattachment point of the recirculation zone. Plot a graph relating the mesh size against the reattachment location, highlighting the most economical mesh for numerical computations (hint, when grid independence is achieved, the reattachment point will not vary with increasing mesh density).
4. Using the mesh determined from step 3, perform simulations for other Reynolds numbers of 50, 150, and 200. Compile your results and create a graph illustrating the relationship between the Reynolds numbers and reattachment points. Explain the phenomenon and provide your own conclusions.

E.2 ASSIGNMENT 2

E.2.1 Background and Aim

One common CFD application is the study of flows over external structures. In the automotive industry it is important to determine the aerodynamic effects of the spacing between adjacent motor vehicles. The Ahmed model is often used in experiments as a representation of the motor vehicle due to its simple geometry and the ease of varying a number of important parameters.

The aims of simulating these models are:

1. To create a CFD simulation of a single Ahmed model and extract meaningful data.
2. To obtain CFD simulations as well as study the effects of spacing between two Ahmed models.
3. To gain an understanding of the model requirements for turbulent flow and the importance of the distribution of y^+ values.
4. To better understand boundary layer flows.
5. To learn how to distinguish and access results with available published experimental results.

E.2.2 Problem Description

E.2.2.1 Single Car Configuration

The geometry of a typical Ahmed configuration is shown below. Students are required to develop a model simulation of a single two-dimensional Ahmed configuration model (Fig. E.2).

The coordinates for the Ahmed model are given as:

x	y
−0.94400	0.00000
0.00000	0.00000
−1.04400	0.10000
0.00000	0.17700
−1.04400	0.18800
−0.94400	0.28800
−0.19226	0.28800

Fig. E.2 A two-dimensional geometry of an Ahmed model.

Note that the characteristic length (taken as the length of the vehicle) of the Ahmed configuration is not equivalent to unity. You will need to take into consideration the length when calculating the Reynolds numbers and the spacing between adjacent vehicles.

Simulations are to be performed in air. The outer domain for the *single model case* should be constructed according to the following Cartesian coordinates below, which should allow any wakes and vortices to be properly resolved within the computational domain.

x	y
−9.39600	−0.05000
−9.39600	4.12600
19.83600	−0.05000
19.83600	4.12600

At the inlet, the velocity should be set according to the *Reynolds number* (with respect to the car length) of 2.3×10^6. The flow *turbulence intensity* is assumed to be *1.8%*.

Instructions

1. Students are required to generate another mesh for a second *vehicle trailing the one created above* (hint, the domain may need to be purposefully extended to accommodate the additional vehicle).
2. Appropriate meshing should be employed, preferably similar to the above. The same meshing methods and boundary conditions as used in the previous section should be used herein.
3. *Vary the distance* between the trailing and leading vehicles. Formulate at least three additional cases. Discuss the flow characteristics and compare the drag and lift coefficients for both the lead and the rear car models (Figs. E.3 and E.4).

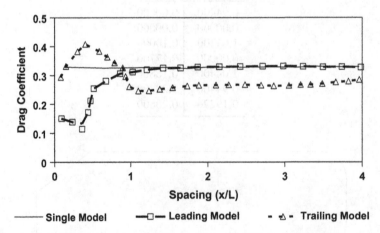

Fig. E.3 Drag coefficient for single Ahmed model and two Ahmed models at different vehicle spacing. *(From Watkins and Vino, 2004.)*

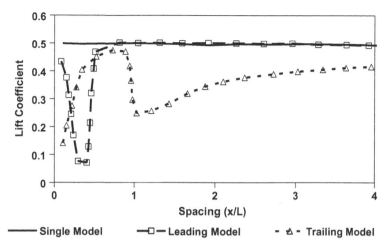

Fig. E.4 Lift coefficient for single Ahmed model and two Ahmed models at different vehicle spacing. *(From Watkins and Vino, 2004.)*

4. Compare the drag and lift coefficients against experimental results (Figs. E.3 and E.4).
5. Discuss the flow physics obtained from the predicted drag and lift coefficients.

E.2.2.2 Drafting Configuration

Instructions

1. Generate a mesh of sufficient quality (keeping in mind the numerical considerations of aspect ratios and grid skewness). Ensure *mesh independence* is reached (hint, use two or three different types of mesh densities).
2. Students must apply the attained knowledge gained from tutorials and the previous assignment to determine *suitable boundary conditions*. (*Note* simulations can be performed by allowing the *floor* to *travel at the same speed as the air*).
3. Discuss the importance of y^+ values in turbulent flows. Students must ensure that acceptable values are achieved. If otherwise, provide an explanation for this.

E.3 ASSIGNMENT 3

E.3.1 Background and Aim

CFD has the ability to model fluid flows coupled with heat transfer. A basic understanding of the thermal and hydrodynamic behavior of fluid within a channel has long been an established area of research. CFD simulations can provide important insights into the flow behavior and heat transfer to improve the heat transfer within a complex channel geometry having a waved-shaped wall (depicted in Fig. E.5), which are increasingly being explored in industrial heat exchangers.

The aims of this assignment are:
1. To create a wavy channel consisting of a sufficient number of complete waves so as to provide developed flow conditions.
2. To create a wavy channel segment using the periodic boundary conditions.
3. To investigate the effect of different turbulence models and wall functions on the solution.
4. To implement a constant surface temperature as well as a constant surface heat flux on the wavy wall and determine the relationship between the Reynolds number and thermal properties.
5. To compare and discuss simulation results with experimental results.

E.3.2 Problem Description

The channel consists of a *repeated section* consisting of a straight wall at the top and a sine wave–shaped wall at the bottom:

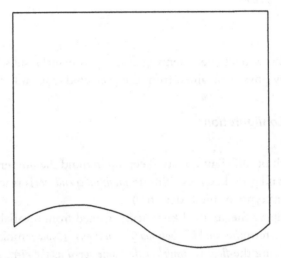

Fig. E.5 A section of the wavy channel.

The *coordinates* for the geometry are given as:

0	0
0.25	0.1
0.5	0
0.75	−0.1
1	0
1	0
0	1
1	1

The *flow properties* of the air are:

Mass flow rate: $\dot{m} = 0.816$ kg/s

Density: $\rho = 1$ kg/m^3

Dynamic viscosity: $\mu = 0.0001$ kg/ms

Bulk fluid temperature: $T_b = 300$ K

Other thermal fluid properties are set as default

The flow is *initialized* according to:

X Velocity = 0.816 m/s

Turbulence Kinetic Energy = 1 m^2/s^2

Turbulence Dissipation Rate = 1×10^5 m^2/s^3

Enhanced wall treatment is applied for all turbulence models (suitable y^+ values are adjusted accordingly as discussed in Chapter 7).

Instructions

1. Create a channel consisting of sufficient numbers of the section given above to provide developed flow conditions (hint, approximately 12 sections).

2. Compare the *NORMALISED* axial-velocity at the crest and the trough of the full model where the flow has become fully developed with the experimental results.

3. Create a section as described above using periodic boundary conditions (see Chapter 8).

4. Compare the *NORMALISED* axial-velocity at the crest ($x = 0.25$) and the trough ($x = 0.75$) of the periodic model with the experimental results. Discuss the relationship between the full model, periodic model, and experimental results.

5. Using the periodic model, investigate the accuracy of using the standard k-ε, RNG, and realizable turbulence models and compare them against the experimental results. Discuss which turbulence model is the most aptly suited for such a complex configuration.

6. Implement a *constant wall temperature of 500K* on the wavy wall surface for a full model and discuss the thermal characteristics. Vary the flow to provide Reynolds numbers of 10,000 and 5400. Compare and discuss the effect of changing the flow rate.

7. Implement a *constant wall heat flux of 1000 W/m^2* on the wavy wall surface for a full model and discuss the thermal characteristics. Use only the initial Reynolds number.

8. Generate other results of interest: velocity, turbulence parameters, temperature distribution (wall and fluid), total heat transfer, and Nusselt number.

E.4 PROJECT GUIDELINE

E.4.1 Aim

The aim of the CFD project is to provide an opportunity for students to demonstrate their understanding of the fundamentals and usage of CFD software as well as introduce them to the numerous applications within the software. Students are allowed to freely determine any topic of interest. It may be desirable that the intended CFD project topic coincides with the undertaking of your final year project, keeping in mind the many constraints and complexity of the flow problem to be simulated. Students are therefore strongly encouraged to consult and discuss the project proposal with the lecturer/instructor before embarking on the next stage of their numerical study.

E.4.2 Objectives

The following abilities should be demonstrated and explained in your project report:

1. The ability to use the commercial CFD software packages
   - Mesh generation and grid quality.
   - Defining the settings of a flow problem, that is, boundary conditions and solver settings.
   - Selecting appropriate CFD models, that is, turbulence models, heat transfer, or other types of simulations.
2. The ability to use CFD as a tool for engineering design
   - Reduction of drag or increase of lift for flow over geometries, for example, car body or aerofoils.
   - Increase of heat transfer for cooling of a car engine, or reduction of heat transfer to prevent heat loss.
   - To create desired flow control, that is, the ability to cause flow to move within a desired region.
3. The ability to apply CFD knowledge to analyze numerical results
   - Discussions of the flow patterns and behaviors (e.g. wake flow, flow separation, boundary layer and convective effects).
   - Discussions of the accuracy of the CFD solution with regard to the mesh quality, flow models chosen, and boundary/domain setting.
   - Assumptions made on the actual model to allow for modeling simplifications to the original geometry.

Note   Not all the items listed above are applicable to all types of CFD problems. Please consult with lecturers or tutors concerning the above, as well as ascertaining the suitability of the project. A brief project proposal outlining aims and scopes of the project, problem description, and objectives of the project should be prepared and submitted to the lecturers, which will account for 5% of your project mark.

Examples of past CFD project topics:
- Effect of vehicle spacing on vehicles in convoy
- CFD analysis of wing in ground effect

- Investigation of the effect of winglet design on the lift and drag performance
- CFD simulation of flow over a bicycle helmet
- Study of the reduction of aerodynamic drag on a car-caravan combination
- Turbulent flow analysis over two two-dimensional wings of variable horizontal separation
- Investigation of flow field in areas of different hydro power plants
- Modeling of car air intake system and comparison of different designs
- Numerical investigation of trailing edge flows
- Comparison of fowler flap systems through CFD

E.5 EXAMPLE – CFD PROJECT PROPOSAL PREPARED BY THE STUDENT

E.5.1 Introduction

Aerofoil design plays a pivotal role in the wing and control surface performance in aerospace engineering. While wind tunnel testing can be rather time consuming and expensive, CFD provides an attractive alternative. Modeling of flow over an aerofoil is an important CFD problem. Among the many aerofoil design features is the flap (Fig. E.6), which when deployed increases the camber of the wing to give increased lift (and drag). This phenomenon is being utilized during take-off and landing in most aircraft as lift-enhancing (and drag-enhancing) devices (Fig. E.7).

Fig. E.6 Plain flap.

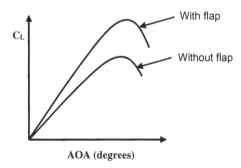

Fig. E.7 Lift curve slope.

E.5.2 Scope

The project will focus on the aerodynamic characteristics of a two-dimensional NACA 23012 aerofoil with a 20% scaled NACA 23012 plain flap. The aerofoil will be modeled at 0° angle of attack with flap settings of 0°, 10°, 20°, 30°, and 40°. The Reynolds number of 1,400,000 helps to study the turbulent characteristics of the combo. The lift and drag coefficient and flow separation will be determined using the standard, RNG k-ε, and realizable k-ε turbulence models to assess the most appropriate model to be applied.

Results will be validated against benchmark experimental data of Carl Wenzinger (*Carl J. Wenzinger, 1937, Pressure distribution over a NACA 23012 Airfoil with a NACA 23012 external-airfoil flap. NACA Report No. 614*).

E.5.3 Objectives

- Create a quality mesh around the aerofoil/flap
- Model the airflow around the aerofoil/flap at $Re = 1,400,000$ and SSL conditions with flap settings varying from 0° to 40° using standard k-ε turbulence model
- Measure the lift and drag coefficient and pressure and velocity distribution of the aerofoil/flap
- Repeat with RNG k-ε and realizable k-ε turbulence models
- Compare and discuss the results of the three turbulence models
- Evaluate the accuracy of the results against experimental data

E.6   OTHER TOPICS FOR CFD PROJECTS

E.6.1 CFD Project A: CFD Simulation of Turbulent Flow Over a Backward-Facing Step

E.6.1.1 Background

The backward-facing step is commonly used as a benchmark for the validation of numerous flow characteristics, including turbulence model, multi-phase flows, and fundamental numerical methods. This model also has applications in industry, such as for combustion and HVAC.

E.6.1.2 Objectives

The aims of this simulation are to:
1. Create a backward-facing step simulation appropriate for turbulence modeling.
2. Determine the effects of different meshing schemes.
3. Simulate turbulent flow and compare with the benchmark data.
4. Determine the effects of different turbulence models.
5. Understand the relevant flow characteristics.
6. Prepare a concise and well-written professional report.

E.6.1.3 Problem Description

Dry air at 27°C flows through a 2D duct with a backward-facing step at a Reynolds number, $Re_h = 5100$, based on the step height, h. The dimensions of this flow configuration are shown below. "No-slip" conditions are applied to the walls and it can be assumed they are perfectly smooth (i.e. roughness height $= 0$ m). The turbulence intensity of the incoming flow can be assumed to be 0.01% and thermal interaction between the walls and the fluid is assumed adiabatic (i.e. no heat transfer). For the purpose of this assignment, it is expected that no vortex shedding will occur (i.e. steady-state analysis). From extremely accurate direct numerical simulations (DNS), it has been determined that re-attachment of the separated boundary layer occurs on the bottom wall at $x = 16.28h$.

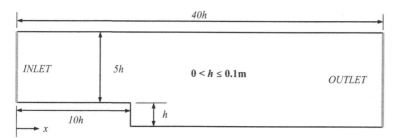

E.6.1.4 Required Discussions

1. Compare the following meshing schemes: (a) uniform structured quadrilateral mesh; (b) non-uniform structured quadrilateral mesh with mesh refinement in appropriate regions (with similar number of cells); and (c) unstructured triangular mesh with refinement in appropriate regions (with similar number of cells). Discuss the differences and the necessity of mesh refinement.

2. Using the best mesh determined from (1), run the case using appropriate settings for (a) a *standard k-ε* turbulence model and (b) a *realizable k-ε* turbulence model. Determine which model gives the better prediction. Discuss the differences between the k-ε models, referring to their characteristic equations.

3. Using the best model determined from (1) and (2), replace the dry air with Ethylene Glycol at 37°C flowing at the same Reynolds number ($Re_h = 5100$). Discuss any similarities or differences with the dry air simulation.

4. Discuss the relevant flow characteristics of the simulation.

E.6.2 CFD Project B: CFD Simulation of Pickup Trucks With Open/Closed Tubs

E.6.2.1 Background

CFD has been a vital tool in studying vehicle aerodynamics. These simulations aim to provide a better understanding of the flow behavior over the complex geometries as well as give insights into methods of increasing vehicle efficiency.

E.6.2.2 Objectives

The aims of this simulation are to:
1. Create two-dimensional models of a pickup truck with the tub open and closed.
2. Create a mesh of sufficient quality to achieve grid independence.
3. Simulate turbulent flow and compare with benchmark data.
4. Understand and discuss the relevant flow characteristics, such as flow velocity and pressure, to design parameters, such as drag and lift forces.
5. Create a trailing pickup truck and investigate the effect of different distances on flow characteristics.
6. Prepare a concise and well-written report.

E.6.2.3 Problem Description

A truck, as shown below (dimensions provided in Addendum), is to be modeled (with tailgate closed). The moving truck is simulated traveling at a Reynolds number of 3.3×10^6 (based on the truck length). The standard k-ε turbulence model and standard wall functions are employed. Students are to create a sufficiently large domain to capture the generated wake. The inlet should have a turbulence intensity of 2% and the lower boundary (parallel to the x-axis) should be set to a non-moving smooth wall, so as to simulate wind-tunnel conditions. The system is assumed to be adiabatic and time independent. For a single pickup truck with a closed tailgate, the drag coefficient was found to be 0.44.

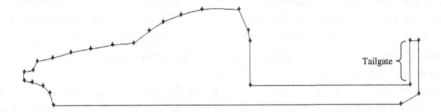

E.6.2.4 Required Discussions

1. Determine appropriate boundary layer settings to provide a mesh of sufficient quality for this simulation, including calculations of the first layer thickness for a standard wall model turbulence case.
2. Compare the simulation drag coefficient with the benchmark data to determine mesh independence.
3. Compare and discuss the flow characteristics of a single truck with the *tailgate open* with one with the *tailgate closed* to determine if this has an effect on the overall drag of the vehicle.
4. Compare and discuss the effect of adding a trailing truck of variable distance (recommended you do not exceed 3 lengths of the truck) on both the leading and trailing vehicles. You may set either truck to have an open or closed tailgate for this investigation. Plot and compare the drag and lift coefficients for both trucks against different separation lengths on a graph and discuss.

Note You may need to extend the length of the domain to incorporate the extra truck.

E.6.2.5 Addendum

Coordinates for the pickup truck with closed tailgate:

x	y
0	0.354
0	0.224
0.0215	0.246
0.0215	0.354
0.045	0.199
0.4226	0.246
0.4226	0.354
0.4264	0.38
0.4504	0.432
0.5458	0.432
0.5981	0.419
0.6438	0.402
0.6795	0.38
0.7185	0.35
0.7743	0.346
0.8284	0.337
0.8792	0.328
0.9263	0.199
0.9263	0.199
0.9267	0.315
0.935	0.229
0.9518	0.246
0.966	0.307
0.9771	0.285
0.9798	0.255
1	0.26
1	0.281

E.6.3 CFD Project C: Investigation of Cooling Electronic Components Within a Computer

E.6.3.1 Background

CFD is a powerful tool for optimizing applications involving heat transfer. The cooling of electronic components within a computer has recently become an important issue. Particularly, increased processing power tends to generate more heat, which may cause hardware damage or failure.

E.6.3.2 Objectives

The aims of this simulation are to:

1. Preparing a suitable mesh of sufficient quality to analyze the thermal and flow characteristics of air cooling within a computer.
2. Understand the requirements for cooling electronic components within the system.
3. Design the system by placing the components in optimal positions to allow operation within the safety range through the understanding of the thermal and flow behavior.
4. Determine the best positions for the electronic components.

E.6.3.3 Problem Description

A system case is shown below with the dimensions listed in millimeters. There are two exhaust openings (50 mm wide) and one air intake (80 mm wide). Air is to enter the system at a temperature of 20°C at a rate of 0.1 m³/s with a turbulence intensity of 5%. The walls of the system are assumed to be adiabatic and smooth. Within the system, add: (1) 1 processor with dimensions 25 mm × 25 mm with a heat output of 800 W/m², (2) 3 RAM modules with dimensions 135 mm × 30 mm with a heat output of 60 W/m², and (3) 3 other components of dimensions 100 mm × 50 mm with a heat output of 150 W/m². Use the standard k-ε turbulence model to simulate the flow turbulence.

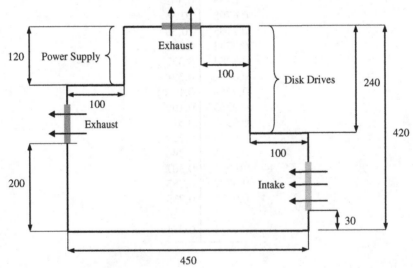

E.6.3.4 Required Discussions

Arrange the components listed above within the system casing with the given boundary conditions in at least three different designs. Additionally, analyze the fluid flow and thermal characteristics to determine the most optimum locations for your components, bearing in mind that no component surface should exceed 80°C.

APPENDIX F

Other Numerical Methods and Techniques

F.1 MULTIGRID METHODS

In Chapter 5 the use of the multigrid method to accelerate the convergence of the iterative process for large systems of non-linear algebraic equations has been discussed, while in Chapter 6, the authors presented the simplest strategy for achieving computational efficiency in solving such a system of equations through the V-cycle with five different grid levels. It is by no means a comprehensive description; the authors have aimed mainly to provide the reader with a bird's eye view of such an approach. However, recognizing its surging importance in areas of modern CFD, the basic philosophy behind this particular method is further explored within this section.

As with the V-cycle, calculations are initially carried out on a fine grid and these results are progressively transferred downward to a series of coarser grids; the results on the coarsest grid are obtained and transferred upward back to the fine grid level. The process is repeated until satisfactory convergence is achieved. On a mathematical basis, the advantage of the multigrid method stems from the consideration of the enhanced damping of numerical errors through the flow field. A whole spectrum of errors can propagate throughout the numerical solution of such a flow field. For a discrete grid increment Δ_m, the *high-frequency errors* can be represented by the smallest value $\lambda_{min} = 2\Delta_m$. The majority of iterative procedures such as the Jacobi and Gauss-Seidel (Chapter 5) are rather efficient in removing the high-frequency errors in a few iterations. It is the removal of the *low-frequency errors* that causes the slow convergence of iterative methods on a fixed grid. Nonetheless, let us imagine that after carrying out a few iterations on the fine grid, the immediate results are transferred to a coarser grid. The *high-frequency errors* are now essentially lost or hidden in the coarse grid and the solution procedure begins to damp at a more rapid rate than would have taken place in the fine grid because of the larger Δ_m. Hence, by progressively moving the intermediate results to coarser grids, the low-frequency errors are essentially damped; when these results are transferred back to the fine grid, the *low-frequency errors* are indeed much smaller than they would have been for an equal number of sweeps that are performed on the fine grid itself.

It is clear from above that the multigrid method is more of a strategy rather than applying a particular solution method to solve the algebraic equations for a given flow field. Currently, ongoing research is still being undertaken to this very day to optimize the solution process particularly on improving the numerical methods in solving the algebraic equations at each grid level. Many challenges remain in designing the "best" multigridding strategy, whether they may be better in ascertaining a suitable combination of simple or advanced iterative methods on fine grids accompanied by direct methods to obtain solutions on the coarse grids or by employing different cycling strategies using either a series combination of the V-cycle with the W-cycle or the implementation of a *full-multigrid*, F-cycle. Extensive efforts are also being heavily invested in extending the multigrid method to parallel computing to achieve quicker computational speed-ups in comparison to traditional single- or dual-processor computations. Interested readers can refer to Mavriplis (1988), Wesseling (1995), Timmermann (2000), and Thomas et al. (2003) for the latest trends and developments in this subject area.

References

Acharya, S., Moukalled, F.H., 1989. Improvements to Incompressible Flow Calculation on a Nonstaggered Curvilinear Grid. Numer. Heat Transf. 15, 131–152.

Aiba, S., Tsuchida, H., Ota, T., 1982a. Heat Transfer Around Tubes in in-Line Tube Banks. Bulletin of the JSME Vol. 25, 919–926.

Aiba, S., Tsuchida, H., Ota, T., 1982b. Heat Transfer Around Tubes in Staggered Tube Banks. Bulletin of the JSME Vol. 25, 927–933.

Alvarez, G., Palacios, M.J., Flores, J.J., 2000. A Test Method to Evaluate the Thermal Performance of Window Glazings. Appl. Therm. Eng. Vol. 20, 803–812.

Álvarez-Farré, X., Gorobets, A., Trias, F.X., 2021. A Hierarchical Parallel Implementation for Heterogeneous Computing. Application to Algebra-Based CFD Simulations on Hybrid Supercomputers. Comput. Fluids Vol. 214, 104768.

American Institute of Aeronautics and Astronautics, 1998. Guide for the Verification and Validation of Computational Fluid Dynamics, *AIAA G-077-1998*. American Institute of Aeronautics and Astronautics, Reston, USA.

Anderson Jr., J.D., 1995. Computational Fluid Dynamics – The Basics With Applications. McGraw-Hill, New York.

Anderson, W., Thomas, J.L., Van Leer, B., 1986. Comparison of Finite Volume Flux Vector Splittings for the Euler Equations. AIAA J. Vol. 24, 1453–1460.

Anderson, J., Lorenz, C., Travesset, A., 2008. General Purpose Molecular Dynamics Simulations Fully Implemented on Graphics Processing Units. J. Comput. Phys. Vol. 227, 5342–5359.

ANSYS Inc., 2009. ANSYS FLUENT 12.0 Theory Guide.

Apsley, D., Chen, W.-L., Leschziner, M., Lien, F.-S., 1997. Nonlinear Eddy Viscosity Modelling of Separated Flows. IAHR J. Hydraulic Research Vol. 35, 723–748.

Arcilla, A.S., Häuser, J., Eiseman, P.R., Thompson, J.F. (Eds.), 1991. Numerical Grid Generation in Computational Fluid Dynamics and Related Fields. In: Proceedings of the Third International Conference on Numerical Grid Generation in Computational Fluid Dynamics and Related Fields, Barcelona, Spain, North-Holland, 3–7 June 1991.

Armengol, J.M., Vicquelin, R., Coussement, A., Santos, R.G., Gicquel, O., 2019. Scaling of heated Plane Jets With Moderate Radiative Heat Transfer in Coupled DNS. Int. J. Heat Mass Transf. Vol. 139, 456–474.

Aufiero, M., Fratoni, M., 2017. A New Approach to the Stabilization and Convergence Acceleration in Coupled Monte Carlo–CFD Calculations: The Newton Method via Monte Carlo Perturbation Theory. Nucl. Eng. Technol. Special Issue on International Conference on Mathematics and Computational Methods Applied to Nuclear Science and Engineering 2017 (M&C 2017). Vol. 49, 1181–1188.

Bahiraei, M., 2014. A Comprehensive Review on Different Numerical Approaches for Simulation in Nanofluids: Traditional and Novel Techniques. J. Dispers. Sci. Technol. Vol. 35, 984–996.

Baker, A.J., 1983. Finite Element Fluid Mechanics. McGraw-Hill, New York.

Balaras, E., Benocci, C., Piomelli, U., 1996. Two-layer Approximate Boundary Conditions for Large-Eddy Simulations. AIAA J. Vol. 34, 1111–1119.

Baldwin, W.S., Lomax, H., 1978. Thin-Layer Approximation and Algebraic Model for Separated Turbulent Flows. In: 16th Aerospace Sciences Meeting. American Institute of Aeronautics and Astronautics.

Bassi, F., Rebay, S., 1997. A High-Order Accurate Discontinuous Finite Element Method for the Numerical Solution of the Compressible Navier–Stokes Equations. J. Comput. Phys. 131, 267–279.

Battan, P., Goldberg, U., Chakravarthy, S., 2004. Interfacing Statistical Turbulence Closures With Large-Eddy Simulation. AIAA J. Vol. 42, 485–492.

Baum, H.R., McCaffrey, B.J., 1989. Fire Induced Flow Field – Theory and Experiment. Fire Saf. Sci. Vol. 2, 129–148.

Beam, R.M., Warming, R.F., 1978. An Implicit Factored Scheme for the Compressible Navier-Stokes Equations. AIAA J. Vol. 16, 393–402.

Belytschko, T., Krongauz, Y., Organ, D., Fleming, M., Krysl, P., 1996. Meshless Methods: An Overview and Recent Developments. Comp. Meth. Appl. Mech. Eng. Vol. 139, 3–47.

Benson, R.A., McRae, D.S., 1991. A Solution Adaptive Mesh Algorithm for Dynamic/Static Refinement of Two and Three Dimensional Grids. In: 3rd Int. Conf. Num. Grid Generation in Computational Fluid Dynamics and Related Fields. Barcelona, Spain.

Benzi, R., Struglia, M.V., Tripiccione, R., 1996. Extended Self-Similarity in Numerical Simulations of Three-Dimensional Anisotropic Turbulence. Phys. Rev. E Vol. 53, 5565–5568.

Bertrand, F., Leclaire, L.-A., Levecque, G., 2005. DEM-Based Models for the Mixing of Granular Materials. Chem. Eng. Sci. Vol. 60, 2517–2531.

Bhatnagar, P.L., Gross, E.P., Krook, M., 1954. A Model for Collision Processes in Gases. I: Small Amplitude Processes in Charged and Neutral One-component System. Phys. Rev. Vol. 94, 511–525.

Bird, G.A., 1988. Direct Simulation of Gas Flows at the Molecular Level. *Commun.* Appl. Numer. Methods Vol. 4, 165–172.

Bird, G.A., 1994. Molecular Gas Dynamics and the Direct Simulation of Gas Flows. Oxford Press, Clarendon, UK.

Bisetti, F., Attili, A., Pitsch, H., 2014. Advancing Predictive Models for Particulate Formation in Turbulent Flames via Massively Parallel Direct Numerical Simulations. Philos. Trans. R Soc. Math. Phys. Eng. Sci. Vol. 372, 20130324.

Bourgoyne, D.A., Ceccio, S.L., Dowling, D.R., Jessup, S., Park, J., Brewer, W., Pankajakshan, R., 2000. Hydrofoil Turbulent Boundary Layer Separation at High Reynolds Numbers. In: 23rd Symp. Naval Hydrodynamics. Val de Reuil, France. September 17–22.

Boussinesq, J., 1868. Mémoire sur l'influence des frottements dans les mouvements réguliers des fluides. J. Math. Pures Appl. 13.

Bradshaw, P., 1994. Turbulence: The Chief Outstanding Difficulty of Our Subject. Exp. Fluids Vol. 16, 203–216.

Briley, W.R., McDonald, H., 1977. Solution of the Multidimensional Compressible Navier-Stokes Equations by a Generalized Implicit Method. J. Comp. Phys. Vol. 24, 372–379.

Brunton, S.L., Noack, B.R., Koumoutsakos, P., 2020. Machine Learning for Fluid Mechanics. Annu. Rev. Fluid Mech. Vol. 52, 477–508.

Cai, W., Ma, H., Wang, Y., Chen, J., Zheng, X., Zhang, H., 2019. Development of POD Reduced-Order Model and Its Closure Scheme for 2D Rayleigh–Bénard Convection. Appl. Math. Model. Vol. 66, 562–575.

Calzolari, G., Liu, W., 2021. Deep Learning to Replace, Improve, or Aid CFD Analysis in Built Environment Applications: A Review. Build. Environ. Vol. 206, 108315.

Canuto, C., Hussaini, M.Y., Quateroni, A., Zang, T.A., 1987. Spectral Methods in Fluid Dynamics. Springer-Verlag, Berlin.

Cao, W., Huang, W., Russell, R.D., 1999. A Moving Mesh Method in Multiblock Domains With Application to a Combustion Problem. Numer. Methods Partial Differ. Equ. 15, 449–467.

Çengel, Y.A., 2003. Heat Transfer: A Practical Approach. McGraw-Hill.

Chalmers, N., Agbaglah, G., Chrust, M., Mavriplis, C., 2019. A Parallel HP-Adaptive High Order Discontinuous Galerkin Method for the Incompressible Navier-Stokes Equations. J. Comput. Phys. X Vol. 2, 100023.

Chang, J.L.C., Kwak, D., 1984. On the Method of Pseudo Compressibility for Numerically Solving Incompressible Flows. In: 22nd Aerospace Sciences Meeting. American Institute of Aeronautics and Astronautics.

Chao, C.Y.H., Wan, M.P., Morawska, L., Johnson, G.R., Ristovski, Z.D., Hargreaves, M., Mengersen, K., Corbett, S., Li, Y., Xie, X., Katoshevski, D., 2009. Characterization of Expiration Air Jets and Droplet Size Distributions Immediately at the Mouth Opening. J. Aerosol Sci. Vol. 40, 122–133.

Chen, J.H., 2011. Petascale Direct Numerical Simulation of Turbulent Combustion – Fundamental Insights Towards Predictive Models. Proc. Combust. Inst. Vol. 33, 99–123.

Chen, S., Doolen, G.D., 1998. Lattice Boltzmann Method for Fluid Flows. Ann. Rev. Fluid Mech. Vol. 30, 329–364.

Chen, L., Tu, J.Y., Yeoh, G.H., 2003. Numerical Simulation of Turbulent Wake Flows Behind Two Side-By-Side Cylinders. J. Fluids Structures Vol. 18, 387–403.

Chen, L., Tu, J., Zhou, Y., Virahsawmy, H., McGillivray, I., 2004. Computation of Flow Behind Three Side-By-Side Cylinders of Unequal/Equal Spacing. ANZIAM J. 46, 672–689.

Cheng, H.K., 1993. Perspectives on Hypersonic Viscous Flow Research. Ann. Rev. Fluid Mech. Vol. 25, 455–484.

Cheng, H.K., Emanuel, G., 1995. Perspective on Hypersonic Nonequilibrium Flow. AIAA J. Vol. 33, 385–400.

Cheung, S.C.P., Yeoh, G.H., Cheung, A.L.K., Yuen, R.K.K., Lo, S.M., 2007. Flickering Behavior of Turbulent Buoyant Fires Using Large-Eddy Simulation. Numer. Heat Transf. Part Appl. Vol. 52, 679–712.

Chew, P., 1989. Guaranteed-Quality Triangular Meshes. Computer Science Technical Reports. TR 89-983.

Chiesa, M., Mathiesen, V., Melheim, J.A., Halvorsen, B., 2005. Numerical Simulation of Particulate Flow by the Eulerian-Lagrangian and the Eulerian-Eulerian Approach with the Application to a Fluidized Bed. Comp. Chem. Eng. Vol. 29, 291–304.

Choi, S.-K., Lin, C.-L., 2010. A Simple Finite-Volume Formulation of the Lattice Boltzmann Method for Laminar and Turbulent Flows. Numer. Heat Transf. Part B Fundam. Vol. 58, 242–261.

Chorin, A.J., 1968. Numerical Solution of Navier-Stokes Equations. Math. Comp. Vol. 22, 745–762.

Choy, Y.-H., Merkle, C.L., 1993. The Application of Preconditioning in Viscous Flows. J. Comp. Phys. Vol. 105, 207–223.

Colburn, A.P., 1933. A Method of Correlating Forced Convection Heat Transfer Data and a Comparison with Fluid Friction, Trans. In: American Institute of Chemical Engineers, Vol. 29. American Institute of Electrical Engineers, New York, pp. 174–210.

Colella, P., Woodward, P., 1984. The Piecewise Parabolic Method for Gas-Dynamical Simulations. J. Comp. Phys. Vol. 54, 174.

Cottet, G.H., Poncet, P., 2004. Advances in Direct Numerical Simulations of 3D Wall-Bounded Flows by Vortex-in-Cell Methods. J. Comp. Phys. Vol. 193, 136–158.

Cox, G., Chitty, R., 1980. A Study of the Deterministic Properties of Unbounded Fire Plumes. Combust. Flame Vol. 39, 191–209.

Crabill, J., Witherden, F.D., Jameson, A., 2018. A Parallel Direct Cut Algorithm for High-Order Overset Methods With Application to a Spinning Golf Ball. J. Comput. Phys. Vol. 374, 692–723.

Crespo, A.C., Dominguez, J.M., Barreiro, A., Gómez-Gesteira, M., Rogers, B.D., 2011. GPUs, a New Tool of Acceleration in CFD: Efficiency and Reliability on Smoothed Particle Hydrodynamics Methods. PLOS ONE Vol. 6, e20685.

Crowe, C., Sommerfield, M., Tsuji, Y., 1998. Multiphase Flows With Droplets and Particles. CRC Press, Florida.

Cundall, P.A., Strack, O.D.L., 1979. A Discrete Numerical Model for Granular Assemblies. Geotechnique Vol. 29, 47–65.

Daru, V., Tenaud, C., 2004. High Order One-Step Monotonicity-Preserving Schemes for Unsteady Compressible Flow Calculations. J. Comp. Phys. Vol. 193, 563–594.

Davidson, L., Peng, S.-H., 2003. Hybrid LES-RANS: A One-Equation SGS Model Combined With a k-w Model for Predicting Recirculating Flows. Int. J. Num. Meth. Fluids Vol. 43, 1003–1018.

De Berg, M., van Kreveld, M., Overmars, M., Schwarzkopf, O., 2000. Computational Geometry: Algorithms and Applications, Second Edition. Springer-Verlag, Berlin.

de Langhe, C., Merci, B., Dick, E., 2005a. Hybrid RANS/LES Modelling with an Approximate Renormalization Group. I. Model Development. J. Turb Vol. 6, 1–18.

de Langhe, C., Merci, B., Lodefier, K., Dick, E., 2005b. Hybrid RANS/LES Modelling with an Approximate Renormalization Group. II. Applications. J. Turb Vol. 6, 1–16.

Deen, N.G., Van Sint Annaland, M., Van Der Hoef, M.A., Kuipers, J.A.M., 2007. Review of Discrete Particle Modeling of Fluidized Beds. Chem. Eng. Sci. Vol. 62, 28–44.

Deen, N.G., Peters, E.A.J.F., Padding, J.T., Kuipers, J.A.M., 2014. Review of Direct Numerical Simulation of Fluid–Particle Mass, Momentum and Heat Transfer in Dense Gas–Solid Flows. Chem. Eng. Sci. Vol. 116, 710–724.

Demirdzic, I., Muzaferija, S., Peric, M., Schreck, E., 1997. Numerical Method for Simulation of Flow Problems Involving Moving and Sliding Grids. In: Proc. 7th Int. Symp. Computational Fluid Dynamics. Int. Academic Publishers, Beijing, China.

Deng, S., Jiang, L., Liu, C., 2007. DNS for Flow Separation Control Around an Airfoil by Pulsed Jets. Comput. Fluids 36, 1040–1060.

Desjardin, P.E., Frankel, S.H., 1999. Two-Dimensional Large Eddy Simulation of Soot Formation in the Near-Field of a Strongly Radiating Nonpremixed Acetylene-Air Turbulent Jet Flame. Comb. Flame Vol. 119, 121–132.

Di Renzo, A., Di Maio, F.P., 2005. An Improved Integral Non-Linear Model for the Contact of Particles in Distinct Element Simulations. Chem. Eng. Sci. Vol. 59, 3461–3475.

Ding, C., Lam, K.P., 2019. Data-Driven Model for Cross Ventilation Potential in High-Density Cities Based on Coupled CFD Simulation and Machine Learning. Build. Environ. Vol. 165, 106394.

Dong, S., Karniadakis, G.E., 2004. Dual-Level Parallelism for High-Order CFD Methods. Parallel Comput. Vol. 30, 1–20.

Dong, J., Shang, Y., Tian, L., Inthavong, K., Qiu, D., Tu, J., 2019. Ultrafine Particle Deposition in a Realistic Human Airway at Multiple Inhalation Scenarios. Int. J. Numer. Methods Biomed. Eng. Vol. 35, e3215.

Dongarra, J., Grigori, L., Higham, N.J., 2020. Numerical Algorithms for High-Performance Computational Science. Philos. Trans. R Soc. Math. Phys. Eng. Sci. Vol. 378, 20190066.

Dopazo, C., 1993. Recent Developments in PDF Methods. Turbul. React. Flows 375, 474.

Drikakis, D., 2002. Embedded Turbulence Model in Numerical Methods for Hyperbolic Conservation Laws. Int. J. Num. Meth. Fluids Vol. 39, 763–781.

Duarte, C.A., Oden, J.T., 1996. An H-P Adaptive Method Using Clouds. Comp. Meth. Appl. Mech. Eng. Vol. 139, 237–262.

Du Fort, E.C., Frankel, S.P., 1953. Stability Conditions in the Numerical Treatment of Parabolic Differential Equations. Math. Tables and Other Aids to Computation Vol. 7, 559–573.

Ellero, M., Kröger, M., Hess, 2002. Viscoelastic Flows Studied by Smoothed Particle Dynamics. J. Non-Newton. Fluid Mech. Vol. 105, 35–51.

Elsen, E., LeGresley, P., Darve, E., 2008. Large Calculation of the Flow Over a Hypersonic Vehicle Using a GPU. J. Comput. Phys. Vol. 227, 10148–10161.

Ewart, T., Firpo, J.L., Graur, I.A., Perrier, P., Méolans, J.G., 2008. DSMC Simulation: Validation and Application to Low Speed Gas Flows in Microchannels. J. Fluids Eng. 131.

Fanelli, M., Feke, D.L., Manas-Zloczower, I., 2006. Prediction of the Dispersion of Particle Clusters in the Nano-Scale, Part I: Steady Shearing Responses. Chem. Eng. Sci. Vol. 61. pp. 473–388.

Farhat, C., 2005. CFD on Moving Grids: From Theory to Realistic Flutter, Maneuvering, and Multidisciplinary Optimization. Int. J. Comp. Fluid Dyn. Vol. 19, 595–603.

Fauci, L.J., McDonald, A., 1994. Sperm Motility in the Presence of Boundaries. Bull. Math. Biol. Vol. 57, 679–699.

Ferrero, A., Iollo, A., Larocca, F., 2020. Reduced Order Modelling for Turbomachinery Shape Design. Int. J. Comput. Fluid Dyn. Vol. 34, 127–138.

Ferziger, J.H., Perić, M., 2002. Computational Methods for Fluid Dynamics. Springer, Berlin, Heidelberg.

Fletcher, C.A.J., 1984. Computational Galerkin Methods. Springer-Verlag, Berlin.

Fletcher, C.A.J., 1991. Computational Techniques for Fluid Dynamics. Volumes I and II. Springer-Verlag, Berlin.

Forum, H.P.F., 1993. High Performance Fortran Language Specification. Sci. Prog. Vol. 2, 1–70.

Freret, L., Ivan, L., Sterck, H.D., Groth, C.P., 2017. A High-Order Finite-Volume Method With Anisotropic AMR for Ideal MHD Flows. In: 55th AIAA Aerospace Sciences Meeting. AIAA SciTech Forum, Grapevine, Texas.

Fujii, K., 2005. Progress and Future Prospects of CFD in Aerospace – Wind Tunnel and Beyond. Prog. Aero. Sci. Vol. 42, 455–470.

Fujii, K., Obayashi, S., 1987a. Navier-Stokes Simulations of Transonic Flows Over a Practical Wing Configuration. AIAA J. Vol. 25, 368–370.

Fujii, K., Obayashi, S., 1987b. Navier-Stokes Simulations of Transonic Flows Over a Wing Fuselage Configuration. AIAA J. Vol. 25, 1587–1596.

Gallis, M.A., Torczynski, J.R., Plimpton, S.J., Rader, D.J., Koehler, T., 2014. Direct Simulation Monte Carlo: The Quest for Speed. AIP Conf. Proc. Vol. 1628, 27–36.

Garnier, E., Mossi, M., Sagaut, P., Comte, P., Deville, M., 1999. On the Use of Shock-Capturing Schemes for Large-Eddy Simulation. J. Comp. Phys. Vol. 153, 273–311.

Gao, Y., Liu, C., 2018. Rortex and Comparison With Eigenvalue-Based Vortex Identification Criteria. Phys. Fluids 30, 085107.

Germano, M., Piomelli, U., Moin, P., Cabot, W., 1991. A Dynamic Subgrid-Scale Eddy Viscosity Model. Phys. Fluids Vol. 3, 1760–1765.

Geurts, B.J., 2001. Modern Simulation Strategies for Turbulent Flow. Edwards, Inc., Philadelphia.

Gharebaghi, M., Irons, R.M.A., Ma, L., Pourkashanian, M., Pranzitelli, A., 2011. Large Eddy Simulation of Oxy-coal Combustion in an Industrial Combustion Test Facility. Int. J. Greenh. Gas Control 5, 100–110.

Giles, M.B., 1990. Nonreflecting Boundary Conditions for Euler Equation Calculations. AIAA J. 28, 2050–2058.

Ghias, R., Mittal, R., Lund, T., 2004. A Non-Body Conformal Grid Method for Simulation of Compressible Flows With Complex Immersed Boundaries. In: 42nd AIAA Aerospace Sciences Meeting and Exhibit. American Institute of Aeronautics and Astronautics.

Godunov, S.K., 1959. A Finite Difference Method for the Numerical Computation of Discontinuous Solutions of the Equations of Fluid Dynamics. Mat. Sb. Vol. 47, 357.

Gorden, W., Thiel, L., 1982. Transfinite Mappings and Their Application to Grid Generation. Appl. Math. Comput. 10–11, 171–233.

Gresho, P.M., 1991. Incompressible Fluid Dynamics: Some Fundamental Formulation Issues. Ann. Rev. Fluid Mech. Vol. 23, 182–188.

Gresho, P.M., Sani, R.L., 1990. On Pressure Boundary Conditions for the Incompressible Navier-Stokes Equations. Int. J. Num. Meth. Fluids Vol. 7, 11–46.

Griffith, B.E., Patankar, N.A., 2020. Immersed Methods for Fluid–Structure Interaction. Annu. Rev. Fluid Mech. Vol. 52, 421–448.

Grotberg, J.B., Jensen, O.E., 2004. Biofluid Mechanics in Flexible Tubes. Ann Rev. Fluids Mech. Vol. 36, 121–147.

Gumbert, C., Lohner, R., Parikh, P., Pirzadeh, S., 1989. A Package for Unstructured Grid Generation and Finite Element Flow Solvers. In: Aerodynamics Conference.

Gunzburger, M.D., Nicolades, R.A. (Eds.), 1993. Incompressible Computational Fluid Dynamics Trends and Advances. Cambridge University Press, Cambridge, UK.

Gustafsson, B., Lotsedt, P., Goran, A., 2002. A Fourth-Order Difference Method for the Incompressible Navier-Stokes Equations. Numerical Simulation of Incompressible Flows. World Scientific Press, Washington.

Hafez, M., 2002. Numerical Simulation of Incompressible Flows. World Scientific Press, Washington.

Hahn, M., Drikakis, D., 2005. Large Eddy Simulation of Compressible Turbulence Using High-Resolution Methods. Int. J. Numer. Methods Fluids 47, 971–977.

Hájek, J., Kermes, V., Stehlík, P., Šikula, J., 2005. Utilizing CFD as an Efficient Tool for Improved Equipment Design. Heat Transf. Eng. Vol. 26, 15–24.

Hagen, G., 1839. Ueber die Bewegung des Wassers in engen cylindrischen Röhren. Ann. Phys. 122, 423–442.

Harlow, F.H., Welch, J.E., 1965. Numerical Calculation of Time-Dependent Viscous Incompressible Flow With Free Surface. Phys. Fluids Vol. 8, 2182–2189.

Harten, A., 1989. ENO Schemes With Subcell Resolution. J. Comp. Phys. Vol. 83, 148.

Hartwich, P., Agrawal, S., 1997. Method for Perturbing Multiblock Patched Grids in Aero-elastic and Design Optimization Applications. In: 13th Computational Fluid Dynamics Conference, Fluid Dynamics and Co-Located Conferences. American Institute of Aeronautics and Astronautics.

Hassan, O., Probert, E.J., 1999. In: Thompson, J.F., Soni, B.K., Wetherhill, N.P. (Eds.), Grid Control and Adaptation, Handbook of Grid Generation. CRC Press, Florida. pp. 35-1 to 35-29.

Hattori, H., Nagano, Y., 2012. Structures and Mechanism of Heat Transfer Phenomena in Turbulent Boundary Layer With Separation and Reattachment via DNS. Int. J. Heat Fluid Flow Vol. 37, 81–92.

Hazel, A.L., Heil, M., 2003. Steady Finite-Reynolds-Number Flows in Three-Dimensional Collapsible Tubes. J. Fluid Mech. Vol. 486, 79–103.

He, X., Doolen, G.D., 1997. Lattice Boltzmann Method on Curvilinear Coordinates System: Vortex Shedding Behind a Circular Cylinder. Phys. Rev. E Vol. 56, 434–440.

He, X., Luo, L.-S., 1997. A Priori Derivation of the Lattice Boltzmann Equation. Phys. Rev. E Vol. 55, 6333–6336.

He, Y., Muller, F., Hassanpour, A., Bayly, A.E., 2020. A CPU-GPU Cross-Platform Coupled CFD-DEM Approach for Complex Particle-Fluid Flows. Chem. Eng. Sci. Vol. 223, 115712.

Heil, M., 1997. Stokes Flow in Collapsible Tubes: Computation and Experiment. J. Fluid Mech. Vol. 353, 285–312.

Heil, M., Pedley, T.J., 1996. Large Post-Buckling Deformations of Cylindrical Shells Conveying Viscous Flow. J. Fluids Struct. Vol. 10, 565–599.

Hertz, H., 1882. Über die Berührung Fester Elatischer Körper. J. Reine Angew. Math. Vol. 92, 156–171.

Hesthaven, J.S., Rozza, G., Stamm, B., 2016. Certified Reduced Basis Methods for Parametrized Partial Differential Equations, SpringerBriefs in Mathematics. Springer International Publishing, Cham.

Hieber, S.E., Koumoutsakos, P., 2005. A Lagrangian Particle Level Set Method. J. Comp. Phys. 210, 342–367.

Hijazi, S., Stabile, G., Mola, A., Rozza, G., 2020. Data-Driven POD-Galerkin Reduced Order Model for Turbulent Flows. J. Comput. Phys. Vol. 416, 109513.

Hill, D.J., Pullin, D.I., 2004. Hybrid Tuned Center-Difference-WENO Method for Large Eddy Simulations in the Presence of Strong Shocks. J. Comput. Phys. Vol. 194, 435–450.

Hirt, C.W., Nichols, B.D., 1981. Volume of Fluid (VOF) Method for Dynamics of Free Boundaries. J. Comp. Phys. Vol. 39, 201–225.

Hiyama, K., Kato, S., Ishida, Y., 2010. Thermal Simulation: Response Factor Analysis Using Three-Dimensional CFD in the Simulation of Air Conditioning Control. Build. Simul. Vol. 3, 195–203.

Höhne, T., Krepper, E., Rohde, U., 2010. Application of CFD Codes in Nuclear Reactor Safety Analysis. Sci Tech. Nucl. Installations. Vol. 2010, 1–8. https://doi.org/10.1155/2010/198758.

Hoomans, B.P.B., Kuipers, J.A.M., Briels, W.J., Swaaij, W.P., 1996. Discrete Particle Simulation of Bubble and Slug Formation in a Two-Dimensional Gas-Fluidized Bed: A Hard-Sphere Approach. Chem Eng. Sci. Vol. 51, 99–118.

Hörschler, I., Meinke, M., Schröder, W., 2003. Numerical Simulation of the Flow Field in a Model of the Nasal Cavity. Comput. Fluids. Vol. 32, 39–45.

Horsfield, K., Dart, G., Olson, D.E., Filley, G.F., Cumming, G., 1971. Models of the Human Bronchial Tree. J. Appl. Physiol. Vol. 31, 207–217.

Hou, S., Zuo, Q., Chen, S., Doolen, G., Cogley, A.C., 1995. Simulation of a Cavity Flow by the Lattice Boltzmann Method. J. Comp. Phys. Vol. 118, 329–347.

Hou, S., Sterling, J., Chen, S., Doolen, G., 1996. A Lattice Boltzmann Subgrid Model for High Reynolds Number Flows. Fields Inst. Comm. Vol. 6, 151–166.

Hu, F.Q., Hussaini, M.Y., Manthey, J.L., 1996. Low-Dissipation and Low-Dispersion Runge-Kutta Schemes for Computational Acoustics. J. Comp. Phys. Vol. 124, 177–191.

Hu, L., Zhou, L., Luo, Y., Xu, C., 2013. Measurement and Simulation of Swirling Coal Combustion. Particuology Vol. 11, 189–197.

Huang, W., 2001. Practical Aspects of Formulation and Solution of Moving Mesh Partial Differential Equations. J. Comp. Phys. Vol. 171, 753–775.

Huang, W., Ren, Y., Russel, R.D., 1994. Moving Mesh Methods Based on Moving Mesh Partial Differential Equations. J. Comp. Phys. Vol. 113, 279–290.

Huang, W., Russel, R.D., 1999. Moving Mesh Strategy Based on a Gradient Flow Equation for Two-Dimensional Problems. SIAM J. Sci. Comp. Vol. 20, 998–1015.

Hubbard, B., Chen, H.-C., 1994. A Chimera Scheme for Incompressible Viscous Flows With Application to Submarine Hydrodynamics. In: Fluid Dynamics Conference, Fluid Dynamics and Co-Located Conferences. American Institute of Aeronautics and Astronautics.

Hubbard, B.J., Chen, H.C., 1995. Calculation of Unsteady Flows Around Bodies with Relative Motion Using a Chimera RANS Method. In: Proc. 10th ASCE Engineering Mechanics Conference. Vol. II. University of Colorado at Boulder, Boulder, Colorado. pp. 7832–785, May 21–24.

Humphrey, J.A.C., Whitelaw, J.H., Yee, G., 1981. Turbulent Flow in a Square Duct With Strong Curvature. J. Fluid Mech. 103, 443–463.

Hurtado Rosas, R., Wang, B.-C., 2022. DNS Study of Turbulent Heat Transfer in an Elliptical Pipe Flow Subjected to System Rotation About the Major Axis. Int. J. Heat Mass Transf. Vol. 184, 122230.

Huynh, H., Wang, Z., Vincent, P., 2014. High-Order Methods for Computational Fluid Dynamics: A Brief Review of Compact Differential Formulations on Unstructured Grids. Comput. Fluids 98, 209–220.

Iaccarino, G., Kalitzin, G., Elkins, C., 2003. Numerical and Experimental Investigation of the Turbulent Flow in a Ribbed Serpentine Passage. Presented at the Center for Turbulence Research Annual Research Briefs 2003, 379–388.

Ince, N.Z., Launder, B.E., 1995. Three-Dimensional and Heat-Loss Effects on Turbulent Flow in a Nominally Two-Dimensional Cavity. Int. J. Heat Fluid Flow Vol. 16, 171–177.

Ingram, C.L., McRae, D.S., Benson, R.A., 1993. Time Accurate Simulation of a Self-Excited Oscillatory Supersonic External Flow with a Multi-Block Solution Adaptive Mesh Algorithm, AUAA 9303387. In: 11th Computational Fluid Dynamics Conference. Orlando, Florida.

Inthavong, K., Tian, Z.F., Li, H.F., Tu, J.Y., Yang, W., Xue, C.L., Li, C.G., 2006. A Numerical Study of Spray Particles Deposition in a Human Nasal Cavity. Aerosol Sci. Tech. Vol. 40, 1034–1045.

Inthavong, K., Ge, Q., Se, C.M.K., Yang, W., Tu, J.Y., 2011. Simulation of Sprayed Particle Deposition in a Human Nasal Cavity Including a Nasal Spray Device. J. Aerosol Sci. 42, 100–113.

Ishii, M., Hibiki, T., 2006. Thermo-Fluid Dynamics of Two-Phase Flows. Springer-Verlag, Berlin.

Issa, R.I., 1986. Solution of the Implicitly Discretised Fluid Flow Equations by Operator-Splitting. J. Comp. Phys. Vol. 62, 40–65.

Jagode, H., Anzt, H., Juckeland, G., Ltaief, H. (Eds.), 2020. High Performance Computing: ISC High Performance 2020 International Workshops, Frankfurt, Germany, June 21–25, 2020, Revised Selected Papers, Lecture Notes in Computer Science. Springer International Publishing, Cham.

Jiang, G.-S., Shu, C.-W., 1996. Efficient Implementation of Weighted ENO Schemes. J. Comp. Phys. Vol. 126, 202–228.

Jiang, L., Choudhari, M., Chang, C.-L., Liu, C., 2006. Numerical Simulations of Laminar-Turbulent Transition in Supersonic Boundary Layer. In: 36th AIAA Fluid Dynamics

Conference and Exhibit, Fluid Dynamics and Co-Located Conferences. American Institute of Aeronautics and Astronautics.

Johansen, S.T., 2003. Mathematical Modeling of Metallurgical Processes. In: 3rd International Conference on CFD in the Minerals and Process Industries. CSIRO, Melbourne, Australia.

Kadanoff, L., 1986. On Two Levels. Phys. Today Vol. 39, 7–9.

Kader, B., 1993. Temperature and Concentration Profiles in Fully Turbulent Boundary Layers. Int. J. Heat Mass Transfer Vol. 24, 1541–1544.

Kallinderis, Y. (Ed.), 2000. Adaptive Methods for Compressible CFD. Comp. Meth. Appl. Sci. Eng. Vol. 189. Preface.

Kamakoti, R., Shyy, W., 2004. Fluid-Structure Interaction for Aeroelastic Applications. Prog. Aero. Sci. Vol. 40, 535–558.

Kampolis, I., Trompoukis, X., Asouti, V., Giannakoglou, K., 2010. CFD-based Analysis and Two-Level Aerodynamic Optimization on Graphics Processing Units. Comput. Methods Appl. Meh. Eng. Vol. 199, 712–722.

Kamyar, A., Saidur, R., Hasanuzzaman, M., 2012. Application of Computational Fluid Dynamics (CFD) for Nanofluids. Int. J. Heat Mass Transf. Vol. 55, 4104–4115.

Kandil, O.A., Chung, H.A., 1988. Unsteady Vortex-Dominated Flows Around Maneuvering Wings Over a Wide Range of Mach Numbers. AIAA Paper No. 88-0317, In: AIAA 26th Aerospace Sciences Meeting & Exhibit. Reno, Nevada.

Kato, M., Launder, B.E., 1993. Three-Dimensional Modelling and Heat-loss Effects on Turbulent Flow in a Nominally Two-Dimensional Cavity. Int. J. Heat and Fluid Flow Vol. 16, 171–177.

Kefayati, G.H.R., Hosseinizadeh, S.F., Gorji, M., Sajjadi, H., 2011. Lattice Boltzmann Simulation of Natural Convection in Tall Enclosures Using Water/SiO$_2$ Nanofluid. Int. Commun. Heat Mass Transf. Vol. 38, 798–805.

Keyhani, K., Scherer, P.W., Mozell, M.M., 1995. Numerical Simulation of Airflow in the Human Nasal Cavity. J. Biomech. Eng. Vol. 117, 429–441.

Khan, W.A., Culham, J.R., Yovanovich, M.M., 2006. Analytical Model for Convection Heat Transfer from Tube Banks. J. Thermophys.Heat Trans. Vol. 20, 720–727.

Kim, S.-E., Choudhury, D., 1995. A Near-Wall Treatment Using Wall Functions Sensitized to Pressure Gradient. Separated Complex Flows. In: Presented at the 1995 ASME/JSME Fluids Engineering and Laser Anemometry Conference and Exhibition. Hilton Head, SC.

Kim, D., Braun, J.E., Cliff, E.M., Borggaard, J.T., 2015. Development, Validation and Application of a Coupled Reduced-Order CFD Model for Building Control Applications. Build. Environ. Vol. 93, 97–111.

Kiris, C., Kwak, D., Rogers, S., 2002. Incompressible Navier-Stokes Solvers in Primitive Variables and their Applications to Steady and Unsteady Flow Simulations. Numerical Simulation of Incompressible Flows, World Scientific Press, Washington.

Kliafas, Y., Holt, M., 1987. LDV Measurements of a Turbulent Air-Solid Two-Phase Flow in a 90° Bend. Exp. Fluids Vol. 5, 73–85.

Knio, O.M., Ghoniem, A.F., 1992. The Three-Dimensional Structure of Periodic Vorticity Layers Under Non-Symmetrical Conditions. J. Fluid Mech. Vol. 243, 353–392.

Kochkov, D., Smith, J.A., Alieva, A., Wang, Q., Brenner, M.P., Hoyer, S., 2021. Machine Learning–Accelerated Computational Fluid Dynamics. Proc. Natl. Acad. Sci. 118, 1–8.

Koelbel, C., Loveman, D., Schreiber, R., Steele, G., Zosel, M., 1994. The High Performance Fortran Handbook. MIT Press, Cambridge, MA.

Kolev, N.I., 2015. Multiphase Flow Dynamics 1: Fundamentals. Springer International Publishing, Cham.

Komatitsch, D., Erlebacher, G., Göddeke, D., Michéa, D., 2010. High-Order Finite-Element Seismic Wave Propagation Modeling with MP on a Large GPU Cluster. J. Comput. Phys. Vol. 229, 7692–7714.

Koobus, B., Tran, H., Farhat, C., 2000. Computation of Unsteady Viscous Flows Around Moving Grids Using k-e Turbulence Model on Unstructured Dynamics Grids. Comp. Meth. Appl. Mech. Eng. Vol. 190, 1441–1466.

Koumoutsakos, P., 2005. Multiscale Flow Simulations Using Particles. Ann. Rev. Fluid Mech. Vol. 37, 457–487.

Koumoutsakos, P., Leonard, A., 1995. High Resolution Simulation of the Flow Around an Impulsively Started Cylinder Using Vortex Methods. J. Fluid Mech. Vol. 296, 1–38.

Krasny, R., 1986. A Study of Singularity Formation in a Vortex Sheet by the Point Vortex Approximation. J. Fluid Mech. Vol. 167, 65–93.

Kruggel-Emden, H., Wirtz, S., Scherer, V., 2009. Applicable Contact Force Models for the Discrete Element Model: The Single Particle Perspective. ASME J. Pressure Vessel Technol. Vol. 131, 1–11.

Ku, D., 1997. Blood Flow in Arteries. Ann. Rev Fluids Mech. Vol. 29, 399–434.

Kurose, R., Makino, H., 2003. Large Eddy Simulation of a Solid-Fuel Jet Flame. Combust. Flame Vol. 135, 1–16.

Kuwabara, G., Kono, K., 1987. Restitution Coefficient in Collision Between Two Spheres. Jpn. J. Appl. Phys., Part 1 Vol. 26, 1230–1233.

Kuzan, J. D. (1986). Velocity Measurements for Turbulent Separated and Near-Separated Flows Over Solid Waves, Ph.D. Thesis, Dept. Chem. Eng., Univ. Illinois, Urbana, Illinois.

Kwak, D., Kiris, C., 1991. Steady and Unsteady Solutions of the Incompressible Navier-Stokes Equations. AIAA J. Vol. 29, 603–610.

Kwak, D., Kiris, C., Kim, C.S., 2005. Computational Challenges of Viscous Incompressible Flows. Comp. Fluids Vol. 34, 283–299.

Labournasse, E., Sagaut, P., 2004. Advance in RAN-LES Coupling, a Review and an Insight on the NLDE Approach. Arch. Comp. Meth. Eng. Vol. 11, 199–256.

Launder, B.E., 1989. Second-Moment Closures: Present and Future? Int. J. Heat Fluid Flow Vol. 10, 282–300.

Launder, B.E., Spalding, D.B., 1974. The Numerical Computation of Turbulent Flows. Comp. Meth. Appl. Mech. Eng. Vol. 3, 269–289.

Launder, B.E., Reece, G.J., Rodi, W., 1975. Progress in the Development of a Reynolds Stress Turbulence Closure. J. Fluid Mech. Vol. 68, Pt. 3. 537–566.

Lax, P.D., Richtmyer, R.D., 1956. Survey of the Stability of Linear Finite Difference Equations. Commun. Pure Appl. Math. Vol. 9, 267–293.

Lebas, R., Menard, T., Beau, P.A., Berlemont, A., Demoulin, F.X., 2009. Numerical Simulation of Primary Break-Up and Atomization: DNS and Modelling Study. Int. J. Multiph. Flow Vol. 35, 247–260.

Lee, J., Herrmann, H.J., 1993. Angle of Repose and Angle of Marginal Stability – Molecular Dynamics of Granular Particles. Phys. Rev. E Vol. 52, 3288–3291.

Lee, K.R., Park, J.H., Kim, K.H., 2011. High-Order Interpolation Method for Overset Grid Based on Finite Volume Method. AIAA J. 49, 1387–1398.

Lei, M., Archie, J., Kleimstreuer, C., 1997. Computational Design of a Bypass Graft that Minimizes Wall Shear Stress Gradients in the Region of the Distal Anastomosis. J. Vasc. Surg. Vol. 25, 637–646.

Leonard, B.P., 1991. The ULTIMATE Conservative Difference Scheme Applied to Unsteady One-Dimensional Advection. Comp. Meth. Appl. Mech. Eng. Vol. 88, 17.

Li, Q., He, Y.L., Wang, Y., Tao, W.Q., 2007. Coupled Double-Distribution-Function Lattice Boltzmann Method for the Compressible Navier-Stokes Equations. Phys. Rev. E Vol. 76, 056705.

Li, D., Da Ronch, A., Chen, G., Li, Y., 2019. Aeroelastic Global Structural Optimization Using an Efficient CFD-Based Reduced Order Model. Aerosp. Sci. Technol. Vol. 94, 105354.

Lilek, Z., Muzaferija, S., Peric, M., Sedil, V., 1997. Computation of Unsteady Flows Using Non-Matching Blocks of Structured Grids. Num. Heat Transfer, Part B: Fund. Vol. 23, 369–384.

Lilly, D.K., 1992. A Proposed Modification of the Germano Subgrid-Scale Closure Model. Phys. Fluid Vol. 4, 633–635.

Lind, S.J., Rogers, B.D., Stansby, P.K., 2020. Review of Smoothed Particle Hydrodynamics: Towards Converged Lagrangian Flow Modelling. Proc. R Soc. Math. Phys. Eng. Sci. Vol. 476, 20190801.

Liseikin, V.D., 1999. Grid Generation Methods. Springer-Verlag, Berlin.

Liu, X., Osher, S., Chan, T., 1994. Weighted Essentially Non-Oscillatory Schemes. J. Comp. Phys. Vol. 115, 200–212.

Liu, C., Gao, Y., Tian, S., Dong, X., 2018. Rortex – A New Vortex Vector Definition and Vorticity Tensor and Vector Decompositions. Phys. Fluids 30, 035103.

Liu, C., Gao, Y., Dong, X., Wang, Y., Liu, J., Zhang, Y., Cai, X., Gui, N., 2019. Third Generation of Vortex Identification Methods: Omega and Liutex/Rortex Based Systems. J. Hydrodyn. 31, 205–223.

Lo, S.H., 1985. A New Mesh Generation Scheme for Arbitrary Planar Domains. Int. J. Num. Meth. Eng. Vol. 21, 1403–1426.

Lockwood, F.C., Naguib, A.S., 1975. The Prediction of the Fluctuations in the Properties of Free, Round-Jet, Turbulent. Diffusion Flames. Combust. Flame 24, 109–124.

Loner, R., Yang, C., Cebral, J., Soto, O., Camelli, F., 2002. On Incompressible Flow Solvers. Numerical Simulation of Incompressible Flows, World Scientific Press, Washington.

Lu, L., Brennan Pecha, M., Wiggins, G.M., Xu, Y., Gao, X., Hughes, B., Shahnam, M., Rogers, W.A., Carpenter, D., Parks, J.E., 2022. Multiscale CFD Simulation of Biomass

Fast Pyrolysis With a Machine Learning Derived Intra-Particle Model and Detailed Pyrolysis Kinetics. Chem. Eng. J. Vol. 431, 133853.

Luan, B., Robbins, M.O., 2005. The Breakdown of Continuum Model for Mechanical Contacts. Nature Vol. 435, 929–932.

Luo, L., 1997. Symmetry Breaking of Flow in 2-D Symmetric Channels: Simulations by Lattice Boltzmann Method. Int. J. Mod. Phys. C Vol. 8, 859–867.

Luo, S., Vellakal, M., Koric, S., Kindratenko, V., Cui, J., 2020. Parameter Identification of RANS Turbulence Model Using Physics-Embedded Neural Network. In: Jagode, H., Anzt, H., Juckeland, G., Ltaief, H. (Eds.), High Performance Computing, Lecture Notes in Computer Science. Springer International Publishing, Cham, pp. 137–149.

MacCormack, R.W., 1969. The Effect of Viscosity in Hypervelocity Impact Cratering. J. Spacecr. Rockets 40, 757–763.

MacCormack, R., Paullay, A., 1972. Computational Efficiency Achieved by Time Splitting of Finite Difference Operators. In: 10th Aerospace Sciences Meeting, Aerospace Sciences Meetings. American Institute of Aeronautics and Astronautics.

Magnussen, B.F., Hjertager, B.H., 1976. On Mathematical Modelling of Turbulent Combustion with Special Emphasis on Soot Formation and Combustion. In: 16th Symp. (Int.) Comb. The Combustion Institute, pp. 1405–1414.

Magolan, B., Baglietto, E., Brown, C., Bolotnov, I.A., Tryggvason, G., Lu, J., 2017. Multiphase Turbulence Mechanisms Identification From Consistent Analysis of Direct Numerical Simulation Data. Nucl. Eng. Technol. Special Issue on International Conference on Mathematics and Computational Methods Applied to Nuclear Science and Engineering 2017 (M&C 2017). Vol. 49, 1318–1325.

Makkawi, Y.T., Wright, P.C., Ocone, R., 2006. The Effect of Friction and Inter-Particle Cohesive Forces on the Hydrodynamics of Gas–Solid Flow: A Comparative Analysis of Theoretical Predictions and Experiments. Powder Technol. Fluidization and Fluid Particle Systems. Vol. 163, 69–79.

Malalasekera, W.M.G., Versteeg, H.K., Gilchrist, K., 1996. A Review of Research and an Experimental Study of the Pulsation of Buoyant Diffusion Flames and Pool Fires. Fire Mater. Vol. 20, 261–271.

Marcum, D.L., Weatherill, N.P., 1985. Unstructured Grid Generation Using Iterative Point Insertion and Local Reconnection. AIAA J. Vol. 33, 1619–1625.

Marcum, D., Weatherill, N., 1995. A Procedure for Efficient Generation of Solution Adapted Unstructured Grids. Comput. Methods Appl. Mech. Eng. 127, 259–268.

MATLAB, 1992. The Student Edition of MATLAB. The Math Works Inc., Prentice Hall, Englewood Cliffs, NJ.

Mattson, T., 1995. Progamming Environments for Parallel and Distributed Computing: A Comparison of p4, PVM, Linda and TCGMSG. Int. J. Supercomputing Vol. 9, 138–161.

Mavriplis, D., 1988. Multigrid Solution of the Two-Dimensional Euler Equations on Unstructured Triangular Meshes. AIAA J. Vol. 26, 824–831.

Mavriplis, D., 1997. Unstructured Grid Techniques. Ann. Rev. Fluid Mech. 29 (1), 473–514.

Maw, N., Barber, J.R., Fawcett, J.N., 1976. Oblique Impact of Elastic Sphere. Wear Vol. 38, 101–114.

McCaffrey, B.J., 1979. Purely Buoyant Diffusion Flames: Some Experimental Results. National Bureau of Standards, Gaithersburg, MD. p. NBS IR 79-1910.

McCaffrey, B.J., 1983. Momentum Implications for Buoyant Diffusion Flames. Combust. Flame Vol. 52, 149–216.

McDonald, P.W., 1971. The Computation of Transonic Flow through Two-Dimensional Gas Turbine Cascades. In: ASME Paper 71-GT-89, Gas Turbine Conference and Products Show. Houston, Texas.

McRae, D.S., 2000. R-Refinement Grid adaptation and Issues. Comp. Meth. Appl. Sci. Eng. Vol. 189, 1288–1294.

Ménard, T., Tanguy, S., Berlemont, A., 2007. Coupling Level Set/VOF/Ghost Fluid Methods: Validation and Application to 3D Simulation of the Primary Break-Up of a Liquid Jet. Int. J. Multiph. Flow Vol. 33, 510–524.

Meneveau, C., Lund, T.S., Cabot, W.H., 1996. A Lagrangian Dynamic Subgrid-Scale Model of Turbulence. J. Fluid Mech. Vol. 319, 353–385.

Menter, F., 1993. Zonal Two Equation k-w Turbulence Models For Aerodynamic Flows. In: 23rd Fluid Dynamics, Plasmadynamics, and Lasers Conference, Fluid Dynamics and Co-Located Conferences. American Institute of Aeronautics and Astronautics.

Menter, F.R., 1994a. Two-Equation Eddy-Viscosity Turbulence Models for Engineering Applications. AIAA J. Vol. 32, 1598–1605.

Menter, F.R., 1994b. Eddy Viscosity Transport Equations and Their Relation to the k-Epsilon Model. NASA STIRecon Tech. Rep. N 95, 14273.

Menter, F.R., 1996. A Comparison of Some Recent Eddy-Viscosity Turbulence Models. J. Fluids Eng. 118, 514–519.

Menter, F., 2018. Stress-Blended Eddy Simulation (SBES) – A New Paradigm in Hybrid RANS-LES Modeling. In: Hoarau, Y., Peng, S.-H., Schwamborn, D., Revell, A. (Eds.), Progress in Hybrid RANS-LES Modelling, Notes on Numerical Fluid Mechanics and Multidisciplinary Design. Springer International Publishing, Cham, pp. 27–37.

Menter, F.R., Egorov, Y., 2010. The Scale-Adaptive Simulation Method for Unsteady Turbulent Flow Predictions. Part 1: Theory and Model Description. Flow Turbul. Combust. Vol. 85, 113–138.

Menter, F., Hüppe, A., Matyushenko, A., Kolmogorov, D., 2021. An Overview of Hybrid RANS–LES Models Developed for Industrial CFD. Appl. Sci. Vol. 11, 2459.

Métais, O., Lesieur, M., 1992. Spectral Large-Eddy Simulation of Isotropic and Stably Stratified Turbulence. J. Fluid Mech. Vol. 256, 157–194.

Mignone, A., Tzeferacos, P., Bodo, G., 2010. High-Order Conservative Finite Difference GLM–MHD Schemes for Cell-Centered MHD. J. Comput. Phys. Vol. 229, 5896–5920.

Mishra, B.K., 2003a. A Review of Computer Simulation of Tumbling Mills by the Discrete Element Method, Part I – Contact Mechanics. Int. J. Miner. Process Vol. 71, 73–93.

Mishra, B.K., 2003b. A Review of Computer Simulation of Tumbling Mills by the Discrete Element Method, Part II – Practical Applications. Int. J. Miner. Process Vol. 71, 95–112.

Mittal, R., Iaccarino, G., 2005. Immersed Boundary Methods. Ann. Rev. Fluid Mech. Vol. 37, 239–261.

Mittal, R., Moin, P., 1997. Suitability of Upwind-Biased Finite Difference Schemes for Large-Eddy Simulation of Turbulent Flows. AIAA J. Vol. 35, 1415.

Mittal, R., Utturkar, Y., Udaykumar, H., 2002. Computational Modeling and Analysis of Biomimetic Flight Mechanisms. In: 40th AIAA Aerospace Sciences Meeting & Exhibit, Aerospace Sciences Meetings. American Institute of Aeronautics and Astronautics.

Mittal, R., Seshadri, V., Udaykumar, H.S., 2004. Flutter, Tumble and Vortex Induced Auto-rotation. Theo. Comp. Fluid Dyn. Vol. 17, 165–170.

Mohammadpour, J., Husain, S., Salehi, F., Lee, A., 2022. Machine Learning Regression-CFD Models for the Nanofluid Heat Transfer of a Microchannel Heat Sink With Double Synthetic Jets. Int. Commun. Heat Mass Transf. Vol. 130, 105808.

Monaghan, J.J., 1988. An Introduction to SPH. Comp. Phys. Comm. Vol. 48, 89–96.

Morawska, L., Johnson, G.R., Ristovski, Z.D., Hargreaves, M., Mengersen, K., Corbett, S., Chao, C.Y.H., Li, Y., Katoshevski, D., 2009. Size Distribution and Sites of Origin of Droplets Expelled From the Human Respiratory Tract During Expiratory Activities. J. Aerosol Sci. Vol. 40, 256–269.

Morris, P., Wang, Q., Long, L., Lockard, D., Morris, P., Wang, Q., Long, L., Lockard, D., 1997. Numerical Predictions of High Speed Jet Noise. In: 3rd AIAA/CEAS Aeroacoustics Conference, Aeroacoustics Conferences. American Institute of Aeronautics and Astronautics.

Morris, P., Long, L., Scheidegger, T., Wang, Q., Pilon, A., 1998. High Speed Jet Noise Simulations. In: 4th AIAA/CEAS Aeroacoustics Conference, Aeroacoustics Conferences. American Institute of Aeronautics and Astronautics.

Morsi, S.A., Alexander, J.A., 1972. An Investigation of Particle Trajectories in Two-Phase Systems. J. Fluid Mech. Vol. 55, 193–201.

Morsi, Y.S., Tu, J.Y., Yeoh, G.H., Yang, W., 2004. Principal Characteristics of Turbulent Gas-Particulate Flow in the Vicinity of Single Tube and Tube Bundle Structure. Chem. Eng. Sci. Vol. 59, 3141–3157.

MPI Forum, 1994. MPI: A Message-Passing Interface Standard (Technical Report CS-93-214). University of Tennessee, USA.

Muntz, E.P., 1989. Rarefied Gas Dynamics. Ann. Rev Fluid Mech. Vol. 21, 387–417.

Murakami, S., 1993. Comparison of Various Turbulence Models Applied to a Bluff Body. J. Wind Eng. Ind. Aerodyn. Vol. 46&47, 21–36.

Naftali, S., Rosenfeld, M., Wolf, M., Elad, D., 2005. The Air-Conditioning Capacity of the Human Nose. Ann. Biomed. Eng. 33, 545–553.

Najm, H.H., Wyckoff, P.S., Knio, O.M., 1998. A Semi-Implicit Numerical Scheme for Reacting Flow: I. Stiff Chemistry. J. Comp. Phys. Vol. 143, 381–402.

Nakayama, Y., Woods, W.A., Engineers, J.S.M., Clark, D.G., 1988. Visualized Flow: Fluid Motion in Basic and Engineering Situations Revealed by Flow Visualization. Pergamon.

Oberkampf, W.L., Trucano, T.G., Hirsch, C., 2004. Verification, Validation, and Predictive Capability in Computational Engineering and Physics. Appl. Mech. Rev. 57, 345–384.

Oran, E.S., Oh, C.K., Cybyk, B.Z., 1998. Direct Simulation Monte Carlo: Recent Advances and Applications. Ann. Rev. Fluid Mech. Vol. 30, 403–441.

Orszag, S.A., Israeli, N., Deville, M.O., 1986. Boundary Conditions for Incompressible Flow. J. Sci. Comp. Vol. 1, 75–111.

Pan, L.S., Liu, G.R., Khoo, B.C., Song, B., 1999. A Modified Direct Simulation Monte Carlo Method for Low-Speed Microflows. J. Micromech. Microeng. Vol. 10, 21–27.

Papadrakakis, M., Stavroulakis, G., Karatarakis, A., 2011. A New Ear in Scientific Computing: Domain Decomposition Methods in Hybrid CPU-GPU Architectures. Comput. Methods Appl. Mech. Enr. Vol. 200, 1490–1508.

Park, N., Yoo, J.Y., Choi, D., 2004. Discretisation Errors in Large-Eddy Simulation on the Suitability of Centered and Upwind-Biased Compact Difference Schemes. J. Comp. Phys. Vol. 198, 580–616.

Patankar, S.V., Spalding, D.B., 1972. A Calculation Procedure for Heat, Mass and Momentum Transfer in Three-Dimensional Parabolic Flows. Int. J. Heat Mass Transfer Vol. 15, 1787–1806.

Patel, V.C., Rodi, W., Scheuerer, G., 1985. Turbulence Model for Near-Wall and Low Reynolds Number Flows: A Review. AIAA J. Vol. 23, 1308–1319.

Patera, A.T., 1984. A Spectral Element Method for Fluid Dynamics: Laminar Flow in a Channel Expansion. J. Comput. Phys. 54, 468–488.

Peaceman, D.W., Rachford Jr., H.H., 1955. The Numerical Solution of Parabolic and Elliptic Differential Equations. J. Soc. Ind. Appl. Math. Vol. 3 (No. 1), 28–41.

Perktold, K., Resch, M., Peter, R.O., 1991. Three-Dimensional Numerical Analysis of Pulsatile Flow and Wall Shear Stress in the Carotid Artery Bifurcation. J. Biomech. Vol. 24, 409–420.

Peskin, C.S., 1973. Flow Patterns Around Heart Valves: A Digital Computer Method for Solving the Equations of Motion. IRE Trans. Med. Electron. BME-20, 316–317.

Peters, N., 2000. Turbulent Combustion. Cambridge University Press, UK.

Piomelli, U., 2008. Wall-Layer Models for Large-Eddy Simulations. Prog. Aerosp. Sci., Large Eddy Simulation – Current Capabilities and Areas of Needed Research Vol. 44, 437–446.

Piomelli, U., Balaras, E., 2002. Wall-Layer Models for Large-Eddy Simulations. Annu. Rev. Fluid Mech. Vol. 34, 349–374.

Piomelli, U., Ferziger, J.H., Moin, P., 1987. Models for Large Eddy Simulation of Turbulent Channel Flows Including Transpiration, Technical Report, Report TF-32. Stanford University, Dept. Mech. Eng.

Piomelli, U., Ferziger, J.H., Moin, P., Kim, J., 1989. New Approximate Boundary Conditions for Large Eddy Simulations of Wall-Bounded Flows. Phys. Fluids A1, 1061–1068.

Poiseuille, J.L., 1844. Recherches experimentales sur le mouvement des liquides dans les tubes de tres-petits diametres. Imprimerie Royale.

Pope, S.B., 1994. Lagrangian PDF Methods for Turbulent Flows. Ann. Rev. Fluid Mech. Vol. 26, 23–63.

Pope, S.B., 1976. The Probability Approach to the Modelling of Turbulent Reacting Flows. Combust. Flame 27, 299–312.

Portscht, R., 1975. Studies on Characteristic Fluctuations of the Flame Radiation Emitted by Fires. Combust. Sci. Tech. Vol. 10, 73–84.

Posner, J.D., Buchanan, C.R., Dunn-Rankin, D., 2003. Measurement and Prediction of Indoor Air Flow in a Model Room. Energy Build. Vol. 35, 269–289.

Prather, K.A., Marr, L.C., Schooley, R.T., McDiarmid, M.A., Wilson, M.E., Milton, D.K., 2020. Airborne Transmission of SARS-CoV-2. Science Vol. 370, 303–304.

Rai, M.M., Moin, P., 1991. Direct Simulation of Turbulent Flow Using Finite-Difference Schemes. J. Comp. Phys. Vol. 96, 15–53.

Rajamani, G.K., 2006. CFD Analysis of Air Flow Interactions in Vehicle Platoons. MEng. Thesis, RMIT University, Australia.

Rappitsch, G., Perktold, K., 1996. Pulsatile Albumin Transport in Large Arteries. J. Biomech. Eng. Vol. 118, 511–519.

Ravindran, A.C., Kokjohn, S.L., 2021. Combining Machine Learning With 3D-CFD Modeling for Optimizing a DISI Engine Performance During Cold-Start. Energy AI Vol. 5, 100072.

Reckinger, S.J., Livescu, D., Vasilyev, O.V., 2016. Comprehensive numerical Methodology for Direct Numerical Simulations of Compressible Rayleigh–Taylor Instability. J. Comput. Phys. Vol. 313, 181–208.

Rhie, C.M., Chow, W.L., 1983. A Numerical Study of the Turbulent Flow Past An Isolated Airfoil With Trailing Edge Separation. AIAA J. Vol. 21, 1525–1532.

Richards, K., Bithell, M., Dove, M., Hodge, R., 2004. Discrete-Element Modelling: Methods and Applications in the Environmental Sciences. Phil. Trans. R Soc. Lond. A Vol. 362, 1797–1861.

Ringstad, K.E., Banasiak, K., Ervik, Å., Hafner, A., 2021. Machine Learning and CFD for Mapping and Optimization of CO_2 Ejectors. Appl. Therm. Eng. Vol. 199, 117604.

Rizzi, A.W., Inuoye, M., 1973. Time Split Finite Volume Method for 3D Blunt Body Flows. AIAA J. Vol. 11, 1478–1485.

Roache, P.J., 1997. Quantification of Uncertainty in Computational Fluid Dynamics. Ann. Rev. Fluid Mech. Vol. 29, 123–160.

Rodi, W., 1980. Turbulence Models and Their Application in Hydraulics – A State of the Art Review. IAHR, Delft, The Netherlands.

Rodi, W., 1991. Experience With Two-Layer Models Combining the k-Epsilon Model With a One-Equation Model Near the Wall. In: 29th Aerospace Sciences Meeting, Aerospace Sciences Meetings. American Institute of Aeronautics and Astronautics.

Rodi, W., 1993. Turbulence Models and Their Application in Hydraulics. Balkema, Rotterdam.

Rogers, S., Kwak, D., 1991. An Upwind Differencing Scheme for the Incompressible Navier–Strokes Equations. Appl. Numer. Math. 8, 43–64.

Roe, P., 1997. A Brief Introduction to High-Resolution Schemes. In: Hussaini, M.Y., van Leer, B., Van Rosendale, J. (Eds.), Upwind and High-Resolution Schemes. Springer, Berlin, Heidelberg, pp. 9–28.

Roe, P.L., 2005. Computational Fluid Dynamics – Retrospective and Prospective. Int. J. Comp. Fluid Dyn. Vol. 19, 581–594.

Roma, A.M., Peskin, C.S., Berger, M.J., 1999. An Adaptive Version of the Immersed Boundary Method. J. Comp. Phys. Vol. 153, 509–534.

Rosa, B., Kopeć, S., Ababaei, A., Pozorski, J., 2022. Collision Statistics and Settling Velocity of Inertial Particles in Homogeneous Turbulence From High-Resolution DNS Under Two-Way Momentum Coupling. Int. J. Multiph. Flow Vol. 148, 103906.

Rosenfeld, M., Kwak, D., Vinokur, M., 1991. A Fractional-Step Method for Unsteady Incompressible Navier-Stokes Equations in Generalized Coordinate Systems. J. Comp. Phys. Vol. 94, 102–137.

Rosenhead, L., 1930. The Spread of Vorticity in the Wake Behind a Cylinder. Proc. R Soc. A Vol. 127. pp. 590–512.

Ruppert, J., 1993. A New and Simple Algorithm for Quality Two-Dimensional Mesh Generation. Proceedings of Fourth ACM-SIAM Symposium Discrete Algorithms, 83–92.

Ruck, B., Makiola, B., 1988. Particle Dispersion in a Single-Sided Backward Facing Step Flow. Int. J. Multiphase Flow Vol. 14, 787–800.

Saad, Y., Schultz, M., 1985. Conjugate Gradient-Like Algorithms for Solving Nonsymmetric Linear Systems. SIAM J. Vol. 44, 417–424.

Sadd, M.H., Tai, Q.M., Shukla, A., 1993. Contact Law Effects on Wave-Propagation in Particulate Materials Using Distinct Element Modeling. Int. J. Non-Linear Mech. Vol. 28, 251–265.

Sagaut, P., 2004. Large-Eddy Simulation for Incompressible Flows – An Introduction. Springer-Verlag, Berlin.

Schroeter, J.D., Kimbell, J.S., Asgharian, B., 2006. Analysis of Particle Deposition in the Turbinate and Olfactory Regions Using a Human Nasal Computational Fluid Dynamics Model. J. Aerosol Med. Off. J. Int. Soc. Aerosols Med. 19, 301–313.

Sethian, J.A., 1996. Level Set Methods: Evolving Interfaces in Computational Geometry, Fluid Mechanics, Computer Vision, and Materials Science, First Edition. Cambridge University Press, Cambridge.

Shadloo, M.S., Oger, G., Le Touzé, D., 2016. Smoothed Particle Hydrodynamics Method for Fluid Flows, Towards Industrial Applications: Motivations, Current State, and Challenges. Comput. Fluids Vol. 136, 11–34.

Shah, K.B., Ferziger, J.H., 1997. A Fluid Mechanisms View of Wind Engineering: Large Eddy Simulation of Flow Over a Cubical Obstacle. In: Meroney, R.N., Bienkiewicsz, B. (Eds.), Computer Wind Engineering. Vol. 2. Elsevier, Amsterdam, pp. 211–236.

Shalaby, H., Thévenin, D., 2010. Statistically Significant Results for the Propagation of a Turbulent Flame Kernel Using Direct Numerical Simulation, Flow Turbul. Combust. Vol. 84, 357–367.

Shan, X., 1997. Simulation of Rayleigh-Bénard Convection Using a Lattice Boltzmann Method. Phys. Rev. E Vol. 55, 2780–2788.

Shan, H., Jiang, L., Liu, C., 2005. Direct Numerical Simulation of Flow Separation Around a NACA0023 Airfoil. Comput. Fluids Vol. 34, 1096–1114.

Shang, Y., Dong, J., Inthavong, K., Tu, J., 2015. Comparative Numerical Modeling of Inhaled Micron-Sized Particle Deposition in Human and Rat Nasal Cavities. Inhal. Toxicol. Vol. 27, 694–705.

Shang, Y., Tao, Y., Dong, J., He, F., Tu, J., 2021. Deposition Features of Inhaled Viral Droplets May Lead to Rapid Secondary Transmission of COVID-19. J. Aerosol Sci. Vol. 154, 105745.

Shariati, V., Ahmadian, M.H., Roohi, E., 2019. Direct Simulation Monte Carlo Investigation of Fluid Characteristics and Gas Transport in Porous Microchannels. Sci. Rep. Vol. 9, 17183.

Shepard, M.S., Georges, M.K., 1991. Three-Dimensional Mesh Generation by Finite Octree Technique. Int. J. Num. Meth. Eng. Vol. 32, 709–749.

Shewchuk, J.R., 2002. Delaunay Refinement Algorithms for Triangular Mesh Generation. Comp. Geo. Vol. 22, 21–74.

Shih, T., Bailey, R., Nguyen, H., Roelke, R., 1991. Algebraic Grid Generation for Complex Geometries. Int. J. Numer. Methods Fluids 13, 1–31.

Shih, T.-H., Liou, W.W., Shabbir, A., Yang, Z., Zhu, J., 1995. A New k-e Eddy Viscosity Model for High Reynolds Number Turbulent Flows. Comp. Fluids Vol. 24, 227–238.

Shu, C.-W., Osher, S., 1988. Efficient Implementation of Essentially Non-Oscillatory Schemes. I. J. Comp. Phys. Vol. 77, 439–471.

Shu, C.-W., Osher, S., 1989. Efficient Implementation of Essentially Non-Oscillatory Schemes. II. J. Comp. Phys. Vol. 83, 32–78.

Simon, H. (Ed.), 1992. Parallel Computational Fluid Dynamics. MIT Press, Cambridge, MA.

Smagorinsky, J., 1963. General Circulation Experiments With the Primitive Equation, Part 1: The Basic Experiment. Mon. Weather Rev. Vol. 91, 99–164.

Smith, R.E., 1982. Algebraic Grid Generation. Appl. Math. Comput. 10–11, 137–170.

Soni, B., Thompson, D., Koomullil, R., Thornburg, H., 2001. In: GGTK: A Tool Kit for Static and Dynamic Geometry-Grid Generation, AIAA 2001-1164. AIAA 39th Aerospace Science Meeting, Reno, Nevada.

Sotiropoulos, F., Yang, X., 2014. Immersed Boundary Methods for Simulating Fluid–Structure Interaction. Prog. Aerosp. Sci. Vol. 65, 1–21.

Spaid, M.A.A., Phelan Jr., F.R., 1997. Lattice Boltzmann Methods for Modeling Microscale Flow in Fibrous Porous Media. Phys. Fluids Vol. 9, 2468–2474.

Spalart, P.R., Jou, W.H., Strelets, M., Allamaras, S.R., 1997. Comments on the Feasibility of LES for Wings, and on a Hybrid RANS/LES Approach. In: Liu, C., Liu, Z. (Eds.), Advances in DNS/LES. Greyden Press, Columbus, OH, pp. 137–148.

Spalding, D.B., 1971. Mixing and Chemical Reaction in Steady Confined Turbulent Flames. In: Thirteenth Symposium (International) on Combustion. Symp. Int. Combust. Vol. 13, 649–657.

Spalding, D.B., 1972. A Novel Finite-Difference Formulation for Differential Expressions Involving Both First and Second Derivatives. Int. J. Num. Meth. Eng. Vol. 4, 551–559.

Spalding, D.B., 1980. Numerical Computation of Multi-Phase Fluid Flow and Heat Transfer. In: Taylor, C., Morgan, K. (Eds.), Recent Advances in Numerical Methods in Fluid. Vol. 1, pp. 139–167.

Speziale, C.G., 1998. Turbulence Modeling for Time-Dependent RANS and VLES: A Review. AIAA J. Vol. 36, 173–184.

Squires, K.D., Constantinescu, G.S., 2003. LES and DES Investigations of Turbulent Flow Over a Sphere at Re = 10,000, Flow. Turb. Comb. Vol. 70, 267–298.

Star, S.K., Spina, G., Belloni, F., Degroote, J., 2021. Development of a Coupling Between a System Thermal–Hydraulic Code and a Reduced Order CFD Model. Ann. Nucl. Energy Vol. 153, 108056.

Stone, H.L., 1968. Iterative Solution of Implicit Approximations of Multidimensional Partial Differential Equations. SIAM. J. Numer. Anal. Vol. 5 (No. 3), 530–558.

Strelets, M., 2001. Detached Eddy Simulation of Massively Separated Flows. In: 39th Aerospace Sciences Meeting and Exhibit, Aerospace Sciences Meetings. American Institute of Aeronautics and Astronautics.

Ström, H., Sasic, S., 2013. A Multiphase DNS Approach for Handling Solid Particles Motion With Heat Transfer. Int. J. Multiph. Flow Vol. 53, 75–87.

Subramaniam, A., Wong, M.L., Lele, S.K., 2019. A High-Order Weighted Compact High Resolution Scheme With Boundary Closures for Compressible Turbulent Flows With Shocks. J. Comput. Phys. Vol. 397, 108822.

Sui, Z., Sui, Y., Wu, W., 2022. Multi-Objective Optimization of a Microchannel Membrane-Based Absorber With Inclined Grooves Based on CFD and Machine Learning. Energy Vol. 240, 122809.

Sun, Q., Fan, J., Boyd, I.D., 2009. Improved Sampling Techniques for the Direct Simulation Monte Carlo Method. Comput. Fluids Vol. 38, 475–479.

Sun, B., Tenneti, S., Subramaniam, S., 2015. Modeling Average Gas–Solid Heat Transfer Using Particle-Resolved Direct Numerical Simulation. Int. J. Heat Mass Transf. Vol. 86, 898–913.

Sunarso, A., Tsuiji, T., Chono, S., 2010. GPU-Accelerated Molecular Dynamics Simulation for Study of Liquid Crystalline Flows. J. Comput. Phys. Vol. 229, 5486–5497.

Sunderam, V., Geist, G., Dongarra, J., Manchek, R., 1990. PVM: A Framework for Parallel Distributed Computing. Concurr. Comput. Prac. Exp. Vol. 2, 315–339.

Suresh, A., Huynh, H.T., 1997. Accurate Monotonicity-Preserving Schemes With Runge-Kutta Time Stepping. J. Comp. Phys. Vol. 136, 83–99.

Sweby, P.K., 1984. High Resolution Schemes Using Flux Limiters for Hyperbolic Conservation Laws, SIAM. J. Num. Anal. Vol. 21, 995–1011.

Sweet, R.A., 1973. Direct Methods for the Solution of Poisson's Equation on a Staggered Grid. J. Comp. Phys. Vol. 12, 422–428.

Tao, Y., Inthavong, K., Tu, J.Y., 2017. Dynamic Meshing Modelling for Particle Resuspension Caused by Swinging Manikin Motion. Build. Environ. Vol. 123, 529–542.

Tao, Y., Zhang, H., Huang, D., Fan, C., Tu, J., Shi, L., 2021. Ventilation Performance of a Naturally Ventilated Double Skin Façade With Low-e Glazing. Energy Vol. 229, 120706.

Taylor, C.A., Draney, M.T., 2004. Experimental and Computational Methods in Cardiovascular Fluid Mechanics. Ann. Rev. Fluids Mech. Vol. 36, 197–231.

Taylor, C.A., Draney, M.T., Ku, J.P., Parker, D., Steele, B.N., et al., 1999. Predictive Medicine: Computational Techniques in Therapeutic Decision-Making. Comp. Aided Surg. Vol. 4, 231–247.

Temmerman, L., Hadziabdic, M., Leschziner, M.A., Hanjalic, K., 2005. A Hybrid Two-Layer URANS-LES Approach for Large Eddy Simulation at High Reynolds Numbers. Int. J. Num. Meth. Fluids Vol. 26, 173–190.

Tennekes, H., Lumley, J.L., 1976. A First Course in Turbulence. MIT Press, Cambridge, Massachusetts.

Tenneti, S., Subramaniam, S., 2014. Particle-Resolved Direct Numerical Simulation for Gas-Solid Flow Model Development. Annu. Rev. Fluid Mech. Vol. 46, 199–230.

Tezduyar, T.E., 2001. Finite Element Methods for Flow Problems With Moving Boundaries and Interfaces. Arch. Comp. Meth. Eng. Vol. 8, 83–130.

Thari, A., Pasquariello, V., Aage, N., Hickel, S., 2021. Adaptive Reduced-Order Modeling for Non-Linear Fluid–Structure Interaction. Comput. Fluids Vol. 229, 105099.

Thomas, J.L., Diskin, B., Brandt, A., 2003. Textbook Multigrid Efficiency for Fluid Simulations. Ann. Rev. Fluid Mech. Vol. 35, 317–340.

Thomasset, F., 1981. Implementation of Finite Element Methods for Navier-Stokes Equations. Springer-Verlag, Berlin.

Thompson, J.F., Warsi, S.U.A., Mastin, C.W., 1982. Boundary-Fitted Coordinate Systems for Numerical Solution of Partial Differential Equations – A Review. J. Comp. Phys. Vol. 47, 1–108.

Thompson, J.F., Warsi, Z.U.A., Mastin, C.W., 1985. Numerical Grid Generation – Foundations and Applications. Elsevier, New York.

Thompson, J.F., Soni, B.K., Weatherill, N.P. (Eds.), 1999. Handbook of Grid Generation, First Edition. CRC Press, Boca Raton, FL.

Thornton, C., 1997. Force Transmission in Granular Media. KONA Powder Part. J. 15, 81–90.

Tian, Z.F., Tu, J.Y., Yeoh, G.H., 2005. Numerical Simulation and Validation of Dilute Gas-Particle Flow Over a Backward Facing Step. Aerosol Sci. Tech. Vol. 39, 319–332.

Tian, Z.F., Tu, J.Y., Yeoh, G.H., 2006. On the Numerical Study of Contaminant Particle Concentration in Indoor Airflow. Build. Env. Vol. 41, 1504–1514.

Tian, W., Sevilla, T.A., Zuo, W.D., 2016. A Systematic Evaluation of Accelerating Indoor Airflow Simulations Using Cross-platform Parallel Computing. J. Build. Perform. Simul. https://doi.org/10.1080/19401493.2016.1212933.

Timmermann, G., 2000. A Cascadic Multigrid Algorithm for Semilinear Elliptic Problems. Numerische Mathematik Vol. 86, 717–731.

Tomas, J. (Ed.), 2003. Mechanics of Nanoparticle Adhesion – A Continuum Approach. In: Particles on Surfaces: Detection, Adhesion and Removal. Volume 8. CRC Press, London, pp. 183–229.

Tong, X., Dong, J., Shang, Y., Inthavong, K., Tu, J., 2016. Effects of Nasal Drug Delivery Device and Its Orientation on Sprayed Particle Deposition in a Realistic Human Nasal Cavity. Comput. Biol. Med. Vol. 77, 40–48.

Toro, E.F., 1997. Riemann Solvers and Numerical Methods for Fluid Dynamics: A Practical Introduction. Springer-Verlag, Berlin.

Trisjono, P., Pitsch, H., 2015. Systematic Analysis Strategies for the Development of Combustion Models From DNS: A Review, Flow Turbul. Combust. Vol. 95, 231–259.

Tsuji, Y., Tanaka, T., Ishida, T., 1992. Lagrangian Numerical Simulation of Plug Flow of Cohesionless Particles in a Horizontal Pipe. Powder Technol. Vol. 71, 239–250.

Tsuji, Y., Kawaguchi, T., Tanaka, T., 1993. Discrete Particle Simulation of Two-Dimensional Fluidized Bed. Powder Technol. Vol. 77, 79–87.

Tu, J.Y., 1997. Computational of Turbulent Two-Phase Flow on Overlapped Grids, Num. Heat. Transfer, Part B, Fund. Vol. 32, 175–195.

Tu, J.Y., Fletcher, C.A.J., 1995. Numerical Computation of Turbulent Gas-Solid Particle Flow in a 90° Bend. AIChE J. Vol. 41, 2187–2197.

Tu, J.Y., Fuchs, L., 1992. Overlapping Grids and Multigrid Methods for Three-Dimensional Unsteady Flow Calculations in IC Engines. Int. J. Numer. Methods Fluids Vol. 15, 693–714.

Tu, J.Y., Abu-Hijleh, B., Xue, C., Li, C.G., 2004a. CFD Simulation of Air/Particle Flow in the Human Nasal Cavity. In: Proc. 5th Int. Conf. Multiphase Flow. Yokohama, Japan.

Tu, J.Y., Yeoh, G.H., Morsi, Y.S., Yang, W., 2004b. A Study of Particle Rebounding Characteristics of Gas-Particle Flow Over Curved Surface. Aerosol Sci. & Tech. Vol. 38, 739–755.

Tu, J., Inthavong, K., Ahmadi, G., 2012. Computational Fluid and Particle Dynamics in the Human Respiratory System. Springer Netherlands, Dordrecht, The Netherlands.

Tu, J., Kiao, I., Wong, K.K.L., 2015. Computational Hemodynamics – Theory, Modelling and Applications. In: Biological and Medical Physics, Biomedical Engineering, first ed. Springer Netherlands, Dordrecht, The Netherlands.

Tucker, P., 2003. Differential Equation Based Length Scales to Improve DES and RANS Simulations. In: 16th AIAA Computational Fluid Dynamics Conference, Fluid Dynamics and Co-Located Conferences. American Institute of Aeronautics and Astronautics.

Tucker, P.G., Davidson, L., 2004. Zonal k-l Based Large Eddy Simulation. Comp. Fluids Vol. 33, 267–287.

Turkel, E., 1999. Preconditioning Techniques in Computational Fluid Dynamics. Annu. Rev. Fluid Mech. 31, 385–416.

Unverdi, S., Tryggvason, G., 1992. A Front-Tracking Method for Viscous, Incompressible, Multi-fluid Flows. J. Comp. Phys. Vol. 100, 25–42.

Valino, L., 1998. A Field Monte-Carlo Formulation for Calculating the Probability Density Function of a Single Scalar in a Turbulent Flow, Flow. Turb. Comb. Vol. 60, 151–172.

Van Doormal, J.P., Raithby, G.D., 1984. Enhancements of the SIMPLE Method for Predicting Incompressible Fluid Flows. Numer. Heat Transfer Vol. 7, 147–163.

Van Doormal, J.P., Raithby, G.D., 1985. An Evaluation of the Segregated Approach for Predicting Incompressible Fluid Flows, ASME Paper 85-HT-9. National Heat Transfer Conference, Denver, Colorado.

Van Driest, E.R., 1952. Investigation of Laminar Boundary Layer in Compressible Fluids Using the Crocco Method. National Advisory Committee for Aeronautics, Washington, D.C.

Van Driest, E.R., 1956. On Turbulent Flow Near a Wall. J. Aero. Sci. Vol. 23, 1007–1011.

Van Leer, B., 1974. Towards the Ultimate Conservative Difference Scheme. II. Monotonicity and Conservation Combined in a Second-Order Scheme. J. Comp. Phys. Vol. 14, 361–370.

Van Leer, B., 1977a. Towards the Ultimate Conservative Difference Scheme. III. Upstream-Centered Finite Difference Schemes for Ideal Compressible Flow. J. Comp. Phys. Vol. 23, 263–275.

Van Leer, B., 1977b. Towards the Ultimate Conservative Difference Scheme. IV. A Second Order Sequel to Godunov's Method. J. Comp. Phys. Vol. 23, 276–299.

Van Leer, B., 1979. Towards the Ultimate Conservative Difference Scheme. V. A New Approach to Numerical Convection. J. Comp. Phys. Vol. 32, 101–136.

Verlet, L., 1967. Computer Experiments on Classical Fluids, I. Thermodynamical Properties of Lennard-Jones Molecules. Phys. Rev. E Vol. 55, 3546–3554.

Versteeg, H.K., Malalasekera, W., 1995. An Introduction to Computational Fluid Dynamics – The Finite Volume Method. Prentice Hall, Pearson Education Ltd., England.

Violeau, D., Rogers, B.D., 2016. Smoothed Particle Hydrodynamics (SPH) for Free-Surface Flows: Past, Present and Future. J. Hydraul. Res. Vol. 54, 1–26.

Vu-Quoc, L., Zhang, X., 1999. An Elastoplastic Force-Displacement Model in the Normal Direction: Displacement-Driven Version, Proc. R Soc. London. Ser. A. Vol. 455, 4013–4044.

Walton, O.R., Braun, R.L., 1986. Computer-Simulation of the Mechanical Sorting of Grains. Powder Technol. Vol. 48, 239–245.

Wang, Z.J., 2011. Adaptive High-Order Methods in Computational Fluid Dynamics, Advances in Computational Fluid Dynamics. Vol. 2 World Scientific, Singapore.

Wang, M., Moin, P., 2002. Dynamic Wall Modeling for Large-Eddy Simulation of Complex Turbulent Flows. Phys. Fluids Vol. 14, 2043–2051.

Wang, B., Wang, J., 2021. Application of Artificial Intelligence in Computational Fluid Dynamics. Ind. Eng. Chem. Res. Vol. 60, 2772–2790.

Wang, Y.Q., Jackson, P., Phaneu, T.J., et al., 2006. Turbulent Flow Through a Staggered Tube Bank. J. Thermophysics & Heat Transfer Vol. 20, 738–747.

Wang, Z.J., Fidkowski, K., Abgrall, R., Bassi, F., Caraeni, D., Cary, A., Deconinck, H., Hartmann, R., Hillewaert, K., Huynh, H.T., Kroll, N., May, G., Persson, P.-O., van Leer, B., Visbal, M., 2013. High-Order CFD Methods: Current Status and Perspective. Int. J. Numer. Methods Fluids Vol. 72, 811–845.

Wang, Z.-B., Chen, R., Wang, H., Liao, Q., Zhu, X., Li, S.-Z., 2016. An Overview of Smoothed Particle Hydrodynamics for Simulating Multiphase Flow. Appl. Math. Model. Vol. 40, 9625–9655.

Wang, S., Xiao, J., Ye, S., Song, C., Wen, J., 2018. Numerical Investigation on Pre-Heating of Coal Water Slurry in Shell-and-Tube Heat Exchangers With Fold Helical Baffles. Int. J. Heat Mass Transf. Vol. 126, 1347–1355.

Wang, Y., Gao, Y., Xu, H., Dong, X., Liu, J., Xu, W., Chen, M., Liu, C., 2020. Liutex Theoretical System and Six Core Elements of Vortex Identification. J. Hydrodyn. 32, 197–211.

Wang, L., Gobbert, M.K., Yu, M., 2020. A Dynamically Load-Balanced Parallel p-Adaptive Implicit High-Order Flux Reconstruction Method for Under-Resolved Turbulence Simulation. J. Comput. Phys. Vol. 417, 109581.

Wang, T., Zhang, F., Furtney, J., Damjanac, B., 2022. A Review of Methods, Applications and Limitations for Incorporating Fluid Flow in the Discrete Element Method. J. Rock Mech. Geotech. Eng. Vol. 14, 1005–1024.

Watkins, S., Vino, G., 2004. On Vehicle Spacing and Its Effect on Drag and Lift. In: Proc. 5th Int. Colloquium Bluff Body Aero. Appl. BBAA5. Ottawa, Canada.

Wesseling, P., 1995. Introduction to Multigrid Methods. Institute for Computer Applications in Science and Engineering, Hampton, Virginia.

Wilcox, D.C., 2004. Turbulence Modeling for CFD. DCW Industries, La Cãnada, California.

Wilson, N.M., 2002. Geometric Algorithms and Software Architecture for Computational Prototyping: Applications in Vascular Surgery and MEMS. Ph.D. Thesis, Stanford University, Stanford.

Witry, A., Al-Hajeri, M.H., Bondok, A.A., 2005. Thermal Performance of Automotive Aluminium Plate Radiator. Appl. Thermal Eng. Vol. 25, 1207–1218.

Wolfshtein, M.W., 1969. The Velocity and Temperature Distribution in a One-Dimensional Flow With Turbulence Augmentation and Pressure Gradient. Int. J. Heat Mass Transfer Vol. 12, 301–312.

Wong, K.K.L., Tu, J., Sun, Z., Dissanayake, D.W., 2013. Methods in Research and Development of Biomedical Devices. World Scientific Publishing Company, Singapore.

World Health Organization, 2014. Infection Prevention and Control of Epidemic- and Pandemic-Prone Acute Respiratory Infections in Health Care. WHO Guidelines Approved by the Guidelines Review Committee, World Health Organization, Geneva.

Wu, J.-S., Lian, Y.-Y., 2003. Parallel Three-Dimensional Direct Simulation Monte Carlo Method and Its Applications. Comput. Fluids Vol. 32, 1133–1160.

Wu, K., Shu, C.-W., 2019. Provably Positive High-Order Schemes for Ideal Magnetohydrodynamics: Analysis on General Meshes. Numer. Math. Vol. 142, 995–1047.

Xiao, J., Wang, S., Ye, S., Wen, J., Zhang, Z., 2020. Multiphysics Field Coupling Simulation for Shell-and-Tube Heat Exchangers With Different Baffles. Numer. Heat Transf. Part Appl. Vol. 77, 266–283.

Xiong, Q., Madadi-kandjani, E., Lorenzini, G., 2014. A LBM-DEM Solver for Fast Discrete Particle Simulation of Particle-Fluid Flows. Contin. Mech. Thermodyn. Vol. 26, 907–917.

Xu, C., Deng, X., Zhang, L., Fang, J., Wang, G., Jiang, Y., Cao, W., Che, Y., Wang, Y., Wang, Z., Liu, W., Cheng, X., 2014. Collaborating CPU and GPU for Large-Scale High-Order CFD Simulations With Complex Grids on the TianHe-1A Supercomputer. J. Comput. Phys. Vol. 278, 275–297.

Xu, X., Waschkowski, F., Ooi, A.S.H., Sandberg, R.D., 2022. Towards Robust and Accurate Reynolds-Averaged Closures for Natural Convection via Multi-Objective CFD-Driven Machine Learning. Int. J. Heat Mass Transf. Vol. 187, 122557.

Yakhot, V., Orszag, S.A., 1986. Renormalization Group Analysis of Turbulence. I. Basic Theory. J. Sci. Comp. Vol. 1, 1–15.

Yakhot, V., Prszag, S.A., Tangham, S., Gatski, T.B., Speciale, C.G., 1992. Development of Turbulence Models for Shear Flows by a Double Expansion Technique. Phys Fluids A: Fluid Dynamics Vol. 4, 1510–1520.

Yan, X., Zhu, J., Kuang, M., Wang, X., 2019. Aerodynamic Shape Optimization Using a Novel Optimizer Based on Machine Learning Techniques. Aerosp. Sci. Technol. Vol. 86, 826–835.

Yanenko, N.N., 1971. The Method of Fractional Steps. Springer-Verlag, Berlin.

Yang, Z.L., Dinh, T.N., Nourgaliev, R.R., Sehgal, B.R., 2000. Numerical Investigation of Bubble Coalescence Characteristics Under Nucleate Boiling Condition by a Lattice Boltzmann Model. Int. J. Therm. Sci. Vol. 389, 1–17.

Yang, Y.X., Zhou, B., Post, J.R., Scheepers, E., Reuter, M.A., Boom, R., 2006. Computational Fluid Dynamics Simulation of Pyrometallurgical Processes, 5th Int. Conf. on CFD in the Minerals and Process Industries, CSIRO, Melbourne, Australia.

Yao, Y., Yeo, K.-S., 2017. An Application of GPU Acceleration in CFD Simulation for Insect Flight. Supercomput. Front. Innov. Vol. 4, 13–26.

Ye, T., Pan, D., Huang, C., Liu, M., 2019. Smoothed Particle Hydrodynamics (SPH) for Complex Fluid Flows: Recent Developments in Methodology and Applications. Phys. Fluids Vol. 31, 011301.

Yeoh, G.H., Tu, J.Y., 2005. Thermal-Hydrodynamics Modelling of Bubbly Flows With Heat and Mass Transfer. AIChE J. Vol. 51, 8–27.

Yeoh, G.H., Tu, J.Y., 2006a. Numerical Modelling of Gas-Liquid With and Without Heat and Mass Transfer. Appl. Math. Model. Vol. 30, 1067–1095.

Yeoh, G.H., Tu, J.Y., 2006b. Two-Fluid and Population Balance Models for Subcooled Boiling Flow. Appl. Math. Model. Vol. 30. pp. 1370–139.

Yeoh, G.H., Tu, J.Y., 2008. Modelling Subcooled Boiling Flows. Nova Science Publishers, Incorporated, New York.

Yeoh, G.H., Tu, J., 2019. Computational Techniques for Multiphase Flows. Elsevier Science & Technology, San Diego, California.

Yeoh, G.H., Behnia, M., de Vahl Davis, G., Leonardi, E., 1990. A Numerical Study of Three-Dimensional Natural Convection During Freezing of Water. Int. J. Num. Meth. Eng. Vol. 30, 899–914.

Yeoh, G.H., Lee, E.W.M., Yuen, R.K.K., Kwok, W.K., 2002. Fire and Smoke Distribution in a Two-Room Compartment Structure. Int. J. Num. Meth. Heat Fluid Flow Vol. 12, 178–194.

Yeoh, G.H., Cheung, C.P., Tu, J., 2014. Multiphase Flow Analysis Using Population Balance Modeling – Bubbles. Drops and Particles, Butterworth-Heinemann, Oxford.

Yerry, M.A., Shepard, M.S., 1984. Three-Dimensional Mesh Generation by Modified Octree Technique. Int. J. Num. Meth. Eng. Vol. 20, 1965–1990.

Yu, C., 2006. An Efficient High-Resolution Shock-Capturing Scheme for Multi-Dimensional Flows I. Hydrodynamics. Chin. J. Astron. Astrophys. Vol. 6, 680–688.

Yu, G., Xhang, Z., Lessmann, R., 1998. Fluid Flow and Particle Diffusion in the Human Upper Respiratory System, Aerosol Sci. Tech. Vol. 28, 146–158.

Yun, G., Kim, D., Choi, H., 2006. Vortical Structures Behind a Sphere at Subcritical Reynolds numbers. Phys. Fluid Vol. 18, pp. 015102:1–14.

Zang, Y., Street, R.L., Koseff, J.R., 1993. A Dynamic Mixed Subgrid-Scale Model and Its Application to Turbulent Recirculating Flows. Phys. Fluids A Vol. 5, 3186–3196.

Zhao, Y., Akolekar, H.D., Weatheritt, J., Michelassi, V., Sandberg, R.D., 2020. RANS Turbulence Model Development Using CFD-Driven Machine Learning. J. Comput. Phys. Vol. 411, 109413.

Zhou, L.X., 2016. A Review for Developing Two-Fluid Modeling and LES of Turbulent Combusting Gas-Particle Flows. Powder Technol. Vol. 297, 438–447.

Žukauskas, A., Ulinskas, R., 1988. Heat Transfer in Tube Banks in Crossflow. Hemisphere, Washington.

Zukoski, E.E., Cetegen, B.M., Kubota, T., 1984. Visible Structure of Buoyant Diffusion Flames. Twentieth Symposium (International) on Combustion. Symp. Int. Combust. 20, 361–366.

Zwartz, G.J., Guilmette, R.A., 2001. Effect of Flow Rate on Particle Deposition in a Replica of a Human Nasal Airway. Inhal. Toxicol. Vol. 13, 109–127.

Further Suggested Reading

Çengel, Y.A., 2003. Heat Transfer: A Practical Approach. McGraw-Hill, New York.

Çengel, Y.A., Turner, R., 2005. Fundamentals of Thermal-Fluid Science, Second Edition. McGraw-Hill, New York.

Chung, T.J., 2002. Computational Fluid Dynamics. Cambridge University Press, Cambridge.

Crowe, T.C., Elger, D.F., Roberson, J.A., 2005. Engineering Fluid Mechanics, Eighth Edition. Wiley, New York.

Drikakis, D., Rider, W., 2005. High-Resolution Methods for Incompressible and Low-Speed Flows. Springer-Verlag, Berlin.

Ferziger, J.H., Perić, M., 1999. Computational Methods for Fluid Dynamics. Springer-Verlag, Berlin.

Patankar, S.V., 1980. Numerical Heat Transfer and Fluid Flow, Hemisphere Publishing Corporation. Taylor & Francis Group, New York.

Roach, P.J., 1998. Fundamental of Computational Fluid Dynamics. Hermosa Publishers, Albuquerque, New Mexico.

Index

Note: Page numbers followed by *f* indicate figures *t* indicate tables and *b* indicate boxes.

Printed in the United States
by Baker & Taylor Publisher Services